Autodesk Inventor 2023 Cookbook

A guide to gaining advanced modeling and automation skills
for design engineers through actionable recipes

Alexander Bordino

BIRMINGHAM—MUMBAI

Autodesk Inventor 2023 Cookbook

Group Product Manager: Rohit Rajkumar

Publishing Product Manager: Kaustubh Manglurkar

Senior Editor: Hayden Edwards

Senior Content Development Editor: Rashi Dubey

Technical Editor: Joseph Aloocaran

Copy Editor: Safis Editing

Project Coordinator: Sonam Pandey

Proofreader: Safis Editing

Indexer: Tejal Daruwale Soni

Production Designer: Alishon Mendonca

Marketing Coordinator: Nivedita Pandey

First published: November 2022

Production reference: 1211122

Published by Packt Publishing Ltd.

Livery Place

35 Livery Street

Birmingham

B3 2PB, UK.

ISBN 978-1-80181-050-0

www.packt.com

To my wonderful wife, Nav, for her love, support, and patience, without which writing this book would not be possible.

For Nonno B.

– Alexander Bordino

Contributors

About the author

Alexander Bordino is an award-winning product design engineer and Autodesk manufacturing technical specialist who specializes in consulting and training across the breadth of the Autodesk design and manufacturing portfolio. He has experience as an Autodesk Inventor Certified Professional, Autodesk Accredited Trainer, Accredited Stratasys FDM, and Polyjet Additive Manufacturing Trainer, and he attained a BSc (Hons) in product design at Nottingham Trent University. He currently works for a leading Autodesk reseller in the UK, specializing in Autodesk design and manufacturing products, empowering design and engineering companies in the UK to design products better, faster, and more efficiently. He has extensive experience of developing innovative product designs for a wide variety of organizations, obtaining industry awards, and patenting technology in his career. In 2017, Alexander was admitted as an Associate of the Worshipful Company of Horners, an ancient guild and livery company in the City of London, representing and supporting the UK plastics industry.

This book would not have been completed were it not for the support and love of my wife, family, and friends. I thank you all for your encouragement throughout. Thank you also to the Packt team, who have made the process of writing a book seamless and efficient. I thank you for your professionalism and valued suggestions throughout this journey. My thanks also go to the fantastic GrabCAD community, where many of the practice files in this book originated or have been inspired by.

About the reviewers

Didi Widya Utama has extensive expertise teaching CAD/CAM/CAE in the mechanical design and robotics fields and is an assistant professor of mechanical engineering at Universitas Tarumanagara, Indonesia. Additionally, he has years of expertise in the industrial sector, including manufacturing, CNC, mold and die making, and more. He is a gold-level Autodesk® Certified Instructor, an Autodesk® Certified Professional for Inventor, Autodesk Certified Professional in Design for Manufacturing, and a Certified Fusion 360® User. He has more than 10 years of experience teaching at the Autodesk® Authorized Training Center on the Autodesk software, including Autocad®, Inventor®, and Fusion 360®, for both the manufacturing industry and educational institutions. He has been involved in several engineering projects for years, including university research, government funding, and projects involving national and international engineering collaboration.

This book, titled Inventor Cookbook 2023, covers all aspects of the Autodesk® Inventor software, from the fundamentals to in-depth expertise. Additionally, the examples in the book are relevant and valuable, and they include additional tips for using the program. This book is excellent for beginners who want to learn about the Inventor software from scratch and more experienced readers who want to refresh their knowledge on the subject. This book can serve as a guide cookbook for those looking to develop and expand their software expertise.

Olanrewaju Sulaimon was born and raised in Lagos, Nigeria. He attended Yaba College of Technology, where he studied mechanical engineering for 2 years. It was during his internship he became interested in CAD in 2017. He is currently studying the same course at the University of Lagos, Nigeria. During a semester break, he started learning how to use Autodesk Inventor as it was one of the courses in the next semester. He found Inventor very interesting and fascinating due to its instant ability to convert from 2D sketches to 3D parts, then to assemblies. He has used Inventor to design several realistic projects that were later manufactured. He is very conversant with other CAD applications, such as Autodesk AutoCAD and SolidWorks.

Table of Contents

5

Advanced CAD Management and Collaboration – Project Files, Templates, and Custom Properties   213

6

Inventor Assembly Fundamentals – Constraints, Joints, and BOMS   253

7

Model and Assembly Simplification with Simplify, Derive, and Model States   327

8

Design Accelerators – Specialized Inventor Tool Sets for Frames, Shafts, and Bolted Connections 351

9

Design Communication – Inventor Studio, Animation, Rendering, and Presentation Files 409

10

Inventor iLogic Fundamentals — Creating Process Automation and Configurations

11

Inventor Stress and Simulation – Workflow and Techniques   499

12

Sheet Metal Design – Comprehensive Methodologies to Create Sheet Metal Products 549

13

Inventor Professional 2023 – What's New? 607

Preface

Autodesk Inventor is an industry-leading, **computer-aided design** (**CAD**) application for 3D mechanical design, simulation, visualization, and documentation. In this book, we aim to bridge the gap between the fundamentals of this software and the more advanced features, workflows, and environments this powerful design solution has to offer.

Using cookbook-style recipes, you will gain a comprehensive understanding and practical experience in creating dynamic 3D parts, assemblies, and complete designs. We will also explore a variety of topics, including automation and parametric techniques, collaboration tools, creating sheet metal designs, design accelerators such as frame generators, surface modeling tools, advanced assembly, and simplification tools, iLogic, and finite element analysis.

By the end of this book, you will not only be able to use the advanced functionality within Autodesk Inventor, but you will also have the practical experience to deploy these techniques in your own projects and workflows.

Who this book is for

This book is aimed at CAD engineers, mechanical/design engineers, and product designers who have a basic understanding and experience of Inventor fundamentals. It aims to guide and coach you beyond the basics and into the advanced functionality of the software and environments within it.

What this book covers

Chapter 1, *Inventor Part Modeling – Sketch, Work Features, and Best Practices*, explains the best practices involved in part modeling with the uses of sketches and all the work features covered. Then combine your knowledge of these best practices to complete a modeling challenge of a complex part.

Chapter 2, *Advanced Design Methodologies and Strategies*, discusses how and when to apply different design methodologies and strategies in Inventor, and the merits of each. You will learn and gain experience with practical examples of how each can be used. This chapter will also cover how to utilize external and non-native CAD data in Inventor.

Chapter 3, *Driving Automation and Parametric Modeling in Inventor*, discusses how to implement levels of automation into both parts and assemblies, by understanding the importance and use of equations and parameters to drive configurations. You can copy features across parts with iFeatures, automate the mating of parts with iMates, and create configurations using iParts and iAssemblies. Finally, you will learn how to link external spreadsheets and parameters to drive models in Inventor.

Chapter 4, Freeform, Surface Modeling, and Analysis, explains how to create surface geometry in various forms and how to use the freeform modeling tools within Inventor to create organic and complex forms. The chapter will break down the various ways surfaces can be created within Inventor. You will also learn about surface validation techniques and how to surface model in the context of an assembly.

Chapter 5, Advanced CAD Management and Collaboration – Project Files, Templates, and Custom Properties, focuses on some of the important admin that's required to deliver successful projects with Inventor. It explains how to manage project files and best practices, how to set up templates and design standards for a company in both parts, assembly and the drawing environment, and how to manage content center libraries. In most cases, these areas are overlooked, but successful data management and organization are essential to delivering projects, on time and within budget. Engineering managers and CAD managers will find this topic of use most of all.

Chapter 6, Inventor Assembly Fundamentals – Constraints, Joints, and BOMS, discusses the fundamentals of successful assembly design and techniques. It will also cover how to effectively use constraints and joints in an assembly and the best practices, and how to use, edit, and customize BOMs in Inventor, including placement in the drawing environment and annotating a GA.

Chapter 7, Model and Assembly Simplification with Simplify, Derive, and Model States, covers techniques for simplifying geometry in a part and assembly to aid in the management of large assemblies and assist in the collaboration of models with external suppliers, customers, and stakeholders. Particular focus will be spent on model states, which were released in 2022 and superseded the traditional level of detail and positional representations of older releases.

Chapter 8, Design Accelerators – Specialised Inventor Tool Sets for Frames, Shafts, and Bolted Connections, explains how to use design accelerators for a range of product types, and how to create frameworks, gears, shafts, and bolted connections within the assembly environment. This is a key skill set in Inventor that allows for automation and extreme efficiency with workflows. The conventional means of creating these types of products or adding them to existing assemblies is very time-consuming. Engineers and designers will greatly appreciate the value that using the design accelerators provides when the situation demands it.

Chapter 9, Design Communication – Inventor Studio, Animation, Rendering, and Presentation Files, explains that successful design is usually measured on how well the benefits and the solution are understood by stakeholders; therefore, being able to demonstrate these in 3D within Inventor is highly important. Designers need to be able to render, animate, and sometimes quickly annotate parts and assemblies to communicate key design changes, updates, and showcase product features. This chapter focuses on how a designer can implement renderings, explode views with presentation files, and create animations. This chapter also focuses on some of the more basic skills of model manipulation.

Chapter 10, Inventor iLogic Fundamentals – Creating Process Automation and Configurations, gives a basic introduction to automation with iLogic and what can be achieved with this. It provides an overview of the environment and practical examples of iLogic in use. You will then progress and create several iLogic rules and use these, before finally creating an iLogic form, controlling the configuration of an assembly. This is a key chapter for enabling you to automate functions, model updates, and changes within the software.

Chapter 11, Inventor Stress and Simulation – Workflow and Techniques, gives an introduction to the stress analysis environment in Inventor, how this works, what is achievable with the standard stress analysis, and workflows to adopt and various types of analysis. It also discusses design and how important it is that designers are able to test and validate their designs, prior to manufacture. Having a basic understanding of the essentials of FEA within Inventor gives the designer the key advantage of being able to iterate and test designers quicker and more efficiently, instead of having to rely on external stress analysis engineers all of the time.

Chapter 12, Sheet Metal Design – Comprehensive Methodologies to Create Sheet Metal Products, provides an introduction to the sheet metal environment within Inventor and a detailed walk-through of the functionality and methods you can use to create sheet metal features. This chapter will also show how you can bring final sheet metal parts into the drawing environment to add detail with bend tables, hole tables, and folded and unfolded flat patterns for export.

Chapter 13, Inventor Professional 2023 – What's New?, talks about what's new in Inventor 2023. You will learn all of the new features and additions to Inventor Professional 2023 in both the part, assembly, and drawing environments.

To get the most out of this book

This book is aimed at advanced users of Autodesk Inventor Professional. It is recommended that you have taken part in an essentials training course at an Autodesk Authorised Training Center or a similar course before proceeding, or that you have sufficient experience with the basic fundamentals of Autodesk Inventor.

You will need a working and licensed copy of Autodesk Inventor Professional 2023 to complete the recipes in this book, and a Windows 10 operating system. Post-2023 versions of Inventor will work with the recipe practice files. Internet access will be required for the initial download of recipe files for this book.

Software/hardware covered in the book	Operating system requirements
Autodesk Inventor 2023	Windows 10

Autodesk Inventor 2023 System Requirements: `https://knowledge.autodesk.com/support/inventor/learn-explore/caas/sfdcarticles/sfdcarticles/System-requirements-for-Autodesk-Inventor-2023.html`

Download the example code files

You can download the Inventor lesson files for this book from `https://packt.link/w6Kik`

Download the color images

We also provide a PDF file that has color images of the screenshots and diagrams used in this book. You can download it here: `https://packt.link/tnAxK`.

Conventions used

There are a number of text conventions used throughout this book.

`Code in text`: Indicates code words in text, database table names, folder names, filenames, file extensions, pathnames, dummy URLs, user input, and Twitter handles. Here is an example: "To begin this recipe, you will need to create a `New Standard (mm) Assembly .iam` file and have it open."

A block of code is set as follows:

```
Parameter("CATL-SIDE PLATE #1:1", "CATH")=CATLL
Parameter("CATL-SIDE PLATE #1:1",
"AMOUNTR")=Ceil((Parameter("CATL-SIDE PLATE #1:1",
"CATH")+32.5)/275)
Parameter("CATL-SIDE PLATE #1:1", "SPACINGR")=Parameter("CATL-
SIDE PLATE #1:1", "CATH")/Parameter("CATL-SIDE PLATE #1:1",
"AMOUNTR")
```

Bold: Indicates a new term, an important word, or words that you see on screen. For instance, words in menus or dialog boxes appear in **bold**. Here is an example: "Select **Finish Sketch** and then repeat this operation on the other side."

> **Tips or Important Notes**
> Appear like this.

Sections

In this book, you will find several headings that appear frequently (*Getting ready*, *How to do it...*, *How it works...*, *There's more...*, and *See also*).

To give clear instructions on how to complete a recipe, use these sections as follows:

Getting ready

This section tells you what to expect in the recipe and describes how to set up any software or preliminary settings required for the recipe.

How to do it...

This section contains the steps required to follow the recipe.

How it works...

This section usually consists of a detailed explanation of what happened in the previous section.

There's more...

This section consists of additional information about the recipe in order to make you more knowledgeable about the recipe.

See also

This section provides helpful links to other useful information for the recipe.

Get in touch

Feedback from our readers is always welcome.

General feedback: If you have questions about any aspect of this book, email us at `customercare@packtpub.com` and mention the book title in the subject of your message.

Errata: Although we have taken every care to ensure the accuracy of our content, mistakes do happen. If you have found a mistake in this book, we would be grateful if you would report this to us. Please visit `www.packtpub.com/support/errata` and fill in the form.

Piracy: If you come across any illegal copies of our works in any form on the internet, we would be grateful if you would provide us with the location address or website name. Please contact us at `copyright@packt.com` with a link to the material.

If you are interested in becoming an author: If there is a topic that you have expertise in and you are interested in either writing or contributing to a book, please visit `authors.packtpub.com`.

Download a free PDF copy of this book

Thanks for purchasing this book!

Do you like to read on the go but are unable to carry your print books everywhere?

Is your eBook purchase not compatible with the device of your choice?

Don't worry, now with every Packt book you get a DRM-free PDF version of that book at no cost.

Read anywhere, any place, on any device. Search, copy, and paste code from your favorite technical books directly into your application.

The perks don't stop there, you can get exclusive access to discounts, newsletters, and great free content in your inbox daily

Follow these simple steps to get the benefits:

1. Scan the QR code or visit the link below

https://packt.link/free-ebook/9781801810500

2. Submit your proof of purchase
3. That's it! We'll send your free PDF and other benefits to your email directly

1

Inventor Part Modeling – Sketch, Work Features, and Best Practices

Sketches and constraints are one of the most important aspects of your Inventor model. A good and well-defined sketch is of paramount importance in ensuring that your model is stable and can be easily edited with design updates or changes with equations and parameters.

A sketch, or your model in this instance, is very much like the skeleton of an organism; if the sketch is not strong, lacks clear definitions, or is overly complex, the rest of the body has issues and problems. You should already have a reasonably good understanding of the basics of sketch geometry in Inventor and already know the fundamentals, but as they are so critical to successful CAD design, it is always worth revisiting and refining these core skills.

In this chapter, before expanding into the practical recipes, you will learn the best practices for applying successful sketches to your models, and additional tips that will complement your existing fundamental knowledge and application of these tools. This chapter then explores the roles of **planes**, **axes**, and **points** (collectively known as **work features**) in the sketch environment, and the best practices for these. Finally, we will explore essential **feature modeling tools**, and combine this knowledge in a part modeling recipe.

Here are the learning objectives for this chapter:

- 2D sketch design – Best practices
- 2D sketch constraints – Best practices
- Work planes – Best practices
- Work axes – Best practices
- Work points – Best practices
- Applying best sketch and work plane practices to model a part

Technical requirements

To complete the recipes, you will require Autodesk Inventor 2023, or a newer version installed on your machine. The recipe files are not backward compatible and will not work with software versions older than the 2023 version. To download and use the practice files that are required to complete the learning objective, type the following address into your internet browser: `https://packt.link/w6Kik`

Downloading your practice files

To make the most of the downloaded practice files' `.zip` folder, you will have to set the Inventor project file up correctly for the practice files. This enables Inventor to quickly access the location of your practice files from your C drive.

To set up the project, follow these steps:

1. Go to your `PC Downloads` folder, and copy the zipped folder titled `Inventor Cookbook 2023 Practice Files`. This is the data you have just downloaded.

2. Paste the zipped folder into **Windows (C:)**. Then, right-click on the folder, select **Unzip**, and the files will unzip into a folder of the same name.

3. Now, start Inventor. From the **Home** screen, in the top right, select **Settings**, as shown in *Figure 1.1*:

Figure 1.1: Browsing the project file

4. Then, in the **Projects** dialog box, select **Browse** from the bottom of the menu:

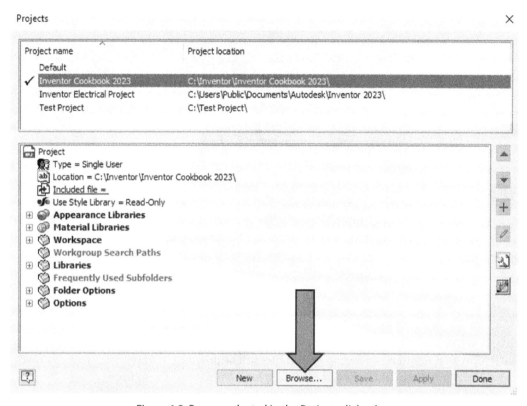

Figure 1.2: Browse selected in the Projects dialog box

5. From here, navigate to the unzipped `Inventor Cookbook 2023 Practice Files` folder and select it. Within the folder, select the `Inventor Cookbook 2023.ipj` file.

6. Select **Open**, followed by **Done**. This will now ensure that Inventor is using the correct project file for the Inventor Cookbook 2023 practice files.

7. You can check this by navigating back to **Home** and hovering the cursor over the active project. The full name of the project will display underneath this, as shown in *Figure 1.3*:

Figure 1.3: Inventor Cookbook 2023.ipj project file successfully loaded

More information regarding managing project files will be provided in *Chapter 5, Advanced CAD Management and Collaboration – Project Files, Templates, and Custom Properties.*

For now, you are all set to begin!

2D sketch design – Best practices

Successful sketch design underpins the entire design. It is the foundation of nearly everything you do in CAD, so reviewing best practices and techniques, even at an advanced level, is worthwhile. We will now begin to focus on some of the important theory behind this before embarking on the recipes.

Here are the best practices you should follow when creating sketches in Inventor.

Start at the origin

Start your initial sketch from the **origin** or, if not possible, create a dimension from one sketch entity to the origin. In any new Inventor file, the only point of reference that is known to Inventor is the origin. By creating your sketch using the origin, you will need to give Inventor less information about your sketch, as it is already working from a known point in time and space.

In *Figure 1.4*, two identical sketches are shown. The right sketch was defined from the origin, while the left was dimensioned to the origin:

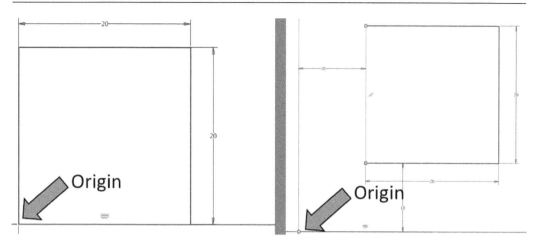

Figure 1.4: Two identical sketches – the right was created using the
origin, while the left was dimensioned to the origin

You can see from here that more dimensions have been added to the left sketch than the right. Both are correct, but the right rectangle has been created in a much more efficient manner with fewer variables. This may seem simple, but with more complex sketches, utilizing the origin becomes of greater significance.

Use sketch constraints often

Effective use of sketch constraint tools allows you to build logic and information about the sketch that dimensions alone cannot provide. If you can fix lines as being horizontal, parallel, or coincident, you will define your sketches much quicker and more efficiently than with just the use of dimension. It also makes editing sketches easier, as there are usually fewer variables to change.

Keep sketches simple

The simpler the sketch, the better. A sketch should be minimal and only contain what is necessary for your next feature command to use. Too often, CAD users will overcomplicate sketches by trying to create complex end features with sketch lines, arcs, and circles. This is not necessary, as it is the feature commands that will do the work and translate the design into something meaningful in 3D.

Before a sketch is created, you should break down what features you will use, two or three steps ahead in your design process, to create the desired outcome. There is no need to put all the information for your complex part in one sketch; be sure to break it down into manageable segments. This makes the modeling process more organized and efficient.

Use keyboard shortcut commands

Much like in Autodesk AutoCAD, the same quick keyboard shortcuts can be used in Autodesk Inventor, such as *L* for line, *C* for circle, and so on. The spacebar will repeat your last used command, and the *Esc* key will exit it.

As shown in the following figure, the *Tab* key allows you to flip between what type of dimension to enter when using a sketch command:

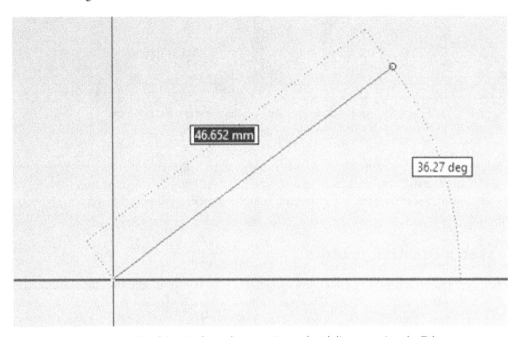

Figure 1.5: Graphics window when creating a sketch line; pressing the Tab
key allows the user to navigate between length and angle

In this instance, a Line command is demonstrated. Pressing the *Tab* key at this stage allows the user to quickly flip between entering a length or angle. The *Tab* key will also lock the dimension value after the value has been added, and then you can switch to another dimension and repeat the process.

All of these shortcuts will save valuable time and clicks. The full list of Autodesk Inventor keyboard shortcuts can be found here: https://damassets.autodesk.net/content/dam/autodesk/www/campaigns/inventor-resource/Inventor-Keyboard-Shortcuts-Guide.pdf.

Always fully constrain sketches

Leaving unconstrained geometry in an Inventor sketch is not a good idea. Later in the design process when you have edits to make to a model or feature, you will find your model will update and change shape in a completely random fashion. This is because elements of the sketch at the model's foundation are undefined, unfixed, and unknown. As you add dimensions and constraints to a sketch, the sketch entities will become defined.

You can tell whether a sketch is fully defined by the color of the sketch lines. By default, in Inventor, unconstrained geometry is black, green, or purple, and constrained geometry is dark blue. In a sketch itself, if you grab an entity and try and move it, a fully constrained sketch will remain in position. Spend time defining sketches, as it pays dividends later in the design process. Always apply geometric constraints before dimensional, as this will prevent your initial sketch from distorting.

You can also use the **Auto Dimension tool**. This will calculate the dimensions required to fully constrain your sketch and will add its own if you select this. Be warned, however, as this will apply dimensions in a random way; if you need a couple of dimensions to fully constrain a sketch, it's worth using Auto Dimension, seeing what dimensions are placed, and then deleting these and adding your own in place. It is not advisable to draw out a whole sketch without dimensions and then default to hitting Auto Dimension.

Create manageable steps

It is tempting to try and get all the model information in one sketch, but this can often cause unnecessary complications and take more time. It is far better to break down the geometry and sketches into manageable sections and let your feature commands do the work. Keep your sketches basic and simple. This also makes future edits much easier to complete, as your Model browser and feature history will become more organized.

Never draw the same thing twice

Always look for lines of symmetry in your design; it is more economical and efficient to draw half of something and then use a Mirror or Pattern command to duplicate it. If you have already drawn a feature that is to be repeated in the design, use Copy or a Pattern/Array tool. This again helps build adaptivity in your sketches. If an element of a repeating pattern must be changed, the other elements will also change and update. Your aim should be to maximize efficiency and productivity at every opportunity!

Those are the best overall practices for a successful sketch design; applying these techniques correctly will have you creating sketches quickly and efficiently as part of your workflow. Sketch constraints are an important aspect of this, and we will now cover the best practices of these.

2D sketch constraints – Best practices

Mastering constraints within a sketch is a common area in which new and experienced users can struggle. Just knowing what each constraint function does and how they can be applied can be challenging enough, and there is also confusion in that constraints can be found in both the 2D sketch and 3D model environments within Inventor. In this section, we will primarily be focusing on 2D sketch constraints in relation to the best practices to follow:

- A 2D sketch constraint constrains sketch geometry in one single sketch plane and, along with applied dimensions, allows you to quantify and fully define the geometry
- A 3D assembly constraint does much the same, only this is used to constrain a 3D part within the context of an assembly

Here are the best practices for working with constraints in Inventor.

Use the Status bar

The Status bar at the bottom right of the graphics window indicates the number of dimensions required to fully constrain a sketch. This will only start to show data once a sketch is created and you are in the sketch environment. In the following example, a simple rectangle was drawn in the sketch environment:

0.000 mm, 0.000 mm   4 dimensions needed   1   1

Figure 1.6: Status bar communicating how many dimensions are required for an unconstrained rectangle

Your goal should be that once you create a sketch, you want to get the dimensions needed down to **0** to ensure that the sketch is fully defined and constrained. As you apply constraints and dimensions, you should see this number reduce. If you are struggling to define a specific sketch, it can be a good indicator as to what you need to achieve a full sketch definition.

Automatic geometric constraints

Geometric constraints are created automatically, through the creation of lines, arcs, and other geometry, as you sketch in the sketch environment. These geometric constraints allow the sketch to be edited with predictable results, and they can also act as guides for aligning geometry and further defining your sketch. As automatic geometric constraints are applied, you will notice that, in the graphics window, alongside your sketch, graphical representations of the sketch will appear, giving you notice of what constraints Inventor has applied for you.

These can of course be manually deleted, and the automatic constraints can be turned off if required, but in most cases, they are useful, as they will limit the required dimensions the user has to input.

In the following figure, you can see that Inventor automatically inferred a **tangential constraint** between the line and arc as it was drawn, and applied a **parallel constraint** to the horizontal line. The visual sketch relationships can also be dragged and moved through the workspace if required:

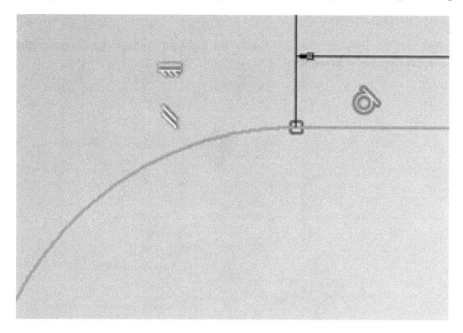

Figure 1.7: Automatically inferred sketch constraint between a line and arc

To access the **Constraint** settings (including turning off this feature), turn off **Constraint Inference**; within a 2D sketch, navigate to the **Sketch** tab, and along the right-hand side of the ribbon, look in the **Constrain** tools, where you will find the **Constraint** settings, shown in *Figure 1.8*.

Figure 1.8: Location of the Constraint settings in the ribbon

---

Tip – Keyboard shortcut

Within a sketch, you can quickly cycle between showing or hiding the graphical representation of the sketch constraints with the *F8* and *F9* keys.

Degrees of freedom

When sketch geometry can change size and shape, this is known as **degrees of freedom**. An example would be that a circle has two degrees of freedom: its center and its radius. Clearly defining both will result in a fully constrained circle sketch.

To display degrees of freedom, with no command active, highlight your sketch geometry, right-click, and select **Display Degrees of Freedom**:

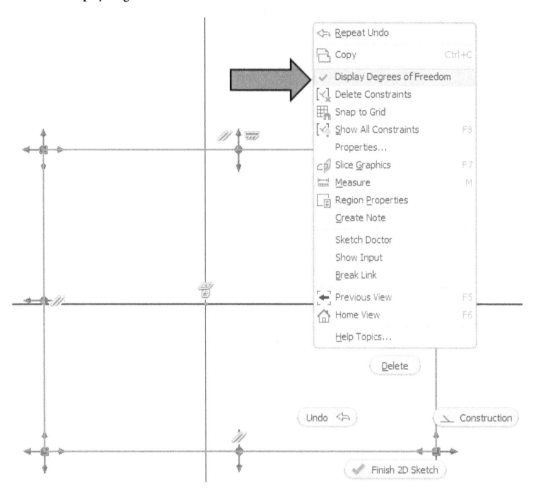

Figure 1.9: Display Degrees of Freedom in a sketch

Make sure to use **Display Degrees of Freedom** in your sketch, as this will notify you, as the user, where you may need to apply additional geometric or dimensional sketches to achieve a fully defined sketch. Your goal with constraints in sketches should be to eliminate the degrees of freedom.

Types of sketch constraint

Having knowledge of what each constraint does and how they can be applied is crucial to successful and efficient sketch design. Quite often, users can get confused about what each constraint means, particularly because some of the graphical representations look quite similar. This summary provides a list of each constraint and what they do, with the corresponding sketch constraint image for ease of reference.

You can locate the sketch constraints in the **Sketch** tab, then **Constrain**:

Figure 1.10: Location and graphical representation of all Inventor sketch constraints in the ribbon

Figure 1.11 gives a breakdown of all the sketch constraints in Inventor and how they are used:

Constraint	Application	Example
Coincident	The **Coincident** constraint allows you to constrain a specific point to another point. An example would be to constrain the center of a circle to the origin. The first image shows the circle with no **Coincident** constraint applied and the circle is unconstrained. The second image shows a single **Coincident** constraint applied to the center of the circle and the origin of the sketch. Because a diameter has already been defined as 10 mm, adding the **Coincident** constraint has fully defined the sketch.	
Collinear	The **Collinear** constraint enables you to constrain two or more line elements to follow the same direction and line. Here, a line on the bottom left of the triangle is made collinear with one of the sides using this constraint.	

Concentric ◎	The **Concentric** constraint is used to define a concentric relationship between circles and arcs so that they share the same center point.	
Fixed 🔒	The **Fixed** constraint fixes geometry to a specific point relative to the sketch coordinate system. Applying a **Fixed** command will usually fully define an element, but be wary of overusing this command, as it will decrease the flexibility of the sketch when it comes to updating it, and parametric changes will be more difficult to realize.	

Parallel	The **Parallel** constraint is used to define a parallel relationship between two lines.	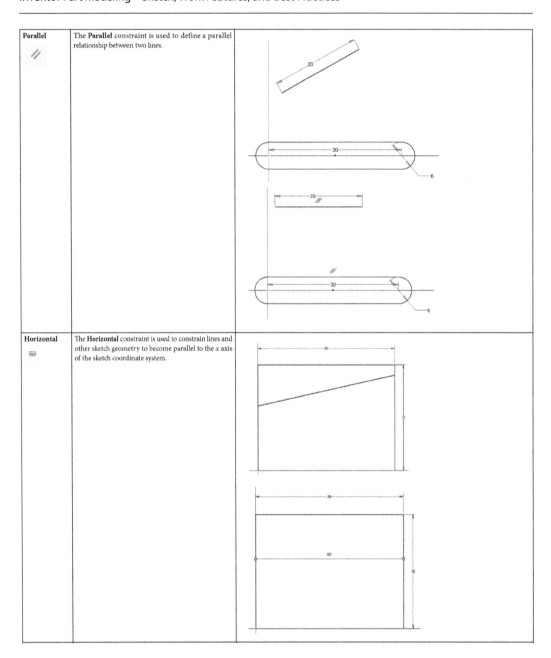
Horizontal	The **Horizontal** constraint is used to constrain lines and other sketch geometry to become parallel to the *x* axis of the sketch coordinate system.	

Vertical	The **Vertical** constraint is used to constrain lines and other sketch geometry to become parallel to the *y* axis of the sketch coordinate system.	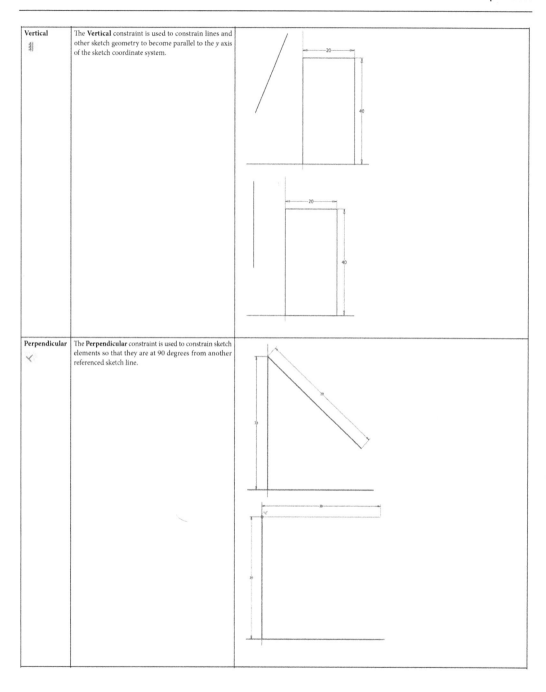
Perpendicular	The **Perpendicular** constraint is used to constrain sketch elements so that they are at 90 degrees from another referenced sketch line.	

Tangent	The **Tangent** constraint is used to create a tangency between splines, circles, arcs, and other geometry within the sketch.	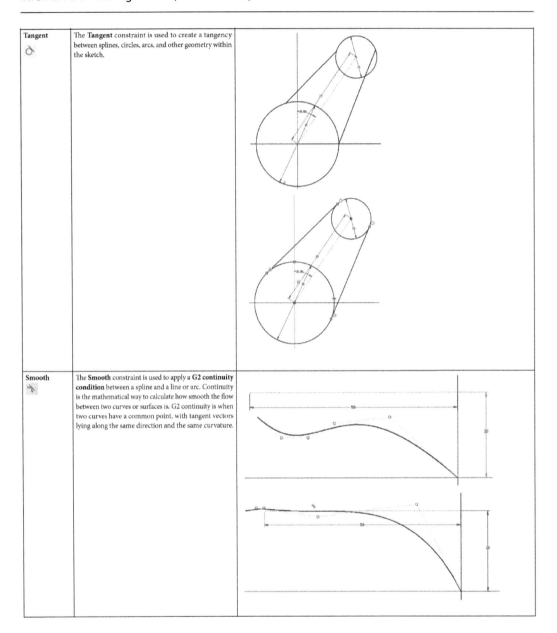
Smooth	The **Smooth** constraint is used to apply a **G2 continuity condition** between a spline and a line or arc. Continuity is the mathematical way to calculate how smooth the flow between two curves or surfaces is. G2 continuity is when two curves have a common point, with tangent vectors lying along the same direction and the same curvature.	

Symmetric [ ]	The Symmetric constraint constrains curves to be symmetrical to other curves about a line of symmetry. This can be a sketch line or axis.	
Equal =	The Equal constraint enables geometry to become equal in size to another piece of geometry.	

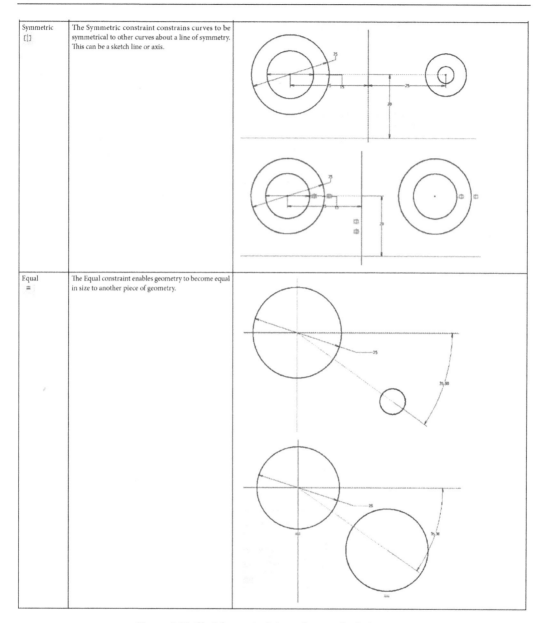

Figure 1.11: Sketch constraints and examples in Inventor

A clear understanding of the constraint options will enable you to create stronger sketches that are more easily defined and can be updated quicker and more efficiently, without causing your sketch to distort when dimensions are changed.

Sketch constraints are important, but a key understanding of planes and their best practices is also necessary to create models effectively in Inventor. The next section covers planes in Inventor, how they work, and the best ways of using them.

Work planes – Best practices

When modeling in Inventor, the correct application of work planes and features is of great importance. When creating features within Inventor, the existing geometry may not possess the required reference to place a new feature or sketch. In this scenario, the creation of a plane is used to create the feature references required.

Several different types of work planes can be applied and having a fundamental knowledge of them is essential to master more advanced functionality.

The full list and an explanation of each one can be found in Inventor. Navigate to the **3D Model** tab, and along the ribbon, you will find the **Work Features** area. Clicking on **Plane** will reveal the drop-down list of all work planes that can be created within Inventor. Hovering the cursor over each **Plane** option shows an explanation of how each one can be used:

Figure 1.12: List of planes from the Plane command

Work planes are non-solids and do not have any mass; they are simply used as a reference from which to create new geometry or features where there normally would not be any. By default, at the origin of a new part, you will have three **origin work planes**.

Figure 1.13 provides a summary of all work planes in Inventor, and examples of how they are used:

Work Plane	Example
Offset from Plane Creates a work plane offset and parallel from an existing face or plane. To apply this, hold the left mouse button and drag away or enter the required distance with the **Offset from Plane** command active. This plane is very useful for creating additional geometry where you do not have any work planes or faces to work from and is probably one of the most used planes in the software. In the recipe later in this chapter, you will get practical experience using this feature.	
Parallel to Plane through Point Creates a work plane that is parallel to another plane or face and passes through an edge or axis. Create by selecting a reference plane and then a point. Note that this does not have to be a point you have placed; it can be a point on a model, for example, a corner of a rectangle. You can always add your own points on the model and within sketches to reference when creating a plane that uses a point as a reference.	
Mid-Plane between Two Planes Generates a work plane that is equally distanced and the center of two existing planes or faces. Simply selecting two existing planes with the command active will generate the mid-plane. This is by far the simplest plane feature to apply and is one of the most important. In most cases, there will be symmetry in a model or a design; by applying a plane through the mid-point, you can efficiently mirror or copy features or even components themselves down a mirror plane. **Mid-Plane between Two Planes** is also useful when splitting solids in a multi-body design workflow.	
Mid-Plane of Torus This will create a work plane through the center plane or mid-plane of a torus. To apply this, select the torus with the command active, and a plane will be placed.	
Angle to Plane around Edge This creates a work plane at an angle to an existing face or plane. Apply this by selecting a plane as a reference in the command, followed by an edge or axis through which the plane is to pass. You can then enter the required angle of the plane. This plane is useful if you require the creation of geometry at a specified angle or direction.	

Three Points Creates a work plane from the reference of three existing points on the model. This can be placed by selecting the command and picking three points of reference on the model. The image shows an example of this with end points, fillets, and edges selected.	
Two Coplanar Edges Creates a work plane from the selection of two coplanar edges on the model. Selecting these two references while in the command will generate the plane.	
Tangent to Surface through Edge Generates a work plane that is tangent to a cylindrical face referenced by a construction line. Select a curved face and then a line or edge on the model to apply it.	
Tangent to Surface and Parallel to Plane Creates a work plane that is tangent to a cylindrical surface or body, and that is parallel to another planar surface. Select the cylindrical face first, followed by the parallel planar surface.	
Tangent to Surface through Point Creates a work plane that is tangent to a cylindrical surface or body, and that is defined by an existing point.	

Normal to Axis through Point Creates a work plane that is perpendicular to an existing axis that passes through a selected point. First, select the axis, and then pick a specific point to place the axis.	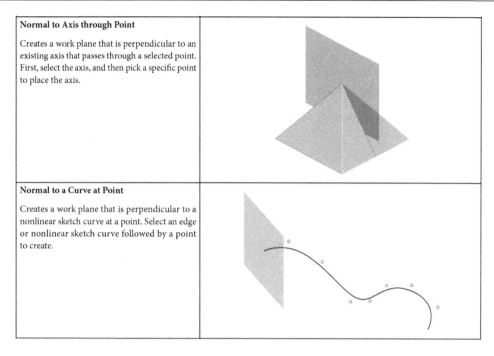
Normal to a Curve at Point Creates a work plane that is perpendicular to a nonlinear sketch curve at a point. Select an edge or nonlinear sketch curve followed by a point to create.	

Figure 1.13: Work planes and examples in Inventor

Now that we have covered all types of work planes, here are some best practices to keep in mind:

- In some instances, you will need to create multiple planes to create and place a plane where it is desired; these can be known as **sacrificial planes**.

- You can rename planes and axes in the **Model** browser. This helps bring organization to your model and allows for faster edits.

- Redefining planes is possible. Right-click on a plane and select **Redefine Feature**. You will then be asked to select a new reference for that plane.

- Work planes are infinite in size although, in the graphics window, they will only display as a small segment. This is so as not to confuse the user or obstruct the view. You can resize and stretch the visible area of any plane.

- Combining axes, points, and planes is encouraged to create the correct references for geometry.

Work planes in Inventor are complementary to the use of work axes; in the next section, we will examine the types of work axes you can apply and how these are used.

Work axes – Best practices

A **work axis** is a construction line of infinite length that is parametrically attached to a part. A work axis compliments the use of work planes and can be used as a reference in the creation of new axes/ planes, or other part modeling features such as a Revolve operation.

A summary of all work axes in Inventor (with examples of how they are used) is shown in *Figure 1.14*:

Axis	Example
On Line or Edge Creates a work axis along an edge or line. Apply by selecting an existing edge or line on the geometry in the **On Line or Edge** axis command.	
Parallel to Line through Point Creates a work axis that is parallel to an edge or line and intersects a point. Apply by first selecting the point and then an edge in this axis command.	
Through Two Points Selecting two work points of a model will generate an axis that intersects them while in this axis command.	
Intersection of Two Planes Selecting two work planes as a reference in this axis command creates an axis located along the intersection of the two selected planes.	

Normal to Plane through Point This creates a work axis that is normal to a surface and goes through a selected point. First, select a point, and then select an existing surface to create within this axis command.	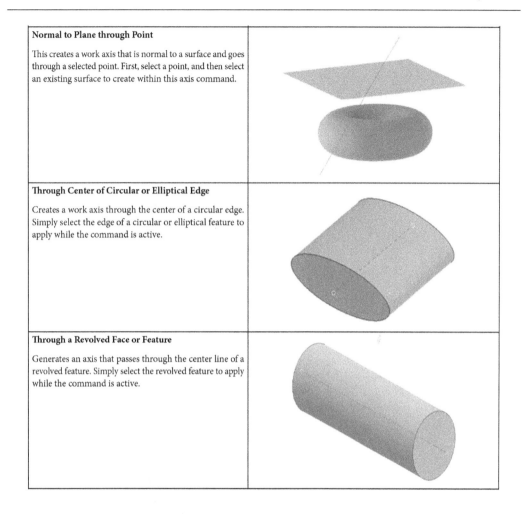
Through Center of Circular or Elliptical Edge Creates a work axis through the center of a circular edge. Simply select the edge of a circular or elliptical feature to apply while the command is active.	
Through a Revolved Face or Feature Generates an axis that passes through the center line of a revolved feature. Simply select the revolved feature to apply while the command is active.	

Figure 1.14: Work axes and examples in Inventor

Now that we have covered all the work axes, here are some best practices to keep in mind:

- Work planes and work points can be created when applying work axes and used as a reference.
- You can redefine references for axes by right-clicking on a work axis and selecting **Redefine Feature**.
- You can change the visibility of work axes in the same way as work planes. Right-click on an axis and select **Visibility**. The global control for this in Inventor can be found by clicking on the **View** tab, then the **Visibility** panel, and selecting **Object Visibility**.

With work axes, quite often, a point of reference needs to be made with a work point. This next section covers the types of work points, how they are used, and the best practices.

Work points – Best practices

Work points are used as references when creating planes and axes. They can also be used in the application of a 3D sketch.

A summary of all work points in Inventor (with examples of how they are used) is shown in *Figure 1.15*:

Work Point	Example
Grounded Point Converts an existing point so that it has all degrees of freedom removed and fixes it in place. It is also useful to move and rotate the object to a desired direction and location.	
On Vertex Creates a work point on a vertex, sketch point, or mid-point.	
Intersection of Three Planes Creates a work point at the intersection of three selected planes.	

Intersection of Two Lines Creates a work point at the intersection of two selected lines, edges, or work axes.	
Intersection of a Plane Creates a point at the intersection of a plane, sketch line, curve, or axis.	
Center Point of Loop Edges When selecting a circle arc or edge, a point is created at the center of the loop.	

Center Point of Torus or Sphere Select a torus or sphere with the command active to place a point at the center point of it.	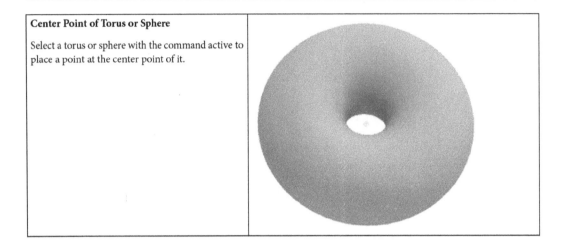

Figure 1.15: Work points and examples in Inventor

Now that we have covered all work points, here are some best practices to keep in mind:

- You can apply work points to a vertex, sketch point, mid-point, and center point of loops and edges, and the intersection of a plane/surface and line point

- Once the work point option is selected, you must then select applicable references to create the point

- **Center Point of Loops and Edges** is useful to act as a reference for a Hole feature that is required to be centered on the referenced geometry, regardless of updates

Now that we have covered the basics of sketches and work features, in the first recipe of the book, you will apply the knowledge learned and create a complex part.

Applying best sketch and work plane practices to model a part

In this exercise, we will combine many of the best practices for sketches, constraints, and planes to model a part most efficiently.

There are numerous ways that this part could be created, and there is no *absolute* correct method, but this method is one of the most efficient and will make use of the skills and topics already covered in this chapter.

Getting ready

To begin, we need to analyze the finished part in *Figure 1.16*. The part is shown from above and underneath. The objective is to recreate this in the most efficient way. No dimensions are given at this stage, as we will go through each step, and the appropriate dimensions will be given at each step.

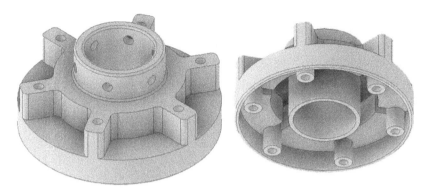

Figure 1.16: Final part – the end objective of this recipe

While this may look like a complex part with a myriad of geometry, using the best practices outlined previously, we can create this in a very efficient way that enables the user to also make quick edits to the part should this be required in the future. All dimensions will be given to you at each step.

There is no requirement to use a practice file with this recipe; you will simply need to create a new part file.

The first step is to look at the model and begin to understand the plan for the features and methodology that will be used to create it. You may just have a hand-drawn sketch or an idea in your head about what is to be created in Inventor, but the same rule applies.

Here are some things that we can decipher from the image of the part in *Figure 1.16*:

- There are revolved profiles that make up the bulk of the model
- It has multiple lines of symmetry, which means that areas of the part can be copied or arrayed
- It has repeating patterns; many of the features in this model are repeating and, therefore, need only to be created once
- The part is hollow, meaning a Shell operation could be used
- The fillets are uniform and can be created in one operation

How to do it...

So, let's combine the best practices for sketches, constraints, and planes to model a part in the most efficient way:

1. Create a new part file. To do this, from **File**, select **New**, then **New** again. Select **Metric**, then `Standard (mm).ipt`.

2. The first sketch can take several forms. To begin, start at the origin, as this will result in fewer dimensions and constraints being required to define the first sketch.

    As the part is made up of revolved geometry, it is logical to proceed with a suitable sketch that can be revolved. The center construction line placed will act as the reference for a **Revolve** operation later. As the part is symmetrical, it also makes sense to only draw half the profile:

    *   So, click on **Start 2D Sketch** and create the sketch shown in *Figure 1.17* on the **XY** plane.

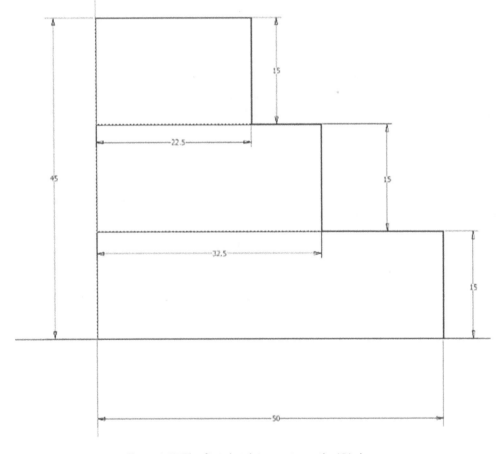

Figure 1.17: The first sketch to create on the XY plane

3. Now that a base sketch has been created and is fully constrained, use the **Revolve** command to revolve this sketch 360 degrees from the axis. This will create the base feature. To do this, select **Finish Sketch**, navigate to the **3D Model** tab, and select **Revolve** (you can also press *R* to access **Revolve** too).

4. Proceed to select the three sketch profiles created, then select **Axis** from the box and pick the construction line to act as an axis for **Revolve**. The preview should look *Figure 1.18*:

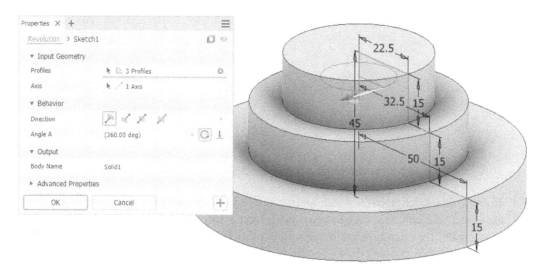

Figure 1.18: Revolve operation of the first sketch

5. Select **OK** to complete the revolved feature.

    Now that the base feature is complete, we can move on to detailing the model further. In the previous step, using a simple sketch and the appropriate feature, we created the base feature of the model efficiently, in a way that is easy to edit and update.

6. Navigate to the top view and select **Start 2D Sketch** on the face highlighted in *Figure 1.19*:

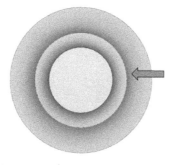

Figure 1.19: Face to create the second sketch

In this next operation, we will now create the arrayed lugs that protrude from the mid-section of the shape. As they are repeated and identical, it is logical to only draw this once and then copy/pattern the feature afterward. The holes will also be applied at this stage.

7.   Create the sketch shown in *Figure 1.20*. Remember to select *F7* while in the sketch to temporarily cut the model, so that you have better visibility of the sketch plane you are working from:

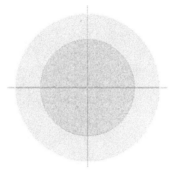

Figure 1.20: Sketch plane with model cut using the F7 command

To create this sketch from the sketch tools, select **Project Geometry** and select the two faces of the model. This will trace the geometry and project a copy of this to your sketch. This allows you to utilize the existing geometry present in other features (see *Figure 1.20*). This results in fewer dimensions and constraints being required to build the sketch.

8.   Then, using the **Line** command, create a line that is joined from the outer profile and inner profile, as shown in *Figure 1.21*:

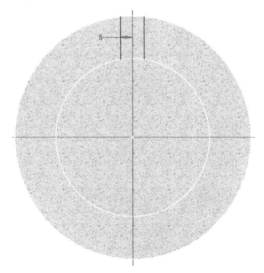

Figure 1.21: The next sketch lines to apply

Then, using **Construction Line**, create a construction line through the center of the part. This is shown by the dotted line in *Figure 1.21*. After that, using **Line**, apply a line from the outer profile to the inner profile. Select **Dimension** and apply a 5 mm dimension between the line that you created previously and the central construction line. Repeat the process for the other side.

9. Select **Finish Sketch** and press *F7* to resume the default view.

10. Select **Extrude** and pick the sketch profile created in *step 8*. Extend this to another face, and select the face, as shown in *Figure 1.22*:

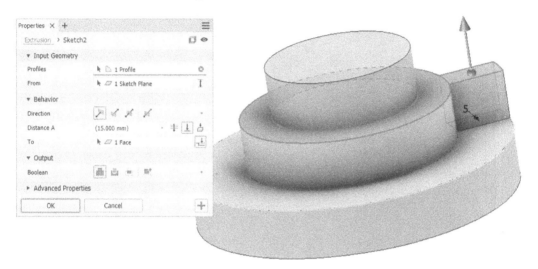

Figure 1.22: Extrusion of newly created lug profile

11. Select **OK** to complete the feature creation.

12. The next step is to create the various hole placements on the model. So, select the top face of the model and select **Start 2D Sketch**.

13. Then, use the **Project Geometry** command to create a center point circle from the sketch origin that is 35 mm in diameter, as shown in *Figure 1.23*:

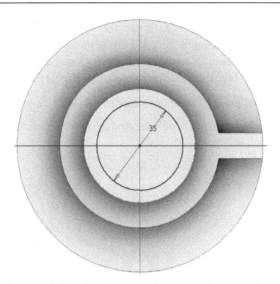

Figure 1.23: The sketch required to create the center bore

14. Select **Extrude** and pick the inner profile of the 35 mm circle. Then, select **Cut**, followed by **Through All**, which creates the central bore of the model:

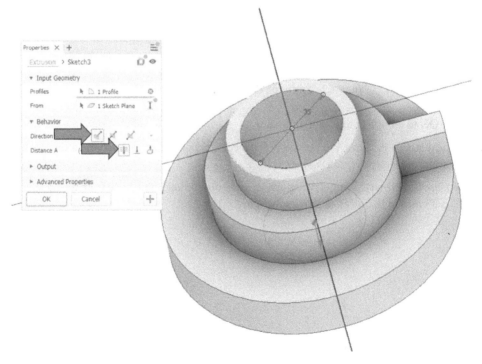

Figure 1.24: Cut operation of the circular sketch to create the bore

15. Create a new sketch on the top face of the lug previously created. Here, we will define the hole placement.

16. Use the **Project Geometry** command to project the geometry of the face and apply construction lines as shown in *Figure 1.25* The construction lines snapped to mid-points will create an intersection where a sketch point can be placed to locate a hole using the **Hole** feature.

17. Add the sketch point to the intersection of the construction lines:

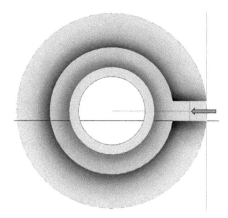

Figure 1.25: Sketch point at the intersection of a sketch curve

18. Create a sketch on the **XY** work plane. If the work plane was not in the correct location, one could be created using an axis through the revolved feature, and then creating a plane referencing this axis and a vertex on the model. Fortunately, in this case, the **XY** plane is in the location required, as we have consistently modeled from the origin throughout.

19. Press *F7* to cut the model to improve visibility and create the following sketch. Use **Project Geometry**, and select the geometry shown in *Figure 1.26*:

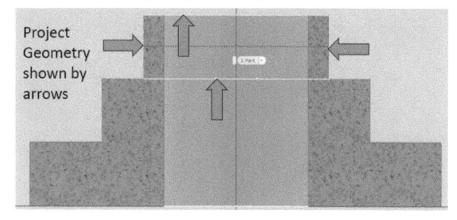

Project Geometry shown by arrows

Figure 1.26: Geometry to project

20. Proceed to create two construction lines that snap to the mid-points of the previously projected geometry. At the intersection of the two construction lines, apply a sketch point to locate the hole, as shown in *Figure 1.27*:

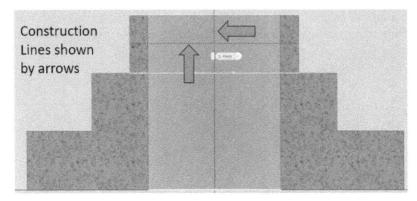

Figure 1.27: F7 cut away in sketch mode required to locate and place the second sketch point

21. Select **Finish Sketch**.

22. Now, we will generate the desired holes. From the **3D Model** tab, in the **Modify** panel, select **Hole** from the ribbon, and select the two sketch points as the center point/locating reference. Only one hole can be created at a time this way, so the same will have to be repeated for the second hole.

23. Apply the settings shown in *Figure 1.28* to detail the hole:

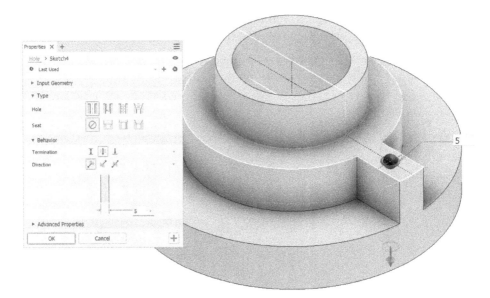

Figure 1.28: Hole command in use, using the sketch point as a reference

You should set the diameter as **5 mm**, the **Hole** type as **Simple Hole**, **Seat** as **None**, and the **Termination** type as **Through All**. Also, ensure the **Direction** setting of the hole is as per *Figure 1.28*.

24. Repeat *steps 22* and *23* for the second hole, but set **Direction** as **Symmetric** instead. Then, select **OK** to complete. The result is shown in *Figure 1.29*:

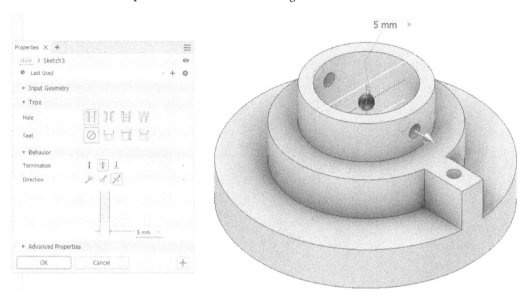

Figure 1.29: Second hole using a second sketch point created as a reference

This allows us to apply two holes in one operation. Exit the **Hole** command.

25. The base features are now created, and we can now focus on creating the circular array of repeating features. Because this has been drawn once, we need to reuse the geometry to be efficient.

26. Begin by selecting the **3D Model** tab, then select **Circular Pattern**.

27. Ensure that the lug extrusion and both holes are selected as features to pattern. For the axis, select the internal cylinder; this will create a work axis within the feature command, from which to revolve **Circular Pattern**. You can also pick the features from the **Model** browser while in the command.

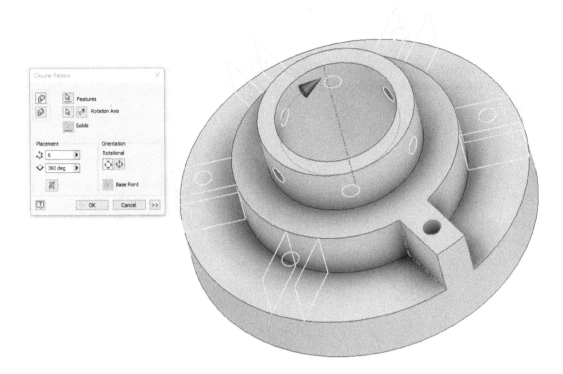

Figure 1.30: Circular Pattern creating the rest of the required geometry

28. Set **Placement** to **6** and ensure the features are patterned 360 degrees. Select **OK**.

29. Now, the fillets on the model must be applied. This could have been done before **Circular Pattern**, but it is best practice to apply fillets after, as this is often quicker and requires less processing time. All the fillets can still be added in one operation.

30. Select **Fillet** from the **3D model** tab, and select the edges shown in *Figure 1.31*. In the **Fillet** command, set the radius of the fillet to **2.5** mm, and for the edges, select **12 Edges**, as shown in *Figure 1.31*:

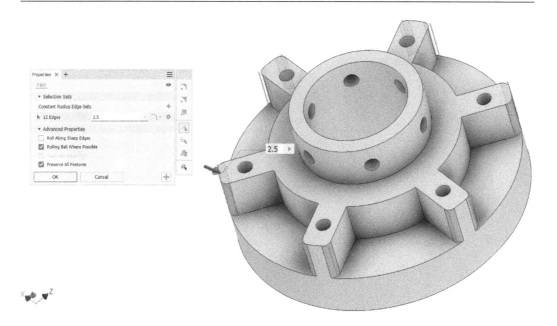

Figure 1.31: First fillet operation and required filleted edges

---

Deselecting a fillet

If you accidentally select the wrong edge and apply a fillet in the **Fillet** command, hold *Shift* and select the incorrect edge to remove the fillet. This keeps you in the command and prevents a restart of the operation.

---

31. *Do not* select **OK** to complete. Instead, hit the green plus icon on the menu to detail another type of fillet on the model. This process can be repeated and enables you to generate many types of fillets in one operation, keeping the **Model** browser and history of features clean.

32. Select the edges shown in *Figure 1.32*, and set the fillet radius to **3** mm:

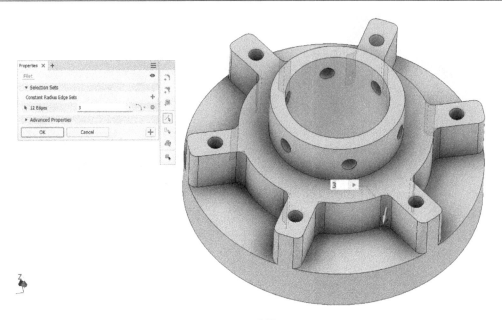

Figure 1.32: Second fillet operation

33. Finally, apply a **1** mm fillet to this edge, as shown in *Figure 1.33*, and select **OK** to complete the operation:

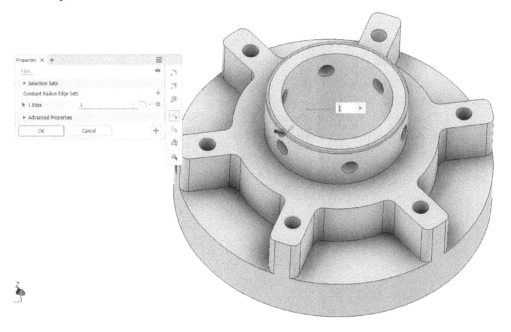

Figure 1.33: Third fillet operation

34. The underside of the model now needs to be completed. While this looks complex, it is actually achieved in one simple **Shell** command. Orbit and rotate the model so that the underside is visible.

35. Navigate to the **3D model** tab in the **Modify** panel, and select the **Shell** command.

36. The model will now preview a shell, select **Remove Face** from the **Shell** command and select the bottom face of the model. This will remove the face, as shown in *Figure 1.34*.

37. Proceed to then change **Thickness** to **1.5** mm.

38. Hit **OK** to complete the operation and the model:

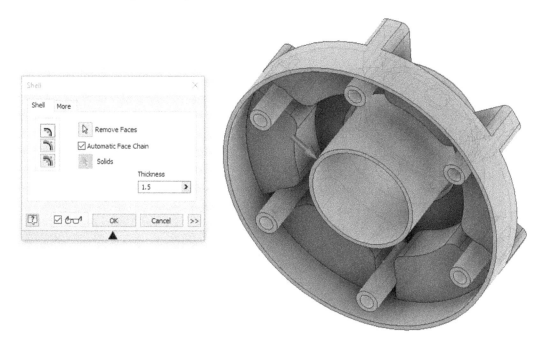

Figure 1.34: Shell command active with bottom face removed

You have now completed the model and used many of the best practices of sketches and work features along the way. By breaking down the tasks into manageable sections, and reusing existing features/geometry, you can create complex shapes with ease.

2

Advanced Design Methodologies and Strategies

Creating parts and assemblies in Inventor can be achieved in a variety of ways. There are multiple strategies that can be employed and combined to create the desired output; all are valid ways of modeling within Inventor, and each strategy has its own advantages and disadvantages. In this chapter, you will learn the differences between these strategies and how to implement them effectively, across a range of scenarios. You will also learn the advantages of each and how and why each strategy should be applied.

Having practical knowledge of all strategies allows you as a designer to make more informed planning decisions when embarking on the design of a product or project within Inventor. The right application of a strategy to a scenario can bring about significant time savings in terms of the amount of rework or updates required that usually must take place as part of an iterative design activity.

In this chapter, we will cover the following topics:

- Understanding the different design processes
- Creating a multi-body part design
- Creating a layout design from imported .dwg CAD data
- Applying and using adaptive modeling

Technical requirements

You will need access to the practice files located in the Chapter 2 folder, within the Inventor Cookbook 2023 folder, to complete the recipes in this chapter.

Understanding the different design processes

Within Inventor, there is a variety of modeling methodologies that can be used and combined. In this section, we will examine each type and how they can be used.

Bottom-up design

The **bottom-up design process** is the traditional way of modeling and creating an assembly within a CAD environment. It is a part-centric process, whereby each part's geometry is created separately in its own independent .ipt part file and then combined into a single top-level assembly .iam file. Independent parts are combined with Mates constraints in the final assembly to create the desired outcome of the final top-level assembly.

The advantages of this methodology are that each part has its own independent file, and therefore its own file structure, which is much easier to understand and organize. Parts can also be updated and changed without influencing related parts in the referenced assembly.

The disadvantages of this structure are that the level of rework can be much higher. Because part files are not linked through geometry, when part A changes in an assembly, the other parts that it may interact with, such as part B or C, will not automatically update. This means that the designer now must spend time manually updating and correcting parts B and C in the assembly to ensure the desired interface and fit between these components is achieved. Depending on the level of change and complexity of the design, this can take a significant amount of time.

Problems can also arise when separate part files are first combined into the assembly. Often, this is due to a lack of engineering information communication, with problems such as misalignment, interference, or an incomplete design.

In *Figure 2.1*, you can see an example of the structure of a bottom-up design methodology:

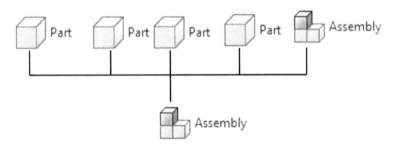

Figure 2.1: Diagram of bottom-up design

Multi-body design

The **multi-body design** methodology places critical design information within a top-level assembly and communicates this information to lower levels of the product structure.

Multi-body design begins with the creation of a single part or solid as a reference model. Additional solids are then created inside the same part file and exported as parts into an assembly environment. Any changes to this methodology, to the original part, are reflected in the linked parts and assembly, resulting in a true parametric methodology.

The advantages of multi-body design are that all the design resides in a single .ipt file, with the solid bodies extracted later to create parts and an assembly. The reliance on Mates and onstraints in the assembly environment is reduced as parts are created in place and *on location*.

With this method, there is usually less reworking required as the parametric link between solids from the initial part file is maintained, so as design changes are made, the linked solids also update to reflect the changes, resulting in less rework.

Finally, the body's visibility can be controlled as a group instead of at an individual feature level, and relationships between bodies can be created and broken freely.

In *Figure 2.2*, you can see an example of the multi-body design methodology:

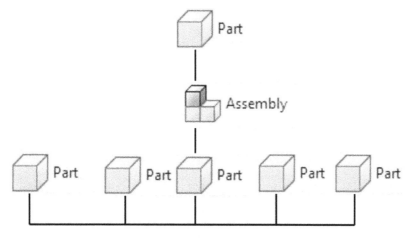

Figure 2.2: Diagram of multi-body design

Layout design

Layout design involves taking an existing arrangement or drawing from a 2D format such as .dwg, importing it into an Inventor sketch environment, and then combining with sketch relationships and feature modeling techniques to derive 3D components from a 2D layout.

The 2D sketch initially contains all the design intent, dimensions, and spatial locations of components within the assembly. Each component is usually broken down into a sketch block, which can be achieved in both Inventor and AutoCAD.

This is widely used in some companies if there is a reliance on the initial design being completed as a 2D layout first, prior to 3D models. It can also be used with other techniques, such as bottom-up or multi-body. An example would be if a supplier of a component could only supply a 2D drawing file, yet the designer required a 3D model to place in the top-level assembly; a 3D variant could be created from the 2D drawing as a reference using the layout design methodology.

With layout design, if the original 2D sketch is updated, these changes are reflected in the 3D model. This means the design can be manipulated from the 2D imported sketch, including the movement and interaction of components.

Figure 2.3 shows the layout design methodology:

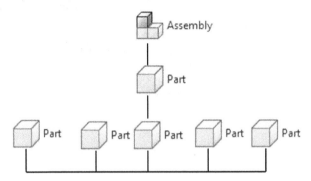

Figure 2.3: Diagram of the layout design process

Adaptive modeling design

Adaptive modeling design uses references from an existing assembly to create parts inside an assembly file, from the projection of existing geometry. Adaptive links are made in this process between sketches and features so that if the parent updates, the child sketch or feature also updates. With an adaptive part, it is intrinsically linked to other parts it shares an adaptive link with.

The advantage of adaptivity is that this link can also be broken if required. It is ideal for determining very specific tolerances in place and on location of where the end part will reside. Because of the adaptive link, the level of rework or updates is further reduced. As parts are created from the assembly environment, the number of mates and constraints needed to position the end components is reduced.

Figure 2.4 shows the adaptive modeling design methodology:

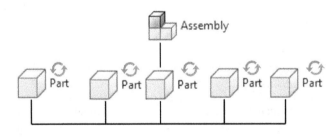

Figure 2.4: Diagram of the adaptive design process

In the next part of the chapter, we will apply this knowledge within Inventor and work with each design methodology.

Creating a multi-body part design

In this recipe, you will create a small subassembly of parts to complete the design of a mountain bike shock absorber. You will learn how to create multiple bodies, in a model, and add features to the bodies. The completed assembly is shown in *Figure 2.5*, which includes the subcomponent that you will design using the multi-body design process. The rest of the components are supplied as separate solids.

Figure 2.5 shows the completed assembly of the shock absorber containing the part we will create:

Figure 2.5: Full assembly of the shock absorber part we will create

Getting ready

Start to familiarize yourself with *Figure 2.5* as it contains the part we will make (dimensions are not given yet but will be given to you in stages throughout this recipe).

The files that you need for this recipe can be found in the Chapter 2 folder of the Inventor 2023 Cookbook folder.

You can also open Shock_Absorber_Assy.iam as a reference, which can be found in the Chapter 2 folder too (although it is not needed directly for the recipe).

How to do it...

To begin the recipe, we will need to create the first component within the assembly:

1. Start the creation of a new part file using the `Standard Metric (mm)` `.ipt` template. To do this, select **File | New | Metric**. Choose `Standard (mm)` `.ipt` for the template, then click **Create** to complete.

2. Select **Start 2D Sketch** and select the XY plane by expanding the **Origin** folder in the **Model Browser**.

3. From the **Sketch** tab, in the **Create** panel, select **Circle** and create an `18` mm diameter circle that is centered on the origin, as shown here in *Figure 2.6*:

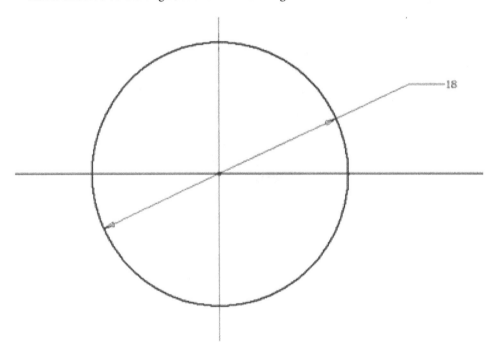

Figure 2.6: The start of the initial sketch

Next, create the rest of the sketch geometry, as shown in *Figure 2.7*. To do this, draw a construction line from the origin of the circle downward by `21` mm. Next, from the end point, create a horizontal `9` mm line from the construction line end point, then a `3` mm vertical line from the end point of the previous line. This is shown in *Figure 2.7*:

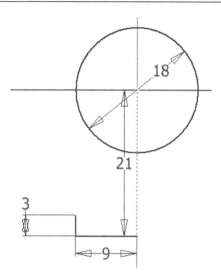

Figure 2.7: The next sketch lines to apply

Ensure the sketch is fully defined before progressing to *step 4*.

4. Next, to create the arc, select the **Three-Point Arc** command. Start at the end point of the line that is 3 mm in length and snap to the edge of the circle. Set the radius of the arc to 15 mm. To fully constrain this arc, you will then need to add a tangent constraint between the arc and the circle, as shown in *Figure 2.8*:

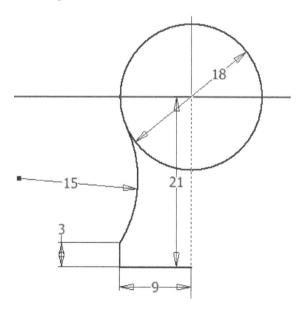

Figure 2.8: 15 mm three-point arc created with a tangency relationship applied between the arc and circle

5. As this is a symmetrical shape, we can use the **Sketch Mirror** command to copy the existing sketch lines about the centerline. This saves time and is the most efficient way to create this sketch. With the **Mirror** command, mirror all geometry, except the circle, using the center construction line as the mirror line.

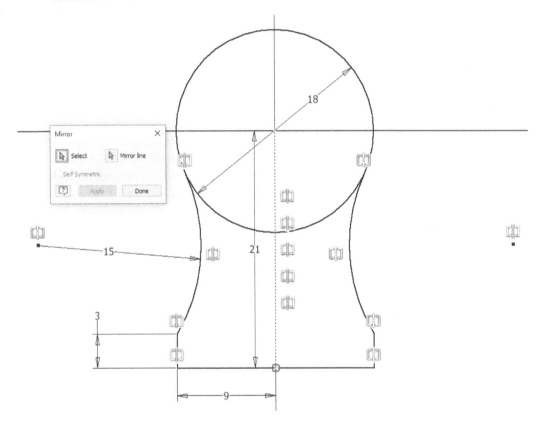

Figure 2.9: The sketch with the mirror applied

6. Complete the sketch by selecting **Finish Sketch**. In the **3D Model** tab, select **Extrude** and select the two closed profiles of the sketch. Enter a value of 14 mm for **Distance A**. Then, ensure **Direction** is set to **Symmetric**, then select **OK**.

    In this case, with 14 mm set as the overall extrude length, this extends the extrusion both ways by 7 mm in each direction.

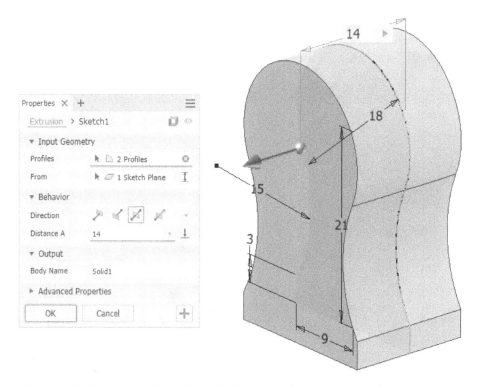

Figure 2.10: Extrusion of the profile applied at 14 mm from the center in both directions

7. We will now create additional features on the model as part of the same solid. Select **Start 2D Sketch** and create a new sketch on the face shown with the following geometry. Select **Project Geometry** and proceed to select the face of the part, before creating the circles, as per *Figure 2.11*:

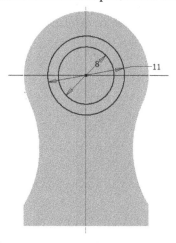

Figure 2.11: The sketch geometry required to create the next features of the part

8. Use the **Extrude** tool on the outer circle to extrude it by 5 mm.

9. We will now cut the center of the circle. Select **Extrude**, then **Cut**, and change **Distance A** to **Through All**. Select **OK**.

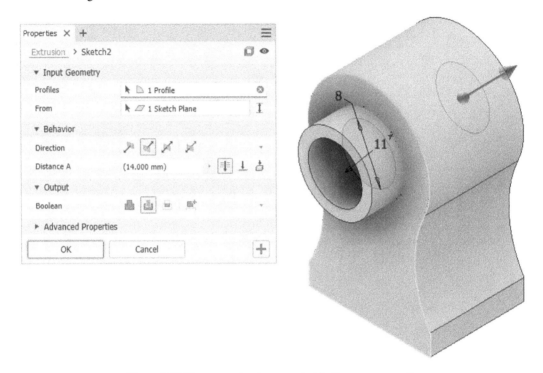

Figure 2.12: The two extrusions created in the same operation

Using the XY plane, in the **3D Model** tab, select **Mirror** and mirror the protruding extrusion from the **Model Browser**. This replicates the extrusion we created previously, but on the other side of the part.

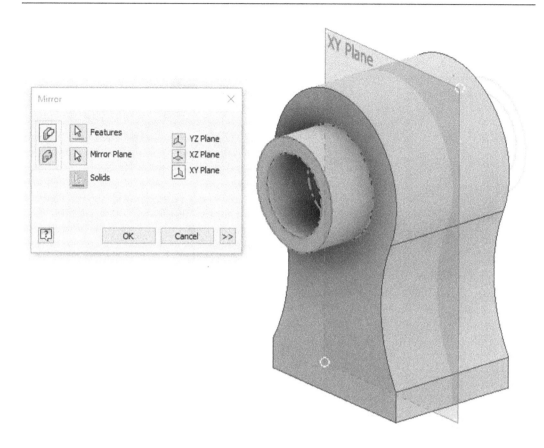

Figure 2.13: Mirror operation in action showing the plane and copied feature geometry

10. Select **Start 2D Sketch** and select the base of the model as the sketch plane. Select **Circle** and create a 10 mm diameter circle central to the base of the model. Once complete, select **Finish Sketch** and then extrude the circle by 50 mm, in the direction away from the model. This is shown in *Figure 2.14*:

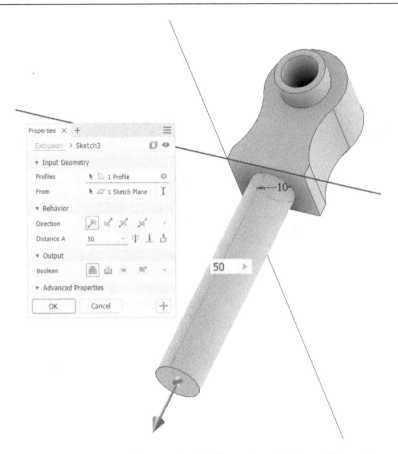

Figure 2.14: 10 mm circle extruded by 50 mm from the base of the part

The first solid body is now complete with additional features added. In the **Model Browser**, you will see the **Solid Bodies** folder now shows **(1)**. The additional solids will now be created within the part file.

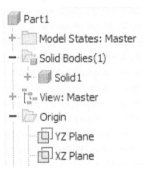

Figure 2.15: Solid Bodies folder of the Model Browser expanded, showing one solid body in the part file

11. Next, select **Start 2D Sketch** and create a sketch of the face shown in *Figure 2.16*. The circle must be 18 mm in diameter.

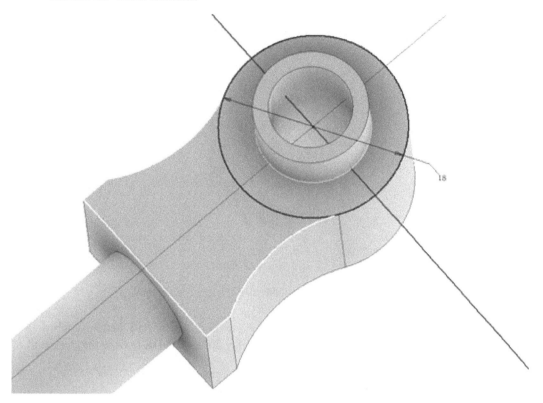

Figure 2.16: Sketch of the circle that must be created, 18 mm in diameter

12. This sketch now becomes the basis for the second solid we will create in this part file. Select **Extrude** and then select the sketch profile created previously in *step 11*. Set **Extrusion Type** as **To**. Select the face of the internal cylinder of **Solid1** as the face to extrude toward.

Crucially, at this stage, open the **Output** area of the **Extrude** menu and set **Boolean** as **New Solid**. This will create the feature as a separate and independent solid from **Solid1**. Select **OK** to complete this.

A second solid has now been created.

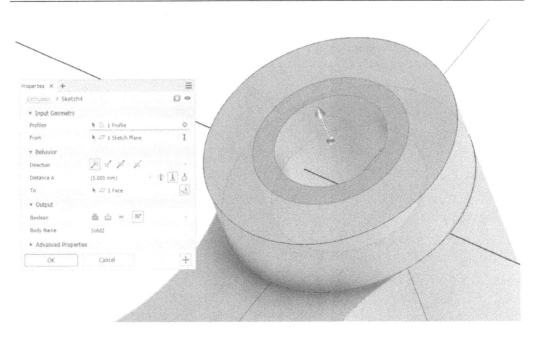

Figure 2.17: Creation of the new solid using the Extrusion command

13. As a new solid has now been created, we will begin to modify it. Select **Fillet** and apply a .5 mm fillet to the edge of **Solid2**, as shown here in *Figure 2.18*. Select **OK** to close the command.

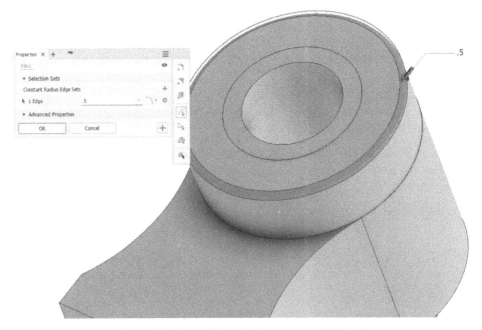

Figure 2.18: .5mm fillet added to the new Solid2 edge

14. A final modification we will make to **Solid2**, independent of **Solid1**, is a property change. In the **Model Browser**, expand the **Solid Bodies** folder, right-click on **Solid2**, and select **Properties**.

15. In the **Body Appearance** drop-down list, select **Smooth Ivory**, followed by **OK**. We have now independently applied an appearance change to **Solid2**.

A second instance of **Solid2** is required on the other side of **Solid1**; rather than creating or modeling this again, we can mirror **Solid2** along the XY plane to create a second identical instance of **Solid2**.

16. To do this, navigate to the **3D Model** panel and find the **Pattern** commands, and select **Mirror**. In the **Mirror** command, rather than using the default **Mirror** feature, select **Mirror Solid**. Then, select **Solid2** from the **Graphics Window**. For **Plane**, use the XY plane. Make sure to also select **New Solid** in the command as the default will be **Join**. Select **OK** to complete. This menu and settings are shown in the top left of the menu in *Figure 2.19*:

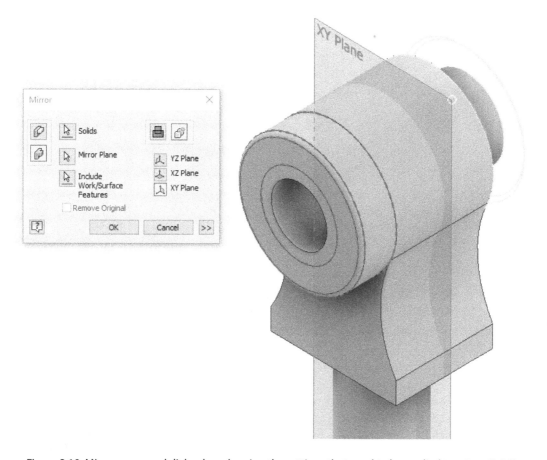

Figure 2.19: Mirror command dialog box showing the settings that need to be applied to mirror Solid2

17. **Solid3** has been created. The appearance has not transferred from the mirror as this is controlled by the solid properties. In the properties of **Solid3**, change the appearance of **Solid3** to **Smooth Ivory** (as you did for **Solid2** previously).

18. To show how all three solids can be edited together, we will now create a small 1 mm hole that goes through all three solids created. Create a new sketch on the face of **Solid2**, as shown in *Figure 2.20*. Add a sketch point where the arrow is shown and then exit the sketch.

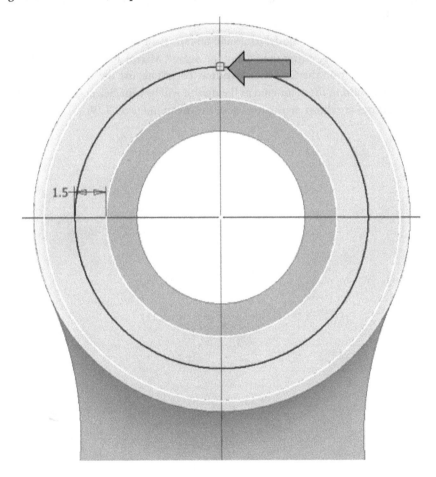

Figure 2.20: Sketch to be created on Solid2 with an arrow indicating the location of the sketch point

19. Select the **Hole** command and pick the sketch point previously placed as the location. Set the **Hole** type as **Simple**, the **Seat** type as **None**, and the **Termination** type as **Through All** with a 1 mm diameter. In the **Solids** area, select all three solids to enable the hole to pass through all three independent solids created. This shows how you can perform feature edits in multi-body design, on multiple solids at once if required. Select **OK** to complete the hole.

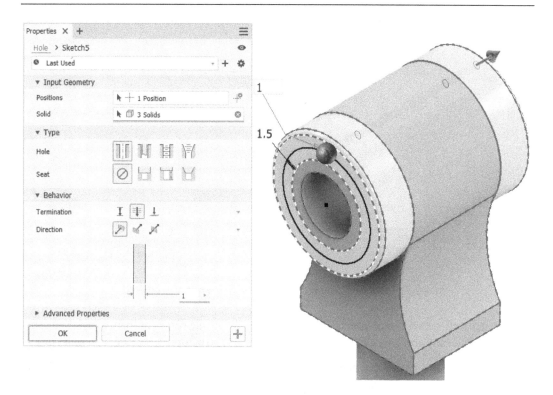

Figure 2.21: Hole command with settings applied in step 19, showing
hole going through all three independent solids

20. Now, you will extract the solid bodies created to create three independent part files; this is the final stage of multi-body part design. To start, select the **Manage** tab, then **Layout** panel. Select **Make Components**. Inventor will prompt you to save the file. Save the file as Part1.ipt in the Shock Absorber folder, within the Chapter 2 folder.

21. Select the three solid bodies from the **Model Browser** to be included in the selection and uncheck **Insert components in target assembly** (this is not required in this case, but would enable you to also build an assembly of solid body parts at the same time). Then, select **Next**.

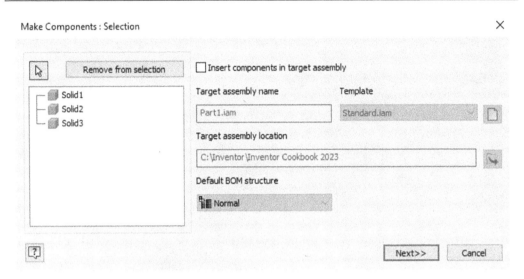

Figure 2.22: Make Components dialog box with the options required

22. You can now rename components and choose the template file you want them to use. You can also include any previous parameters and edit the **BOM** structure.

Select **Solid 1** and change the name to SA_MALE_1.ipt. Repeat this for the two other solids, as shown in *Figure 2.23*, with sequential numbers, in the **Component Name** column.

Then, select the **Page** button above the **Template** column. Navigate to the **Metric** tab and select the Standard (mm) .ipt template to use. Then, click **OK**.

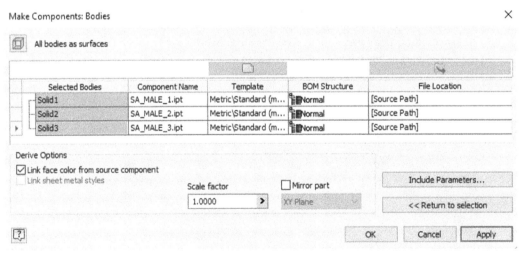

Figure 2.23: Make Components: Bodies dialog box with component
names changed and templates redefined

23. Inventor will now create the solids as separate `.ipt` files. To review the results, click **OPEN** and navigate back to the `Inventor 2023 Cookbook` folder, where the solid parts will be located. This is shown in *Figure 2.24*:

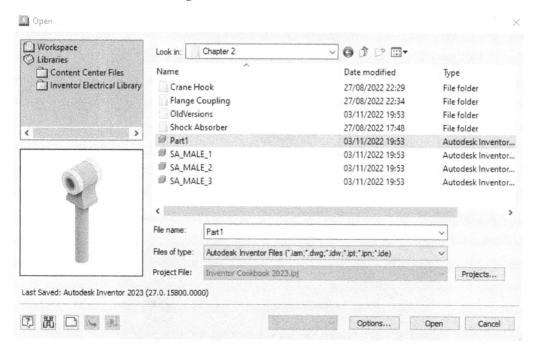

Figure 2.24: Newly created separate .ipt part files from the solids initially created

In this recipe, you have successfully created three independent solids as part of a multi-body design workflow to create a subassembly of parts for a shock absorber. You have then exported each solid as a separate part file. The parent multi-body part file is associative with the exported independent parts, which means that if we change the parent part, the child (exported) parts also change.

Creating a layout design from imported .dwg CAD data

In this recipe, you will create an assembly of part of a crane hook from an existing `.dwg` file using the top-down methodology of layout design. Any changes made to the sketch will be reflected in the 3D model. The steps to create this are, firstly, to create the layout itself in 2D, import the 2D layout into Inventor, define the sketch blocks, and, finally, make the parts and components using the feature commands in Inventor.

You can see the final assembly of the completed crane hook in *Figure 2.25*:

Figure 2.25: The final assembly of the crane hook you will create in this recipe

The original 2D AutoCAD drawing from which this was created is shown in *Figure 2.26*:

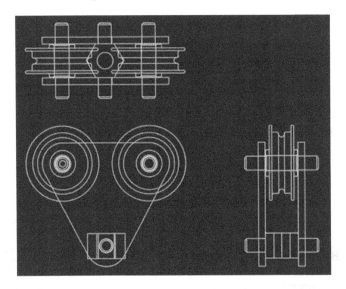

Figure 2.26: The original 2D layout shown in AutoCAD that we will create the 3D assembly from

Getting ready

Navigate to the Chapter 2 folder in Inventor Cookbook 2023 and open the Crane Hook folder. Then, select the 459-25.iam assembly file to look at the completed design in 3D.

The 2D drawing we will create this from can be found in the Crane Hook folder, filename: Crane_Hook.dwg. *Figure 2.26* shows the original 2D layout we will use.

To begin, you will need to open a new `Metric Standard (mm).ipt` file. Have this open and ready before starting the recipe. The original drawing is already drawn to scale 1:1 and does not need dimensional changes when creating the 3D parts; this is simply for reference. The reason for using this methodology is that quite often, at the concept stage, designers create an overall 2D layout in a 2D CAD package, such as AutoCAD, and then require a 3D design for detailed design and manufacturing drawings. This is not the only way to use .dwg files within Inventor.

In this recipe, the original drawing is supplied as a 2D AutoCAD `Crane_Hook.dwg` file. *You do not need AutoCAD to complete this recipe; only Inventor will be used.*

How to do it...

To start, we will need to open Inventor and import the original 2D CAD data:

1. Create a new `Metric Standard (mm).ipt` file.

2. Then, create a new sketch on the XY plane. This will become the plane to which we will import the existing .dwg 2D drawing of the layout, which will form the basis of the assembly. Because the parts are already drawn to scale and in place, minimal constraining and dimensional input are required, as this has already been completed in the AutoCAD `Crane_Hook.dwg` drawing file.

3. With the new sketch active, navigate to the ribbon. In the blank area after **Finish Sketch** (shown in *Figure 2.27*), right-click so that a menu pops up. Then, select **Show Panels**, then **Layout**, so that a *tick* is applied. This loads the **Layout** commands to the ribbon, which will be required for the next steps in this recipe.

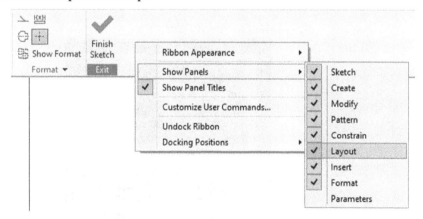

Figure 2.27: Layout panel added to the ribbon

4. Next, navigate to the ribbon. In the **Insert** tab, select the **ACAD** button.

5. You will now import the original 2D AutoCAD drawing into Inventor so that you can use the existing geometry and create the model. Select **ACAD**, then browse to the `Crane Hook` folder and select the `Crane_Hook.dwg` file. Then, click **Open**.

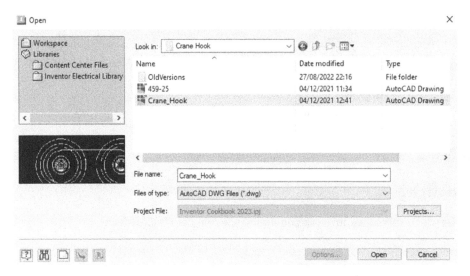

Figure 2.28: Crane_Hook.dwg location shown

6. The **Layers and Objects Import Options** tab has now opened, which allows you to include or exclude certain AutoCAD geometry from the .dwg file for import. Untick the **All** box and click the selection arrow. Then, select the geometry highlighted in *Figure 2.28*. Now, you can click **Finish**. This will import only this highlighted geometry from the model. Then, select **Next**, followed by **Finish**.

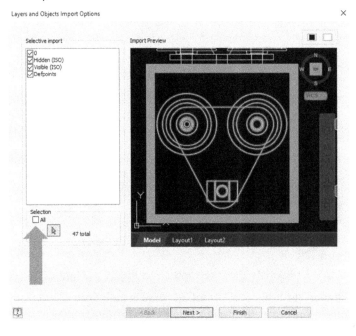

Figure 2.29: Selected geometry we require

Now, the selected geometry will appear in the sketch.

To move the imported geometry to the correct position, we first need to create this as a sketch block. Select **Create Block** from the recently added **Layout** tools in the ribbon.

7. Select all the geometry that was imported and then select the insertion point (*Figure 2.30*) as the center of the bottom circle. Click **OK** to complete the creation of the sketch block. You have now created a sketch block that can be constrained.

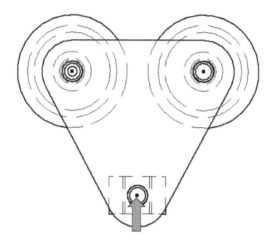

Figure 2.30: Insertion point for the sketch block

8. Next, use **Coincident Constrain** to constrain the center of the insertion point of the sketch block to the origin of the file, then select **Finish Sketch**.

Figure 2.31 shows the newly created sketch block constrained to the origin:

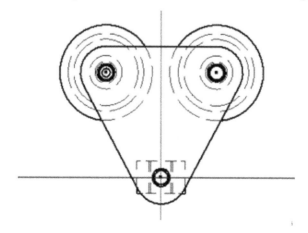

Figure 2.31: Sketch block of imported geometry constrained to origin

9. Turn on the visibility of the YZ plane by right-clicking the YZ plane in the **Model Browser** and selecting **Visibility**; this turns the visibility of the plane on.

10. Select **Plane** and then select the YZ plane as the first reference. For the second reference, select the center of the top-left circle of the sketch block, as shown in *Figure 2.32*:

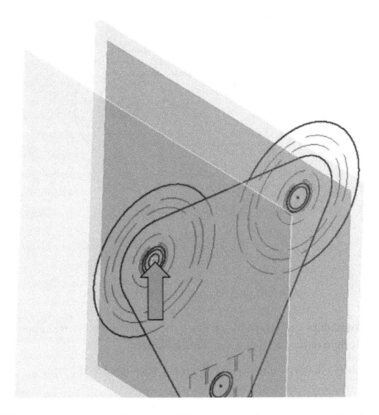

Figure 2.32: Plane creation using the existing YZ plane and reference point on the sketch block

11. The newly created **Workplane 1** will act as the reference plane for the next sketch and the next piece of imported geometry. Select **Start 2D Sketch** on **Workplane 1**.

12. Select the **ACAD** import button and import the Crane_Hook.dwg file. This time, select only the geometry shown in *Figure 2.33* and click **Next** followed by **Finish** to import the geometry into the sketch:

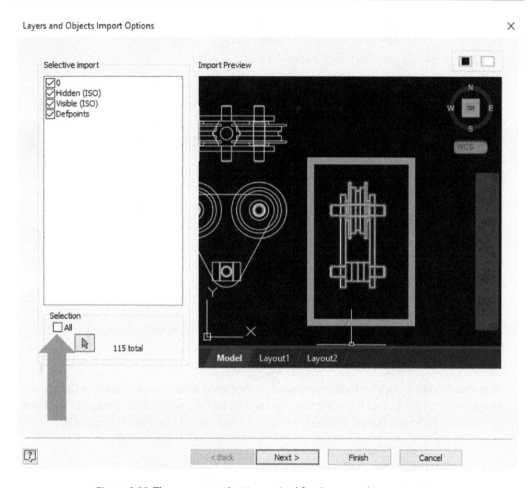

Figure 2.33: The geometry that is required for the second import option

13. Create a sketch block of the newly imported geometry with the same process previously shown in *steps 7* and *8*.

14. Upon completion of these two sketch blocks, the **Blocks** folder in the **Model Browser** should show two sketch blocks: **Block1** and **Block2**.

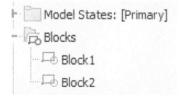

Figure 2.34: Block1 and Block2 in the Model Browser

15. Edit **Sketch2** from the **Model Browser**, that contains **Block2**, select the **Sketch** tab, and select **Rotate** from the **Modify** tools to rotate the sketch block to the orientation shown in *Figure 2.35*. Ensure your sketch matches that of *Figure 2.35*:

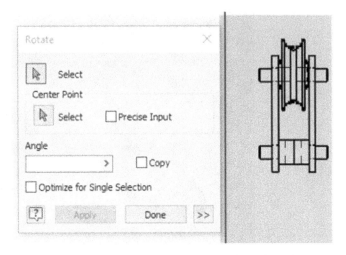

Figure 2.35: Rotate used to rotate the sketch block in sketch mode to the orientation shown

16. Save the file as `Crane_Hook_Layout.ipt`.

17. Now that the sketch block has been created and is in the correct orientation, it needs to be constrained in context with the first sketch block that was created. Edit the sketch that contains **Block2** that you rotated. Select the **Project Geometry** command and project the point shown in *Figure 2.36*:

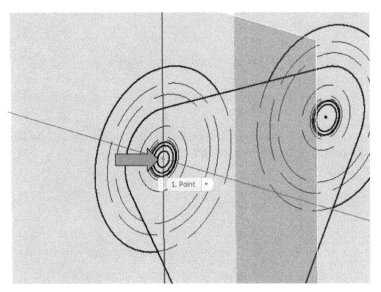

Figure 2.36: Point to project for the second imported sketch

18. In the same sketch, select the coincident constraint. Pick the mid-point shown in *Figure 2.37* as the first reference:

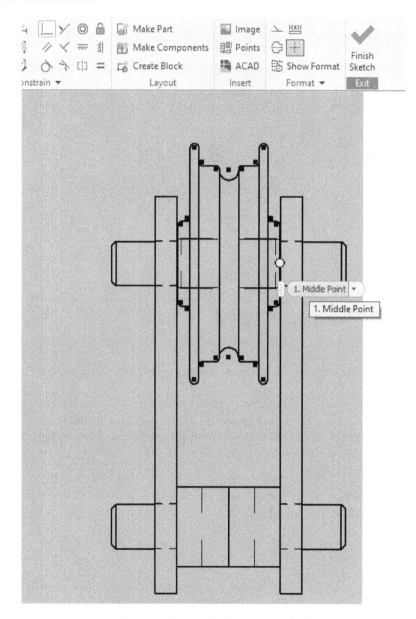

Figure 2.37: The first reference to select

19. For the second reference, select the projected point created in *step 17*. The sketch block should then be constrained in the position shown in *Figure 2.38*:

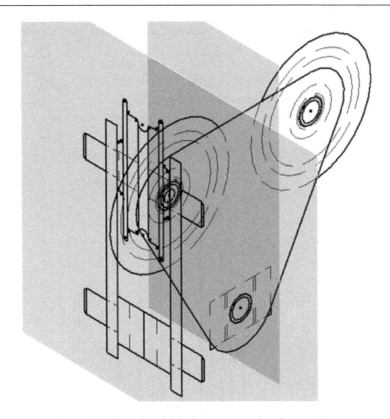

Figure 2.38: Two sketch blocks constrained and in position

Now that the sketch blocks are in position, we will explode them. This is to separate the individual sketch elements so that individual components can be made from them. The reason they were converted to sketch blocks was to enable us to constrain them first to the origin and then to each other in the correct location. Doing this allows us to utilize all aspects of the 2D views of the crane hook provided.

20. In **Model Bowser**, expand the **Sketch2** node and right-click on **Block2:1**, then select **Explode**. Then, select **Finish Sketch** from the ribbon.

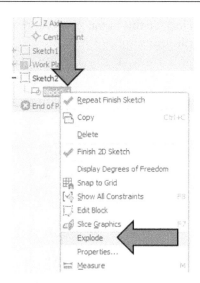

Figure 2.39: Sketch2 expanded and Block2:1 exploded

21. In the **Model Browser**, click on the **Block1** name and rename it to CH_Plate.

22. Select **Edit Sketch** for **Sketch2**. Select **Create Block**, then select the geometry shown in *Figure 2.40*. Set the name of the block to CH_Pulley. Click **OK**.

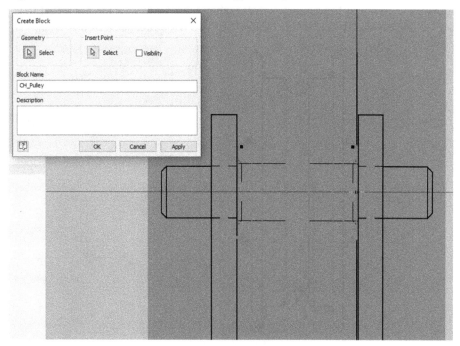

Figure 2.40: CH_Pulley block created

23. Create another new sketch block; in this instance, select the geometry shown in *Figure 2.41* and set the name as CH_HookHousing:

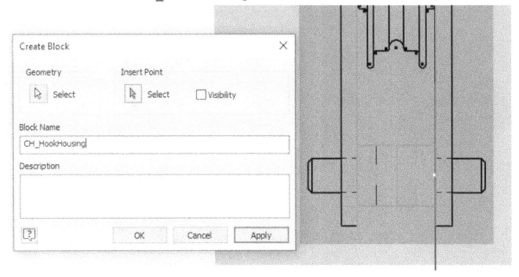

Figure 2.41: CH_HookHousing sketch block created

For the final sketch block for the shaft, edit **Sketch2** and draw additional lines with the **Line** command to complete the geometry. This is shown in *Figure 2.42*. Do not forget to apply a horizontal line across the centerline of the shaft. This is required for a revolve later.

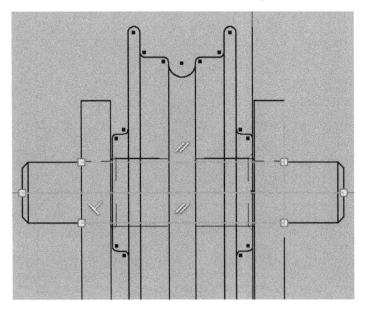

Figure 2.42: Additional two sketch lines created on Sketch2

24. Create a third sketch block, as shown in *Figure 2.43*:

Figure 2.43: Third sketch block to be created from Sketch2

25. All sketch blocks required have now been created. These sketch blocks now need to be exported as components. Exit all sketches in the part file and select **Make Components** from the **Manage** tab. If you have not saved the file yet, you will be prompted to do so. Save the file in the Chapter 2 | Crane Hook folder.

26. Expand the **Blocks** node in the **Model Browser** and select all created sketch blocks (apart from **Block 2**) to be created as new components, then select **Next**.

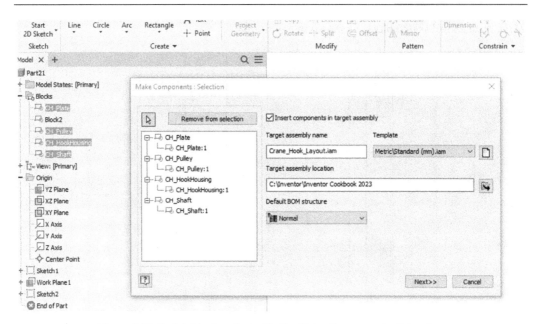

Figure 2.44: Sketch blocks selected in the Make Components dialog box

Note that the target assembly name is the same as the name of the original part file, but with the .iam extension. This is the assembly file Inventor will place the newly created components within. Under **Template**, select the page icon button. Then, navigate to the **Metric** tab and select Standard (mm) .iam, as shown in *Figure 2.45*. Then, select **OK**. Select **Next**.

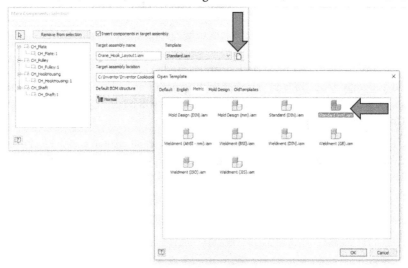

Figure 2.45: Template for the assembly

27. Select the first sketch block, **CH_Plate**, from the list. Select the page icon above **Template**.

28. Then, navigate to the **Metric** tab and select the `Standard (mm) .ipt` template for each sketch block, as shown in *Figure 2.46*:

Figure 2.46: Template for the parts

29. Repeat *steps 27 and 28* for each sketch block in the list, until the list resembles *Figure 2.47*. Select **OK** to complete.

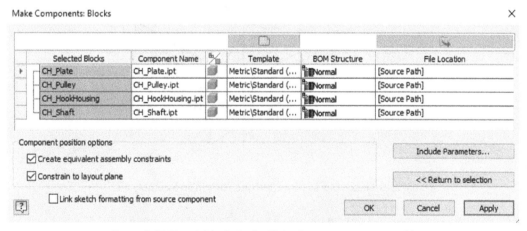

Figure 2.47: Sketch blocks in the Make Components menu, with templates configured to Standard (mm) .ipt

---

Tip – templates

Many of the manual operations in the selecting of templates in these functions can be automatically set once a part file or assembly file template has been created.

30. The newly created components now open in a new assembly `.iam` file named `Crane_Hook_Layout.iam`. Note how in the **Model Browser**, each sketch block has now been created as a component or part within the assembly. The sketch layout geometry is still present in the **Graphics Window**, and we can now use this to add 3D geometry to the individual components.

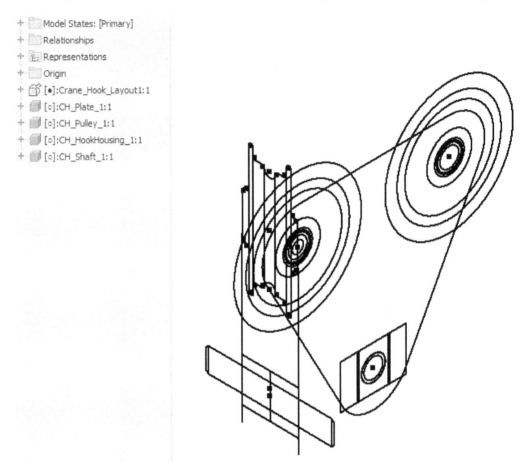

Figure 2.48: Sketch blocks converted into parts in an assembly file

31. Move the cursor over the **CH_Plate** component, right-click, then select **Edit**.

32. Select **Extrude** and pick the profiles of **CH_Plate** to create the end part. Ensure you do not pick the holes for the shafts. (If you do accidentally fill them in, they can be added later as an **Extruded Cut** operation.) Leave **Direction** as the default and set the depth of the extrusion (**Distance A**) to 31.75 mm. This is shown in *Figure 2.49*. Then, select **OK**.

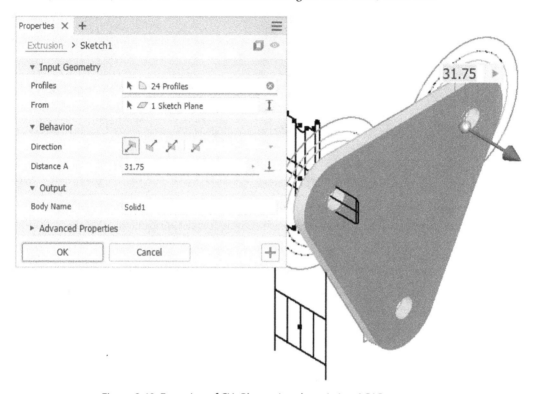

Figure 2.49: Extrusion of CH_Plate using the existing ACAD geometry

33. Select **Return** from the ribbon to return to the assembly and navigate to the other side of the newly created extrusion feature. Under the **Plane** command, select **Offset from Plane** and create a new offset plane from the XY plane at the top level of the assembly, which is a distance of 76.2 mm from it, as shown in *Figure 2.50*:

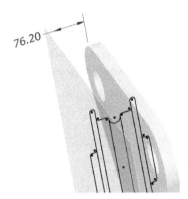

Figure 2.50: The placement of an additional work plane

34. In the **Assemble** tab, select **Mirror** and mirror the newly extruded **CH_Plate** component using the recently created **Offset Work Plane 1** as the mirror plane.

The correct settings to apply in the **Mirror Components** menu are shown in *Figure 2.51*. In the **Mirror** command, select the green icon to activate **Reuse Component**.

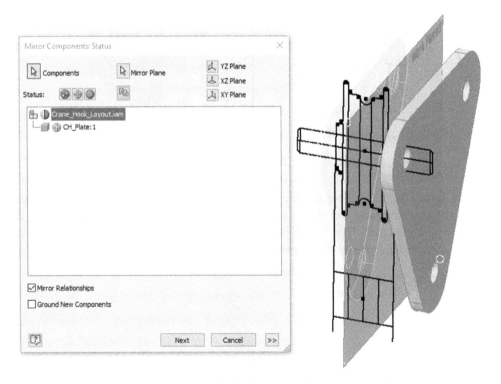

Figure 2.51: Settings applied in the Mirror Component window
to apply and the new component in preview

35. Select **Next** and then **OK**. You have created the first components from the original AutoCAD data that was imported.

36. Next, right-click **CH_Pulley** from the **Model Browser** and then select **Edit**. Then, right-click on **Sketch1** in the **Model Browser** and select **Edit Sketch** on **Sketch1**. Draw the additional sketch line, as shown in *Figure 2.52*. This will act as an axis to revolve the new component.

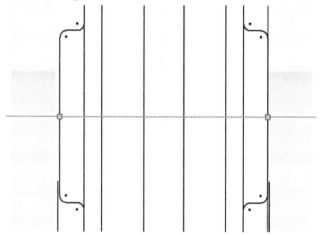

Figure 2.52: Line applied to the sketch to act as an axis for a revolve operation

37. Select **Finish Sketch** and then **Revolve** from the **3D Model** tab.

38. Select the geometry shown in *Figure 2.53* for the profile, and then pick the axis as the newly created sketch line. Then, hit **OK**.

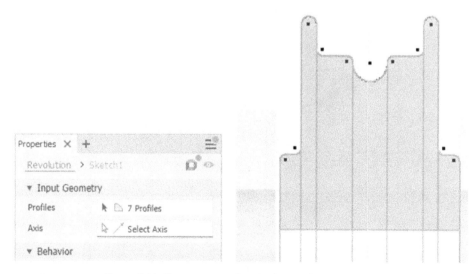

Figure 2.53: Geometry to select in the revolve operation

The resulting feature should resemble *Figure 2.54*:

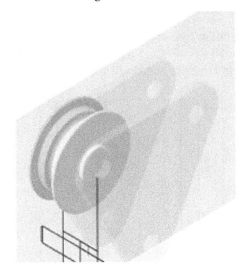

Figure 2.54: Completed revolve operation shown

39. Select **Start 2D Sketch** and select the face shown in *Figure 2.55*. Proceed to select **Project Geometry** and select the geometry shown in *Figure 2.55* from CH_Plate. Select **Finish Sketch**.

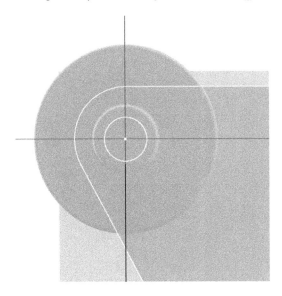

Figure 2.55: Geometry to project

40. Select **Extrude** to create a hole from the circle in *Figure 2.55*. *Figure 2.56* shows the extruded cut and the settings to apply. Select **OK** to complete.

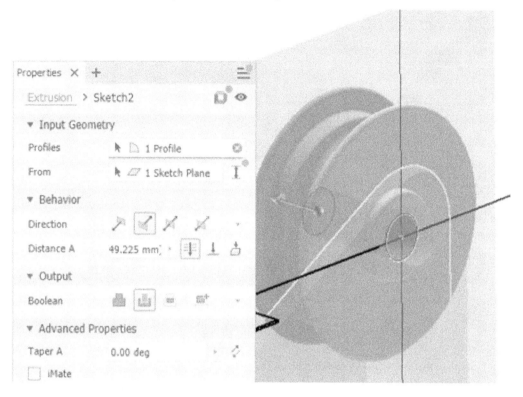

Figure 2.56: Extruded cut deployed on the newly created feature, using projected geometry from CH_Plate

41. Hit **Return** in the ribbon to return to the top level of the assembly after completion.

42. The **CH_Pulley** component must now be mirrored using the same methodology as **CH_Plate**. To start, navigate to the top level of the assembly and mark the YZ plane as **Visible**.

43. Select **Mirror** from the **Assemble** tab and select **CH_Pulley** as the component and the YZ plane as the mirror plane. Select **Reuse** for the **Status** option, as shown in *Figure 2.7*:

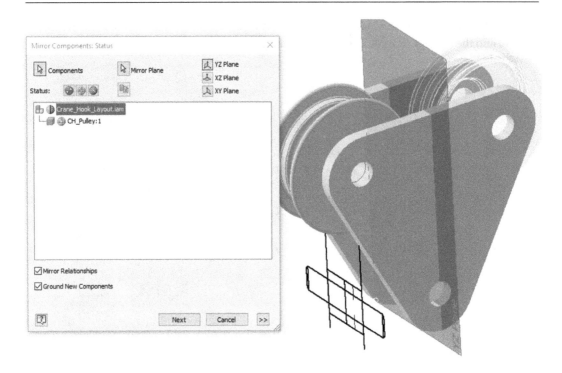

Figure 2.57: Mirror Components menu with a preview of the new CH_Pulley instance

44. Complete the operation by clicking **Next**, then **OK**.

45. For the crane housing component, the imported geometry is not in the correct place, but we can still utilize it. Edit **CH_HookHousing** and in the **Model Browser**, expand the **Origin** folder under **CH_HookHousing**. Select **Start 2D Sketch** on the XY plane of **CH_HookHousing**.

46. Select **Project Geometry** and project the lines shown in *Figure 2.58*. If you cannot project the lines, you can also draw a rectangle over the top; it should snap to the end points, so you know the geometry is correct. The result should be a sketch that is a rectangle.

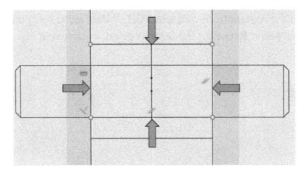

Figure 2.58: Projected lines

47. Select **Finish Sketch** and then click **Extrude**. Pick the profiles shown in *Figure 2.59* and extrude a distance of 500 mm; this will become **CH_HookHousing**.

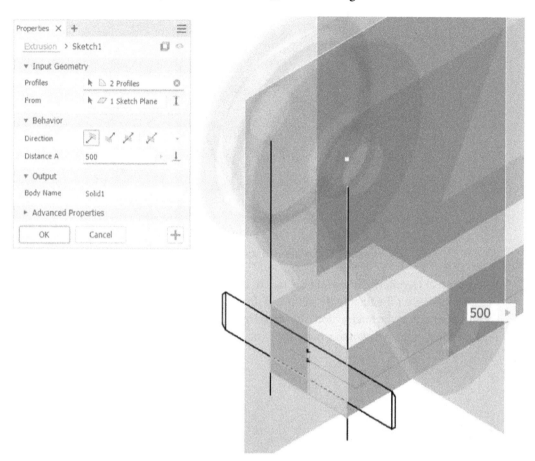

Figure 2.59: 500 mm extrusion to perform

48. Select **OK** to complete.

49. Now, select **Start 2D Sketch** on the underside of the newly created extrusion.

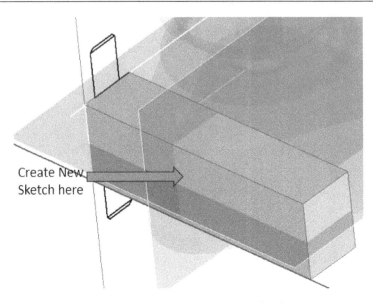

Figure 2.60: Location to start a 2D sketch

Select **Project Geometry** and select the face of the extrusion and the YZ work plane of the assembly file. Select the **Rectangle** command dropdown and select sketch **Polygon**. Create a six-sided polygon at the intersection, shown in *Figure 2.61*. Then, apply a horizontal constraint to the top line of the polygon to straighten it.

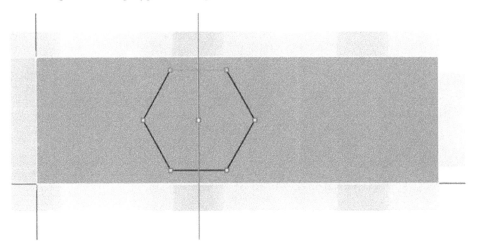

Figure 2.61: Projected sketch lines and polygon

50. To constrain the polygon to the correct size, add the following dimensions to the sketched polygon shown in *Figure 2.62*. Then, sketch a circle from the center point of the polygon with a diameter of 77 mm.

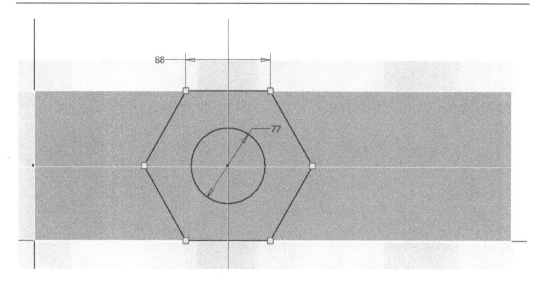

Figure 2.62: Dimensions required and additional sketched circle

51. Select **Finish Sketch**.

52. Select **Extrude**. Then, select **Intersect** for the **Boolean** option.

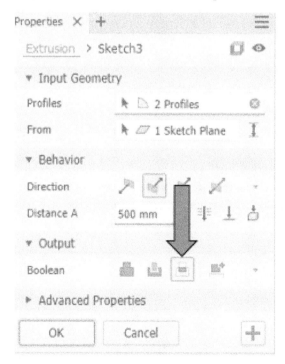

Figure 2.63: Intersect selected in Extrude

Select the two profiles of the polygon, as shown in *Figure 2.64*, with the **Through All** option selected. This will maintain the polygon shape and remove all excess material.

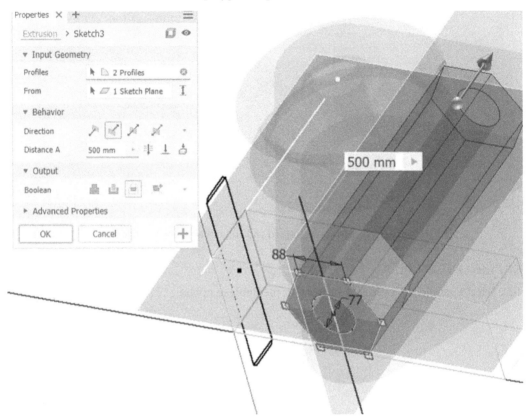

Figure 2.64: Extrude menu shown with profiles selected for the intersect operation

53. Complete the operation with **OK**.

54. Select **Start 2D Sketch**, on the face shown in *Figure 2.65*, then select **Project Geometry** of **CH_Plates Hole** and select **Extrude** as a **Cut**, selecting **Through All**. **CH_HookHousing** is now complete.

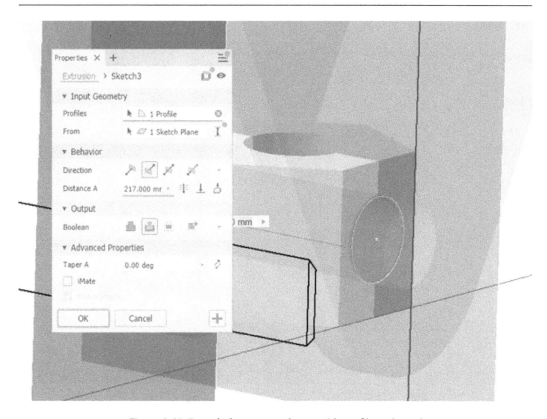

Figure 2.65: Extruded cut menu shown with profiles selected

55. Return to the top level of the assembly. The final stage is to create the **CH_Shaft** component from the original 2D layout.

56. Navigate to the the **Model Browser**, right-click **Crane_Hook_Layout1**, and select **Visibility** to turn the original layout on.

57. Move the cursor over the **CH_Shaft** component in the **Model Browser** and select **Edit**.

58. Select **Revolve**, then select the profile shown in *Figure 2.66*. For the axis, select the centerline of the sketch. The preview should resemble *Figure 2.66*. Select **OK** to complete the revolve operation.

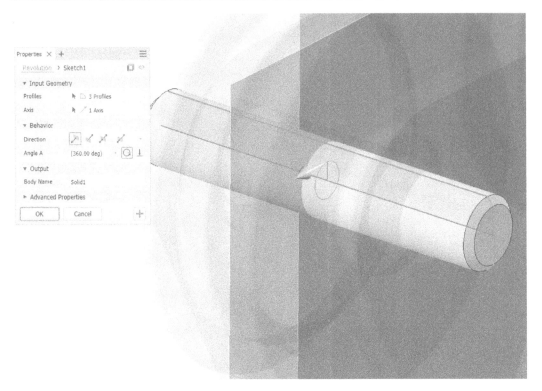

Figure 2.66: Revolve to complete the CH_Shaft component

59. Select **Return** to go back to the top-level assembly.

60. All components have now been created from the initial layout design. The only remaining task is to mirror or copy **CH_Shaft** to the other two locations.

61. *Optional*: Select the **Copy** command and assembly constraints to create the final instance of **CH_Shaft**. The final assembly should resemble *Figure 2.67*.

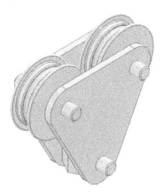

Figure 2.67: Completed assembly of the crane hook with three instances of CH_Shaft applied

You have successfully imported and used an existing 2D AutoCAD sketch to create 3D geometry in Inventor as part of a layout design workflow.

Applying and using adaptive modeling

Adaptive modeling involves creating parts within the context of an assembly. An initial part is created and then placed and grounded within an assembly file. Other parts are created in place from the original part in the assembly file.

In this recipe, you will take an imported flange coupling assembly and create a mating plate and gasket that are adaptively linked. This means that as changes are made in one component, they will automatically be reflected in the linked components, without manual design changes. This is known as an **adaptive link**.

Getting ready

In the Chapter 2 folder, select the Flange Coupling folder. Then, open Flange Coupling. iam.

How to do it...

To begin, we will open the existing assembly. Then, we will create new components within the context of it:

1. With the Flange Coupling.iam assembly open, navigate to the top left of the screen, to the **Assemble** tab, and select **Create**. The menu shown in *Figure 2.68* will appear.

2. Enter Gasket for **New Component Name** and set **Template** to **Metric\ Standard (mm). ipt**. Ensure **New File Location** is set to C:\Inventor\Inventor Cookbook 2023\ Chapter 2\Flange Coupling\.

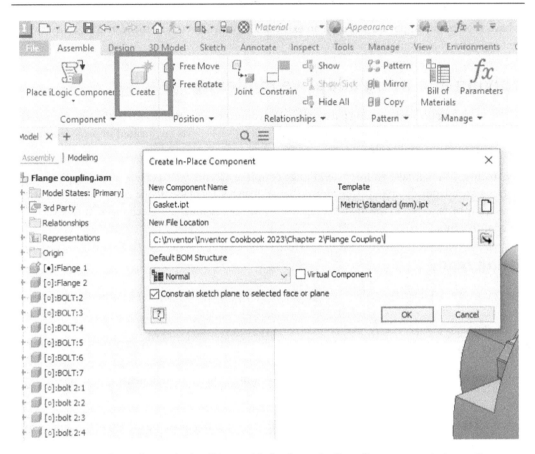

Figure 2.68: Create button in the ribbon, with the Create In-Place Component window active

3.   Select **OK**. Inventor then prompts you to pick a face or plane to build the new component. Select the face highlighted in *Figure 2.69*:

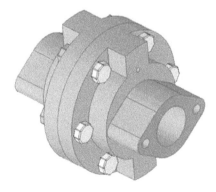

Figure 2.69: Face selected

4. The model will now turn transparent. Select **Start 2D Sketch** and pick the same face previously selected, in *step 3*.

5. In the new sketch, select the **Project Geometry** command, and then select the face shown in *Figure 2.70* to project all the edges shown:

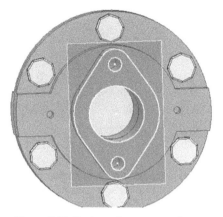

Figure 2.70: Projected geometry shown

6. Select the **Extrude** command and select the profile previously created in *step 5*. Extrude the gasket by 1.5 mm and select **OK**.

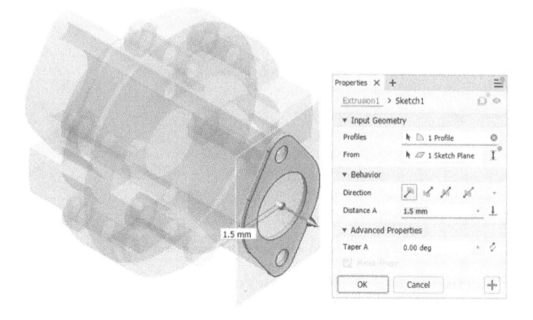

Figure 2.71: Extrusion of previously created projected geometry

7. Select **Return** to complete the gasket. Notice that in the **Model Browser**, next to the recently created gasket, blue and red arrows are placed next to some features. This indicates that **adaptivity** has been created through the projection of geometry (if edits are made to the parent part, then these will be reflected in the child parts). *Figure 2.72* shows the adaptive symbols in the **Model Browser**:

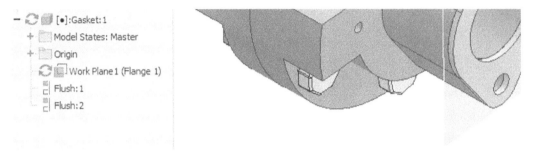

Figure 2.72: Adaptive symbols in the browser shown after the creation of the gasket

8. In the top level of the assembly, create a new in-place component using **Create** as before, but this time select the top face of the recently created gasket to build the part on. Call the new component `Mating Plate` and set **Template** to `standard (mm) .ipt`.

9. Select **Start 2D Sketch** on the top face of the gasket and select **Project Geometry**, as shown in *Figure 2.73*:

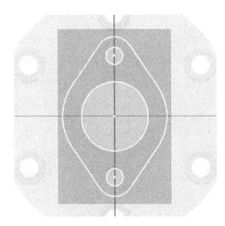

Figure 2.73: Projected sketch on top of the gasket for the creation of the mating plate

10. Select **Finish Sketch**, and then select **Extrude**. Extrude the profiles shown in *Figure 2.74* by `3.5` mm. Select **OK** to complete the operation.

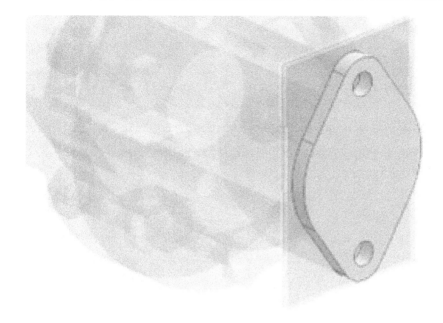

Figure 2.74: Extrusion of the mating plate

11. All additional adaptive parts have now been created. To observe the adaptivity, we will make changes to the original flange coupling components and observe how the newly created parts adapt:

    • Right-click on **Flange 1** from the **Model Browser** and select **Edit**.

    • Expand the **Extrusion 2** node and right-click on **Sketch2**, then select **Edit Sketch**.

    • Change the dimension of the small hole to 2 mm; simply double-click **Dimension** and type the new value. The sketch holes update on completion.

12. Select **Exit Sketch**, then select **Return** to return to the top-level assembly.

    The Mating Plate and Gasket holes have now automatically updated to show the holes as 2 mm. No manual design work was required to complete this.

13. In the **Model Browser**, click **Flange 1** | **Extrusion 2**, right-click **Sketch2**, then edit the value back from 2 mm to 5 mm. Exit and return to the top-level assembly and the parts will update, demonstrating adaptivity again.

This recipe is complete. You have created adaptive parts from an existing part file in the context of an assembly using the adaptive part modeling methodology.

Model credits

The model credits for this chapter are as follows:

- **Shock Absorber** (by Suriya SB) `https://grabcad.com/library/shock-absorber-370/details?folder_id=10235721`

- **Crane Hook** (by Lou Kaminski) `https://grabcad.com/library/crane-hook-225`

- **Flange Coupling** (by Vũ Chí Tường) `https://grabcad.com/library/flange-coupling-144`

3
Driving Automation and Parametric Modeling in Inventor

Automating aspects of the CAD design process should always be your target as an Inventor user. With automation techniques, the utilization of existing geometry and parameters, model updates, and design changes can be accomplished much more efficiently. By applying design intent with these features and methodologies early in the process, models will update and change as expected, without the need for extensive rework.

In this chapter, you will cover how **equations** and **parameters** are used to drive design intent in a model, and how to apply **iMates**, **iFeatures**, **iParts**, and **iAssemblies** to reuse existing digital features, parts, and assembly data throughout your designs.

In this chapter, we will cover the following topics:

- Reviewing equations and parameters in Inventor
- Creating equations and driving a model
- Creating parameters and driving a model
- Reusing design feature data with iFeatures
- Creating automatic mates using iMates
- Creating table-driven master parts to configure sizes and states with iParts
- Building an assembly of iParts with variations such as different sizes, shapes, and content using iAssemblies

Technical requirements

To complete this chapter, you will need access to the `Chapter 3` folder within the `Inventor Cookbook 2023` folder. This is the location of all recipe practice files for this chapter.

Reviewing equations and parameters in Inventor

Inventor equations help to enable the designer to incorporate intelligence and an element of design intent into the model. This ensures that as design changes are made to the design, the model updates and behaves as expected. Inventor equations are established through the creation of mathematical relationships between existing dimensions in a sketch or feature. They can also be established with parameters, which can also be user defined.

Equations sound much more difficult than they are; most of the time, a simple set of basic mathematical functions is required to produce the desired output. For more advanced automation and logic-based rules, **iLogic** would be incorporated. iLogic is the use of `VB.NET` code to run rules within Inventor to automate tasks and model updates. An understanding of the basic equations and parameters is fundamental to understanding iLogic. This chapter will focus solely on equations and parameters; more details about the workings of iLogic will be covered in *Chapter 10, Inventor iLogic Fundamentals – Creating Process Automation and Configurations*.

As you create and add sketches or features to a model, dimensions are generated. With each dimension that is created, Inventor automatically assigns a unique name to each one. The format of this is *d*, followed by a sequential number, starting from *0*. Note that this resets with every new file that is created. Each time you create a new file, the first dimension name that Inventor will apply will always be *d0* by default.

Equations use these *d#* dimension names to establish the desired relationships in the part. Once a value of one dimension is based on the function of another dimension in the model, the relationship is then known as an equation.

In general, the following steps are required to create an equation in a model:

1. A base sketch or model is created, and dimensions are set to be displayed.

2. Equations and relationships are created and established.

3. The model is *flexed* to test the output. Flexing means changing parameters and dimensional values to see whether the resultant changes in the model's behavior are as required.

4. Edits are made, as required.

In the following recipe, you will apply parameters and equations within the context of an Inventor model.

Creating equations and driving a model

For this recipe, you will take an existing model, show the expressions and values already contained within it, and then add and create new equations to facilitate the required changes. The equations that you create will enable you to drive the model (make changes to it with the equations) to the desired outcome.

Getting ready

To complete this recipe, you will need to open the Housing.ipt file from the Chapter 3 folder.

To begin with, we will show the dimensions in the model at present, and then add equations to control the geometry to the desired output.

How to do it...

To begin this recipe, we will first look at what dimensions are present in Housing.ipt and examine the structure of the model. We will then proceed to apply equations. With Housing.ipt open, carry out the following steps:

1. Right-click in any space within the **Graphics Window** and select **Dimension | Display | Name**. Then, right-click on **Extrusion 1** and select **Show Dimensions**. This reveals the dimensions used to create this feature in the model. At present, just the default dimension names are shown, but we will now change this so that the dimension name is converted into a dimension expression.

2. Click the left mouse button to clear the current selection. Then, right-click in the **Graphics Window** and select **Dimension Display | Expression**. The expressions for each dimension are now clearly shown, as per *Figure 3.1*:

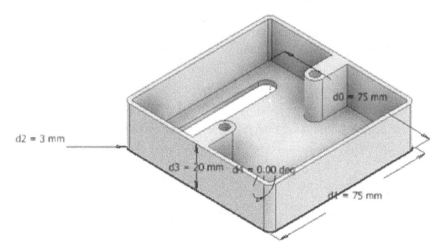

Figure 3.1: Dimensions as expressions shown on Extrusion 1 of Housing.ipt

3. We will now add an equation so that the height of Housing.ipt is always 50 mm shorter than its width. Double-click on the **d3** dimension. Enter the equation d1-25mm and hit *Enter*. Select **Update** if required. Select **d1** and change the value to 85 mm. Select **Update**. The height of the Housing.ipt will change to 60 mm.

4. Double-click on **d1** and change the value to 45 mm. Update the model and see the changes. The equation in **d3** is used again; this is shown in *Figure 3.2*:

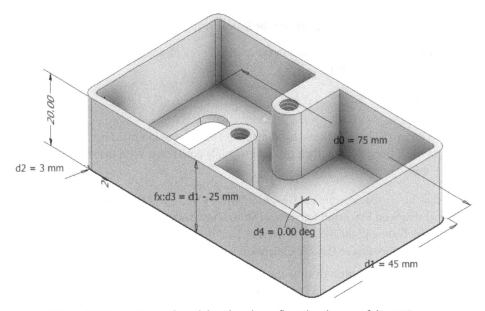

Figure 3.2: Expression and model updated to reflect the change of d1 to 45 mm

5. We will now add an equation to drive the thickness of the Housing.ipt material. Right-click on **Shell** in **Model Browser** and select **Show Dimensions**. At this point, the default value for the thickness is 2 mm.

6. Double-click the **d5** dimension and add the following: d2-1. This controls the thickness of the part so that it is always the value of the radius of the model's fillet minus 1.

7. Now, we will flex the model to test this. Show the dimensions of **Extrusion 1**, double-click **d2**, and change this value to 4. Select **Update** and the thickness of the part will change.

> **Important note**
>
> Adding equations can be made as simple or as complex as desired. The full list of functions and controls can be found here: https://knowledge.autodesk.com/support/ inventor-products/learn-explore/caas/CloudHelp/cloudhelp/2014/ ENU/Inventor/files/GUID-69D04C7D-E195-49C0-B1B2-03C88A806D9D-htm.html.

8. Viewing the range of equations in the model space can be quite limiting; fortunately, this can be managed far easier in the **Parameters** panel. Select the **Manage** tab, and in the **Parameters** panel, click the **Parameters** button.

The **Parameters** panel now opens, which contains a comprehensive list of all equations and dimensions in the model. Equations can be created here too.

There is nothing to change at present, but you can see that the equations added previously are now visible in the **Parameters** panel.

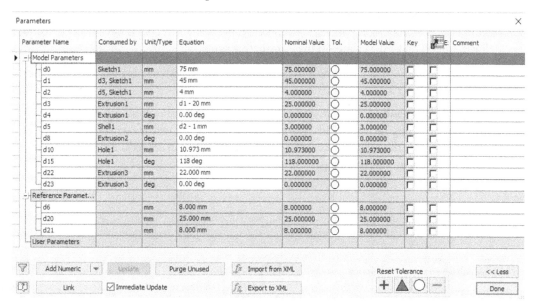

Parameter Name	Consumed by	Unit/Type	Equation	Nominal Value	Tol.	Model Value	Key	E.	Comment
▶ − Model Parameters									
d0	Sketch1	mm	75 mm	75.000000	○	75.000000			
d1	d3, Sketch1	mm	45 mm	45.000000	○	45.000000			
d2	d5, Sketch1	mm	4 mm	4.000000	○	4.000000			
d3	Extrusion1	mm	d1 - 20 mm	25.000000	○	25.000000			
d4	Extrusion1	deg	0.00 deg	0.000000	○	0.000000			
d5	Shell1	mm	d2 - 1 mm	3.000000	○	3.000000			
d8	Extrusion2	deg	0.00 deg	0.000000	○	0.000000			
d10	Hole1	mm	10.973 mm	10.973000	○	10.973000			
d15	Hole1	deg	118 deg	118.000000	○	118.000000			
d22	Extrusion3	mm	22.000 mm	22.000000	○	22.000000			
d23	Extrusion3	deg	0.00 deg	0.000000	○	0.000000			
− Reference Paramet...									
d6		mm	8.000 mm	8.000000	○	8.000000			
d20		mm	25.000 mm	25.000000	○	25.000000			
d21		mm	8.000 mm	8.000000	○	8.000000			
User Parameters									

Figure 3.3: Parameters panel with new equations visible in the Equation column

The equations have now been created, so you can save and close the model.

You have now successfully created and applied equations to drive automatic changes in an Inventor part file.

Creating parameters and driving a model

In this recipe, you will use the **Parameters** panel to create equations and user-defined parameters. You will learn how to rename and organize parameters within a model and drive dimensions with them. You will also learn how to export parameters for use in another model as a reference parameter.

Getting ready

For this recipe, you will need to open the Heatsink.ipt file from Chapter 3 of the Inventor 2023 Cookbook project folder.

How to do it...

To start, we will examine an existing model that has no defined equations or parameters. The first task will be to access the **Parameters** panel and create these. To do this, take the following steps:

1. Open Heatsink.ipt in Inventor. This part has been created and currently has no equations or user-defined parameters.

2. Select **Parameters** from the **Manage** tab.

    Here, we will now rename some of the existing parameters. You can also do this upon placing the dimension in the model space, but in this case, because the model is already created and defined, we will make the change in the **Parameters** panel.

3. Select the **d0** parameter and rename it to Length (be careful with naming parameters as they are *case sensitive*).

4. Do the same for **d1** and rename it to Width. Select the current value of 50 mm for Width and change this by typing Length. This now means that the width will always equal the value of the length. Because the values are currently the same, that is, both are set at 50 mm, the model will not make any geometric changes.

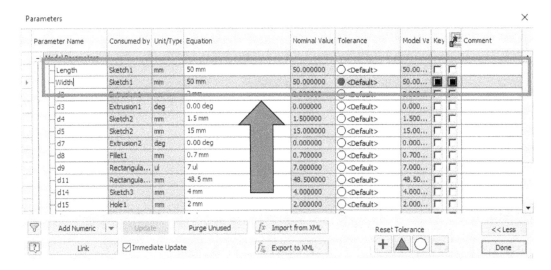

Figure 3.4: Width and Length parameters

5. Exit the **Parameters** panel and select **Start 2D Sketch** on the face shown in *Figure 3.5*. Apply the sketch geometry shown in *Figure 3.5*, including the two sketch points spaced 6.4 mm from each other and the edge of the part. Then, click **Finish Sketch**.

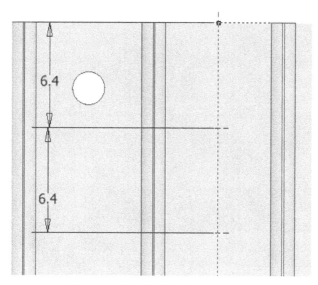

Figure 3.5: The sketch to be created

6. Select the **Parameters** panel again.

7. Rename the **d37** parameter to `Hole_Spacing`. Then, select the tick box for **Export of Parameter**. This allows you to link this parameter to another Inventor model. Change the **d38** parameter value to `d37`.

Ensure your equations match that of *Figure 3.6* before clicking **Done**:

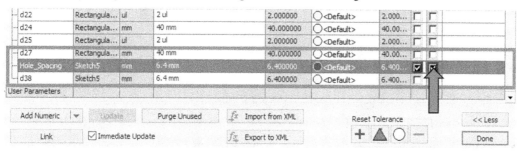

Figure 3.6: Renaming a parameter and selecting Export of Parameter

9. Open `Heatsink2.0.ipt` from the `Chapter 3` folder. Here, we will reference a parameter from the original `Heatsink.ipt` model.

10. Right-click on **Sketch5** in the **Model Browser** and select **Edit Sketch**. Add a dimension between all the preexisting sketch points (don't worry about adding any specific values). An example is shown in *Figure 3.7*:

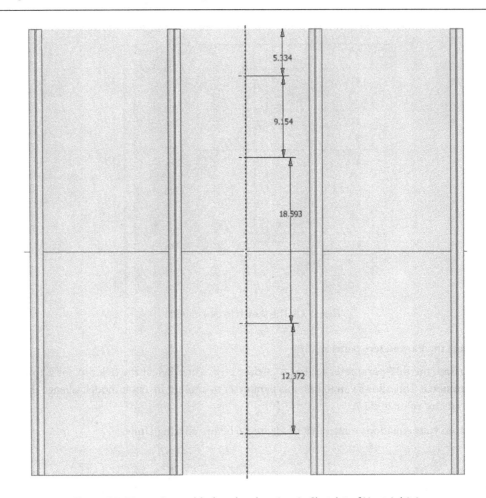

Figure 3.7: Dimensions added to sketch points in Sketch5 of Heatsink2.0

11. Select the **Parameters** panel and click **Link** at the bottom left of the menu.

12. Browse for the Chapter 3 folder and select Heatsink.ipt as the reference file.

13. Scroll down the parameters to the **Hole_Spacing** option. Change the negative symbol to a positive one by selecting it. This notifies Inventor to import this parameter, as shown in *Figure 3.8*:

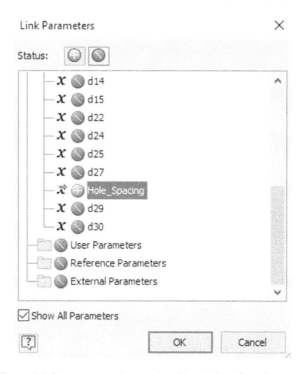

Figure 3.8: Parameters to import into Heatsink2.0 from heatsink

14. Select **OK**. The imported parameters now appear in the **Parameters** dialog.

15. Change the **Equation** value of dimensions **d37** to **d40** to **Hole_Spacing**, as per *Figure 3.9*:

d11	Rectangular...	mm	48.5 mm	48.500000	○	48.500000	☐
d37	Sketch5	mm	Hole_Spacing	5.000000	○	5.000000	☐
d38	Sketch5	mm	Hole_Spacing	5.000000	○	5.000000	☐
d39	Sketch5	mm	Hole_Spacing	5.000000	○	5.000000	☐
d40	Sketch5	mm	Hole_Spacing	5.000000	●	5.000000	■
:\Users\A_Bordin...							
Hole_Spacing	d40, d39, d...	mm	5.000 mm	5.000000	○	5.000000	☐

Add Numeric ▼   Update   Purge Unused   *fx* Import from XML   Reset Tolera

Figure 3.9: Linked parameters changed in the Heatsink2.0 file

16. Select **Done**. The sketch points created are now showing the same hole spacing dimension and value as the original `Heatsink.ipt` file.

17. Select the **Hole** command from the 3D model. Then, apply a hole as per the specifications in *Figure 3.10* to the sketch points in **Sketch5**:

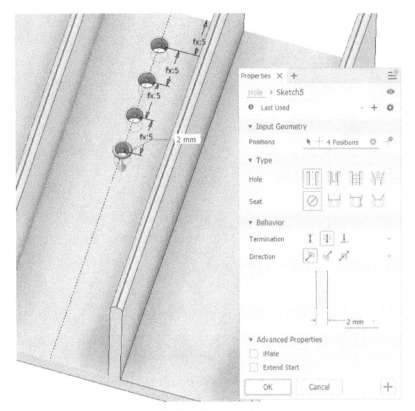

Figure 3.10: Hole feature to be created

18. Select **OK** to complete the holes.

    Now, we will make a change in the Heatsink.ipt file and see how the parameter in Heatsink2.0.ipt is updated.

19. Open Heatsink.ipt. Then, in the **Manage** tab, select **Parameters**.

20. Change the **Hole_Spacing** value from 5 mm to 10 mm. Select **Done**.

21. Open Heatsink2.0.ipt and review the changes that have been made from the original file where the parameter was exported. You may have to select **Local Update** to review the changes. Once the change is complete, you will see how the hole spacings in Heatsink2.0.ipt have also increased to 10 mm.

Heatsink2.0.ipt will show the same parameter changes as **Heatsink**, demonstrating the linking and exporting of a parameter function. Save both files as they are now both linked.

You have created a custom parameter to drive hole spacing in one part, and then exported a link to this parameter to control the hole spacing of another independent part.

Reusing design feature data with iFeatures

iFeatures are a method of copying and reusing design features between files. This duplication technique is an effective way of reusing existing feature design data throughout your models, without needing to remodel features each time. iFeatures can be added to a single part or across multiple parts. Once created, an iFeature is cataloged in a library for future use. iFeatures can be fully parametric with both the size and position driven by parameters.

Using iFeatures effectively enforces consistency in designs and saves time. The typical process of creating an iFeature is as follows:

1. The iFeature is created in a part file.

2. Features are selected for iFeature export.

3. Parameters are established.

4. The iFeature's position is defined.

5. The iFeature is saved in the relevant library.

6. The iFeature is imported, configured, and placed in a model.

In this recipe, you will create a new iFeature of a plastic boss for use in another plastic casing product to complete the part. The iFeature will be defined with a configurable table of options. It will then be exported from the parent file, and then imported, placed, and configured in the new independent model.

Getting ready

For this recipe, you will need to open the `Casing.ipt` file and the `Boss.ipt` file from the `Chapter 3` folder in the `Inventor 2023 Cookbook` folder.

How to do it...

We will start with an existing file that contains the feature that we want to export into another model. Then, we will create an iFeature of this for exporting and importing into the desired target location. To do this, take the following steps:

1. Open the `Boss.ipt` file. You will see that this feature has already been created for you. This is the feature we will convert into an iFeature and export for use in another model, `Casing.ipt` in this example.

2. Navigate to the **Manage** tab and find the **Author** panel. Then, select **Extract iFeature**.

3. The **Extract iFeature** menu now appears on screen. This prompts you to select the features from the **Model Browser** to include as part of the iFeature. Select **Extrusion 1**.

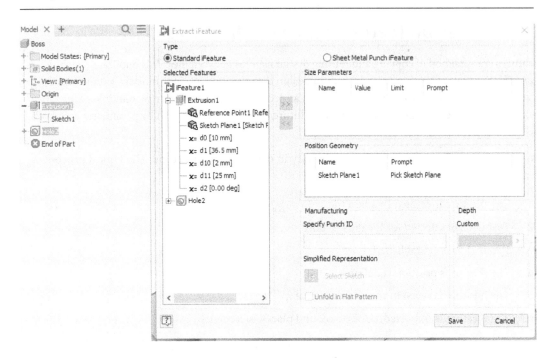

Figure 3.11: Extract iFeature dialog

Extrusion 1 is added, as are all the child features of **Extrusion 1**. This is shown in *Figure 3.11*. Note that the custom parameters that have been created for **Height** and **Diameter** are also pulled across and can be configured here too.

4. The custom parameters have already been defined. We will now set various values and limits as well as prompts so that the placement of the iFeature for future users is clear. The purpose of this is so that the user will simply have to click a face to place the iFeature on it.

5. Click the **iFeature(#)** text in the **Extract iFeature** window and rename it to `iFeatureBoss`. This will make finding and reusing the iFeature in the future much easier, as it now has a meaningful name, as opposed to the default one.

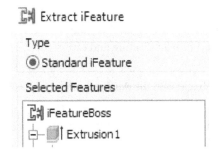

Figure 3.12: Newly created iFeature renamed

6. Select **Save**. You will be prompted to save in the `Catalog` folder. Create a new folder here called `iFeature`. Save your iFeature in the `iFeature` folder as `iFeatureBoss.ide`.

7. Then, navigate to the `Catalog` folder: `C:\Users\Public\Public Documents\ Autodesk\Inventor2023\Catalog\<yourname>\folder`.

8. Find the `iFeatureBoss.ide` file just created and open it. From here, we will now configure the options available to the user upon placement of the iFeature in a model.

9. Select the **iFeature Author Table** icon in the ribbon.

10. You are now prompted to configure the parameters you would like to drive the iFeature and the configurations that will be available to the user upon placement. In this example, we will create three variations of the boss with various sizes of **Height** and **Diameter**. Right-click on the first row of the table and select **Insert Row**. Repeat this again so there is a total of three rows. This is shown in *Figure 3.13*:

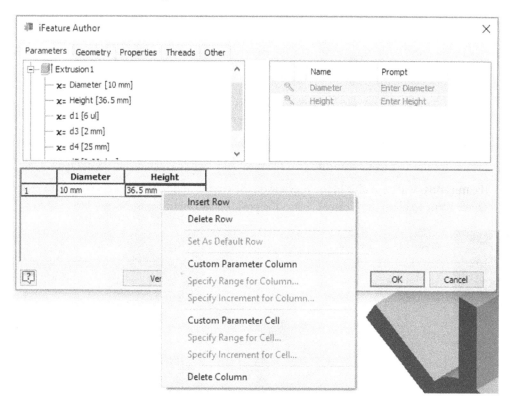

Figure 3.13: Insert Row

11. Click the key icon in the right-hand side box and change both keys so they are blue and active.

12. Change the **Prompt** text to `Select Diameter` and `Select Height`, as shown in *Figure 3.14*:

Figure 3.14: Prompts changed and key selected

13. Select the text in the tables and change the options so they match those in *Figure 3.15*. These will be the **Height** and **Diameter** options the user can pick and select upon placement of **iFeatureBoss**.

	🔑 Diameter	🔑 Height
1	10 mm	25
2	12mm	30
3	14mm	36.5 mm

Figure 3.15: Diameter and Height options required

14. Select **Verify** to check that the values entered are compatible. Select **OK**, save the `iFeatureBoss.ide` file, and close the file.

15. Open the `Casing.ipt` file. We will now import the newly created iFeature and place a boss feature to improve the structure and provide a fixing point to the plastic casing design.

16. In the **Manage** tab, navigate to the **Insert** environment and select **Insert iFeature**.

17. Browse for the `iFeatureBoss.ide` file that was recently created and select **Open**.

18. **iFeatureBoss** will be visible in the preview on the cursor, and Inventor prompts you to pick a sketch plane to place the iFeature. Select the position shown in *Figure 3.16*:

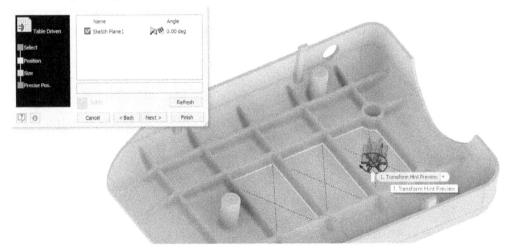

Figure 3.16: iFeature placed on Casing.ipt

19. Once placed, click **Next**.

20. Then, click on the 10 mm value of **Diameter** and change it to 12mm. Upon changing **Diameter**, the iFeature updates **Height** to 30 mm as per the values originally detailed in the table in *Figure 3.15*.

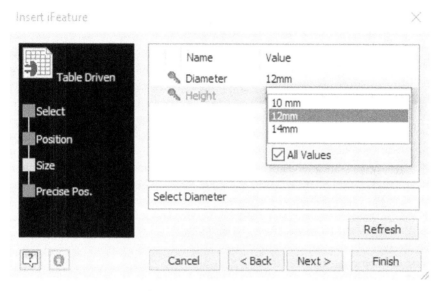

Figure 3.17: iFeature configured and selected, driven with values from the table

21. Select **12mm** as the diameter. Then, select **Next** and then **Finish**. The table-driven configured iFeature is placed on the part.

22. Complete the model as per *Figure 3.18* by adding two more identical iFeatures to the locations shown. Repeat the **Insert iFeature** steps to accomplish this.

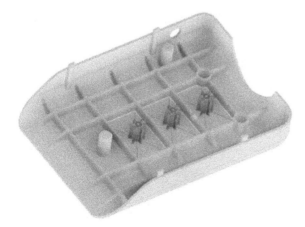

Figure 3.18: Completed model with duplicate iFeatures added and the locations illustrated

You have successfully created and exported a table-driven parametric iFeature, selected a configuration, and then imported this into a new model. `Casing.ipt` does not link to `iFeatureBoss.ide`. These are separate independent files and are not linked.

Creating automatic mates using iMates

iMates are used in Inventor to work more efficiently in part designs when putting them together into a top-level assembly. The iMate functionality enables you to place and mate components in place automatically in an assembly. This is very useful if there are several repeating components in an assembly when identical mates are required. iMates have the constraint information stored within, so manual mates are no longer required with that component. iMates can also be combined into an **iComposite** so that parts that require multiple Mate constraints can be assembled automatically.

This also makes it easy to create interchangeable components within an assembly. If the same iMates are configured for each component within the assembly, they can be swapped out automatically and changed with the iMate functionality.

In this recipe, you will complete an assembly of a drone using iMates to automatically complete the assembly. iComposites will be created, configured, and applied to components and then automatically applied to build and complete the drone.

Getting ready

To begin this recipe, you will need to browse and access the `Inventor Cookbook 2023` folder | `Chapter 3` | `Drone`, and then open the `DRONE.iam` file.

The completed model, once the iMates have been applied, is shown in *Figure 3.18*:

Figure 3.19: The completed DRONE.iam assembly

How to do it...

We will start with an incomplete drone assembly, and then proceed to build in iMates to complete the assembly as per *Figure 3.19*. To do this, take the following steps:

1. Open the DRONE.iam file; *Figure 3.20* shows what it looks like. The assembly of the drone is complete, except the **Motor** and **Blade** components are missing from the model. We will use iComposites/iMates to automatically configure this upon placement.

Figure 3.20: The DRONE.iam file open with missing components

2. We will now begin to configure the first composite iMate. A matching iComposite must also be applied to the other target component for the mate to automatically constrain and place. Start by navigating to the browser, then right-click on the spider arm:1 file and select **Edit**. You are now editing the part within the context of the assembly.

3. Navigate to the **Manage** tab and select the **iMate** button from the **Author** panel.

4. The **Create iMate** window is now active. Next, we must specify the first iMate to apply. Select **Insert** as the type of constraint and select the geometry shown in *Figure 3.21*. No offset is required in this instance.

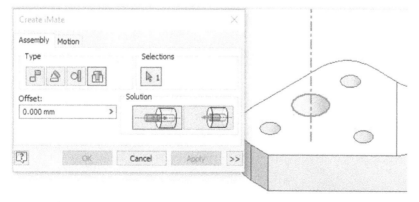

Figure 3.21: Mate constraint and geometry required for this iMate

5. Select **Apply**. The **insert iMate** has now been created. Note in the **Model Browser** on the left-hand side that an `iMates` folder has now appeared. Expand this by clicking the + icon next to it. You will see the iMate has been configured. By clicking on the default iMate name, the name can be changed, but in this instance, leave it as the default.

6. Right-click the iMate in the browser and select **Create Composite**. An instance of **iComposite 1** is created. (If more iMates were required, these could also be added to the iComposite or *group*. iComposites can also be renamed if required.)

    Now that this iComposite has been created, we will need to replicate this in the target component that we want to mate with the spider arm. *Figure 3.22* shows the iComposite after creation:

Figure 3.22: iComposite iMate created in the spider arm file

7. Now the iComposite iMate has been created for the spider arm. There are multiple instances of the spider arm in the assembly. Because of this, the iMates created in the previous step have been generated in the duplicates also, as they share the same source file. Select **Return** from the far right of the ribbon to return to the assembly and exit **Edit Part** mode.

8. Save the assembly at this point.

9. Open the MOTOR.ipt file from the same folder as DRONE.iam. The corresponding iComposite/iMate needs to now be created.

10. Select the **Manage** tab, then **iMate**, and place an insert iMate on the geometry of MOTOR.ipt, as shown in *Figure 3.23*:

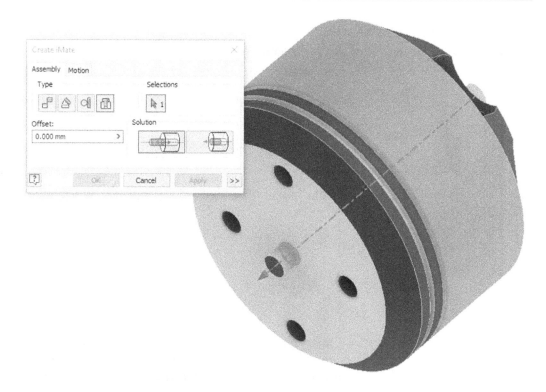

Figure 3.23: Insert iMate to be created on MOTOR.ipt

11. In the **Model Browser**, under the iMates folder, right-click on the insert iMate you just created and select **Create Composite**. The iMate is converted into an iComposite with the *same name* as the spider arm component, as shown in *Figure 3.24*. This is crucial as Inventor uses iMate and iComposite names to match and create the automatic mating.

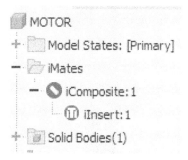

Figure 3.24: Insert iMate as part of an iMate iComposite in MOTOR.ipt

12. Save and close the MOTOR.ipt file.

13. Open DRONE.iam.

14. Select **Place** from the **Assemble** tab.

15. Pick **MOTOR.ipt** and ensure that the **Interactively place with iMates** icon is active. See *Figure 3.25*. If this is not active, the component will not be placed using iMates or iComposites.

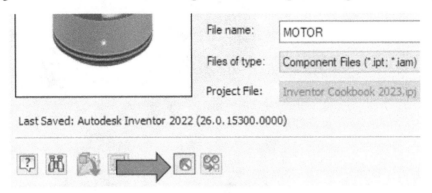

Figure 3.25: Interactively place with iMates glyph selected

16. Select **Open**.

17. MOTOR.ipt will automatically mate with the spider arm, as per the iComposite defined previously. Note the green iComposite glyphs in the **Graphics Window** that are highlighted. To place the part automatically on location with the **iMate**, left-click in the **Graphics Window**.

18. MOTOR.ipt has now been automatically constrained with the iMate. This will cause you to be zoomed into the model by default. Zoom out from the model with the mouse wheel without deselecting anything. Note that the second instance of the motor on the corresponding arm is already in place to be added to the assembly.

19. Continue to left-click until all four instances of MOTOR.ipt have been added to the assembly and placed with the iMates. *Figure 3.26* shows the completed stage:

Figure 3.26: All four MOTOR.ipt parts placed in the assembly with iComposite1

We will now repeat this process for the remaining components.

20. Open Spinner.ipt from the Drone folder.

21. Create another insert iMate, as shown in *Figure 3.27*:

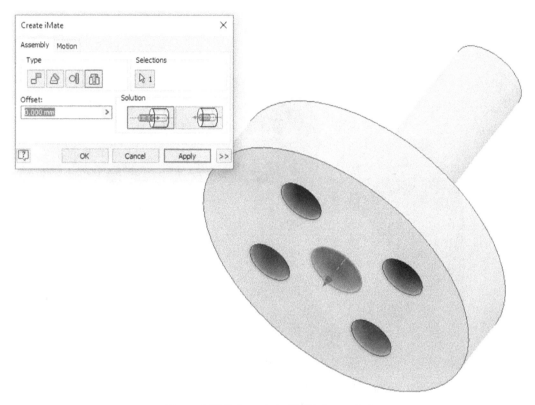

Figure 3.27: Spinner.ipt with iMate created

22. Once created, right-click on the iMate in the browser and then click **Create as an iComposite**. Rename the iComposite created to iComposite:2 and the insert iMate to Insert:2, as shown in *Figure 3.28*:

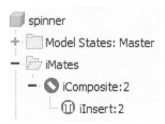

Figure 3.28: iComposite:2 created

23. Save and close Spinner.ipt.

24. Open DRONE.iam

25. In the **Model Browser**, right-click MOTOR.ipt and select **Open**. Create an insert iMate, as shown in *Figure 3.29*:

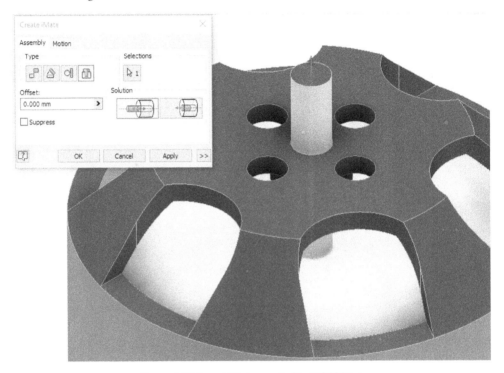

Figure 3.29: Insert iMate created in MOTOR.ipt

26. Once created, create an iComposite from the iMate in the previous step and rename it iComposite:2. Then, rename the insert iMate to Insert:2. The text must match the previous iComposite/iMate to function. This should result in two iComposites in the MOTOR.ipt file. Save MOTOR.ipt.

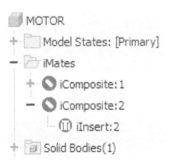

Figure 3.30: Two iComposites in MOTOR.ipt

27. In the **Assemble** tab, select **Place** and then select `Spinner.ipt`. Ensure the iMate glyph is selected.

28. Left-click and place the four instances of `Spinner.ipt` on `DRONE.iam`. The completed step is shown in *Figure 3.31*:

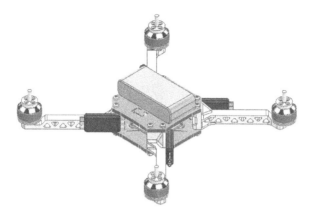

Figure 3.31: Four instances of Spinner.ipt added to DRONE.iam with iComposites

29. Save the assembly.

30. Open `Fan.ipt` from the `Drone` folder.

31. Create the insert iMate as shown in *Figure 3.32*:

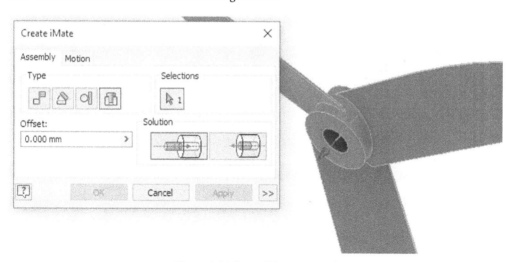

Figure 3.32: Insert iMate required

32. Create an iComposite of the iMate and rename it as per *Figure 3.33*:

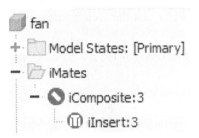

Figure 3.33: iComposite renaming required

33. Save and close `fan.ipt`.

34. Open `DRONE.iam`.

35. In the **Model Browser**, right-click one of the **spinner** components and select **Edit** to edit the part inside of the assembly.

36. Create and apply the following iMate shown in *Figure 3.34*:

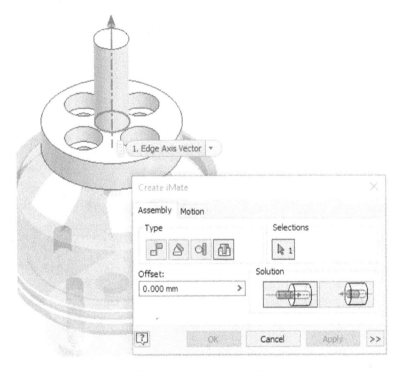

Figure 3.34: iMate to be created on Spinner.ipt

37. Create an iComposite of the iMate just created and rename it `iComposite:3`. Rename the iComposite and insert iMate to `iComposite:3` and `iInsert:3`, respectively. Select **Save** and then return to the assembly.

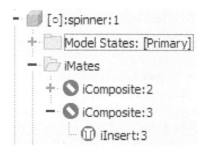

Figure 3.35: Third iComposite to create

38. Place the instances of fan.ipt using the iComposites. *Figure 3.36* shows the result:

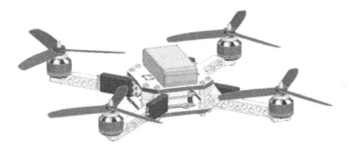

Figure 3.36: All fan.ipt instances added to DRONE.iam

39. To complete the model, add a final **iComposite 5** with a single **Insert** on the prevailing torque hex nut_am_AM-M4-N.ipt. The end assembly should look as in *Figure 3.37*:

Figure 3.37: Completed DRONE.iam

The assembly is now complete. You have successfully created a range of iMates and iComposites to automatically apply the remaining components to complete the assembly of a drone.

Creating table-driven master parts to configure sizes and states with iParts

iParts enable you to create variations in part designs quickly and efficiently. In some cases, you may have a part that is similar to other parts you manufacture, yet only with slight variations. The iPart functionality creates variations or members within the same part file, which can be selected and generated. This means that similar parts do not need to be repeatedly created from scratch. Once a part has been created with iPart members, these members can be documented in a drawing file as a table.

For this iPart recipe, we will create an iPart of an existing bracket already modeled. The iPart will enable the creation of several variations of the bracket.

Getting ready

To begin this recipe, you will need to browse and access the Inventor Cookbook 2023 folder | Chapter 3 and then open the Bracket.ipt file.

How to do it...

For this iPart recipe, we will create an iPart of an existing bracket already modeled. The iPart will enable the creation of several variations of Bracket.ipt. To do this, take the following steps:

1. Open Bracket.ipt. This is a simple bracket containing a central bore, supporting ribs, and four fixing holes. Open the **Parameters** panel, as shown in *Figure 3.38*:

Parameter Name	Consumed by	Unit/Type	Equation	Nominal Value	Tol.	Model Value	Key	Export	Comment
Model Parameters									
Diameter	Fixing_Hole_Cent...	mm	100 mm	100.000000	O	100.000000			
Thickness	Bore_Height, Ext...	mm	Diameter / 10 ul	10.000000	O	10.000000			
d2	Extrusion1	deg	0.00 deg	0.000000	O	0.000000			
Bore_Dia	Hole_Dia, Sketch2	mm	Diameter / 2 ul	50.000000	O	50.000000			
Bore_Height	Extrusion2	mm	Thickness + 5 mm	15.000000	O	15.000000			
d5	Extrusion2	deg	0.00 deg	0.000000	O	0.000000			
Hole_Dia	Hole1	mm	Bore_Dia / 2 ul	25.000000	O	25.000000			
d7	Sketch4	mm	2 mm	2.000000	O	2.000000			
d8	Extrusion3	deg	0.00 deg	0.000000	O	0.000000			
d9	Chamfer1	mm	15.000 mm	15.000000	O	15.000000			
Rib	d13, Circular Patt...	ul	4 ul	4.000000	O	4.000000			
d11	Circular Pattern1	deg	360 deg	360.000000	O	360.000000			
Fixing_Hole_Dia	Hole2	mm	5 mm	5.000000	O	5.000000			
d13	Circular Pattern3	ul	Rib	4.000000	O	4.000000			
d14	Circular Pattern3	deg	360 deg	360.000000	O	360.000000			
Fixing_Hole_Centre	Sketch7	mm	Diameter - 25 mm	75.000000	O	75.000000			
d16	Fillet1	mm	2 mm	2.000000	O	2.000000			
User Parameters									

Figure 3.38: Parameters for Bracket.ipt shown

The model parameter names have already been changed, and the relevant equations have been applied. This is so that the model updates as required. To test the equations and parameters already applied, change the value of **Diameter** to 150 mm and then 200 mm, and observe how the model updates. Note how the diameter changes in the model.

2. Now change the **Diameter** parameter back to 100 mm. The part updates and the original design configuration is restored, but with iParts, we can enable greater differentiation between size variants and set these as absolute values that can be picked by the user. This can be achieved without the need to manually edit the **Parameters** table.

3. In the **Manage** tab, open the **Author** panel and select **Create iPart**.

    The **iPart Author** menu now opens. Note all the custom parameters have automatically been included in the iPart.

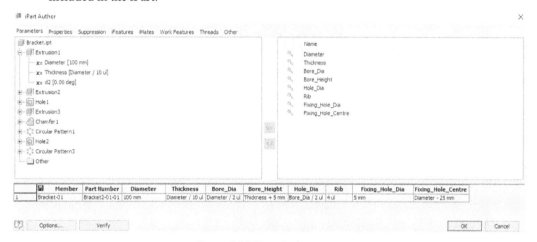

Figure 3.39: iPart Author menu

4. As we want to create five variants (or members) within this iPart, move the cursor over the first line of the table, right-click it, and select **Insert Row**. Build five rows in the table, which will form the basis of five iPart members.

5. Now, we will differentiate between the members. Change the values in the **Diameter**, **Rib**, and **Fixing_Hole_Dia** columns by clicking on the text in the table and typing so that they match those shown in *Figure 3.40*:

	Member	Part Number	Diameter	Thickness	Bore_Dia	Bore_Height	Hole_Dia	Rib	Fixing_Hole_Dia	Fixing_Hole_Centre
1	Bracket-01	Bracket2-01-01	100 mm	Diameter / 10 ul	Diameter / 2 ul	Thickness + 5 mm	Bore_Dia / 2 ul	4 ul	5 mm	Diameter - 25 mm
2	Bracket-02	Bracket2-01-02	150	Diameter / 10 ul	Diameter / 2 ul	Thickness + 5 mm	Bore_Dia / 2 ul	6	5 mm	Diameter - 25 mm
3	Bracket-03	Bracket2-01-03	160	Diameter / 10 ul	Diameter / 2 ul	Thickness + 5 mm	Bore_Dia / 2 ul	6	10	Diameter - 25 mm
4	Bracket-04	Bracket2-01-04	180 mm	Diameter / 10 ul	Diameter / 2 ul	Thickness + 5 mm	Bore_Dia / 2 ul	8	12	Diameter - 25 mm
5	Bracket-05	Bracket2-01-05	200 mm	Diameter / 10 ul	Diameter / 2 ul	Thickness + 5 mm	Bore_Dia / 2 ul	8	15	Diameter - 25 mm

Figure 3.40: iPart member changes required

6. In the **Key** area of the menu, select the following parameters as key parameters to update and change the iPart. You can add up to nine keys in an iPart configuration. The purpose is to allow for easier organization and a logical/hierarchical structure for the configurations. Ensure your menu on screen matches that of *Figure 3.41*:

Figure 3.41: Keys selected

7. With the keys now assigned, select **Verify**. The iPart author checks the values added and ensures your model can compute them.

8. Select **OK**. The **iPart Author** menu closes and you are returned to the base part.

9. In the **Model Browser**, you will notice a **Table** icon has now appeared. Expand this, as in *Figure 3.42*:

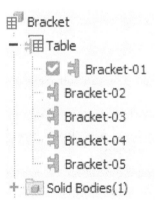

Figure 3.42: iPart table expanded in Bracket.ipt

Note that the iPart members created are now visible. The active one in this instance is the member with a 100 mm diameter. The active member is visible in the **Model Browser** as it has a tick mark next to it, as shown in *Figure 3.42*.

10. Double-click the other members and observe how the model updates. The bracket will update based on the values you entered in the iPart creation. Various configurations can now be generated automatically.

11. Additional features are required on **Bracket-05**. Double-click on **Bracket-05** to activate it. Then, in the **Manage** tab, open the **Author** panel and select **Edit Member Scope**. We can now make changes to this individual member of the iPart.

12. Add a 2 mm fillet to the edge of the central hole. Then, add a thread, as detailed in *Figure 3.43*, to the same hole:

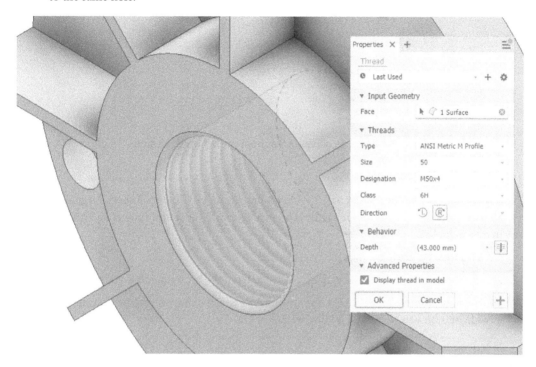

Figure 3.43: A 2 mm fillet and thread added to Bracket-05, in Edit Member Scope

The changes have only been made to **Bracket-05** as desired.

> **iPart changes using Excel**
>
> You can also create changes and manipulate an iPart table using Excel. By right-clicking on an iPart table, you have the option to edit via a spreadsheet.

13. Activate the **Bracket-01** member.

14. We will now generate separate part files for each iPart that has been defined. Hold *Shift* and select all the iPart members. Then, right-click and choose **Generate Files**.

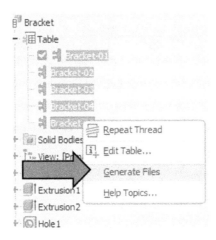

Figure 3.44: How to generate files from the members created

Save the files if prompted.

15. Select **Open** within the Chapter 3 folder. A new Bracket folder has been created to store all of the separate .ipt files from the iPart generation. We will now import and use the iPart in an assembly.

16. Create a New Standard (mm) .iam assembly file.

17. Select **Place**, then select the Bracket.ipt file from the browser, followed by **Open**.

18. You are now prompted to select the iPart member you wish to import into the assembly. Choose a configuration you would like to import. Select **OK**.

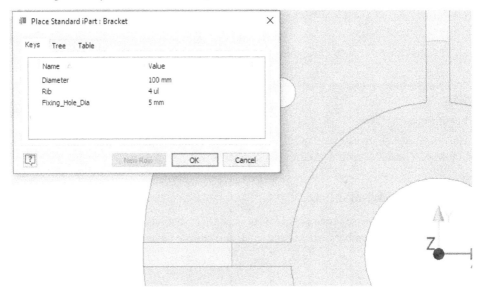

Figure 3.45: Placement of a standard iPart in the assembly

The configured iPart member is imported into the assembly.

19. Close and do not save the assembly file.

You have successfully created an iPart with several configurations, edited an individual member, published the configurations as a table-driven iPart, and imported this into an assembly for use.

Building an assembly of iParts with variations such as different sizes, shapes, and content using iAssemblies

iAssemblies are used to create variations of assemblies and subassemblies that make up your designs. They are an efficient way to create configurations of assemblies, in a single file, without the need to recreate and duplicate work. iAssemblies work and are configured in a similar way to iFeatures and iParts.

Configurable attributes in an iAssembly can consist of suppressing or computing features/parts, constraint values, materials, appearance, and even the BOM structure of components. In most cases, the following steps are used to create an iAssembly:

1. Standard parts are assembled into an assembly.
2. iParts are configured from the parts in the assembly (this can be done prior to being placed into the assembly).
3. The creation of an iAssembly is initiated.
4. Configurable attributes and iAssembly members are specified from the iParts or sub-iAssemblies.
5. The iAssembly is verified.

If the iAssembly contains a subassembly, this will also need to be created as its own independent iAssembly.

In some cases, when creating an iAssembly from existing assemblies, normal parts must be replaced with their iParts. Sometimes, existing mate constraints can become unresolved and will either need to be recovered or reapplied. iAssemblies can also be placed into other assembly files for future use and, once placed, can be configured on demand.

In this recipe, you will create an iAssembly of a premodeled screwdriver so that you can vary the length of the handle and shaft. This is a simple take on an iAssembly but will reveal the steps required to create and configure one. (**iAssemblies** can become much more complex if required.) To start with, you will create two iParts from existing components and then combine the two into the iAssembly.

Getting ready

To begin this recipe, you will need to browse and access the `Inventor Cookbook 2023` folder | `Chapter 3` and then open the `ScrewDriverAssy.iam` file.

How to do it...

Within the assembly, we will first have to make separate iParts, before proceeding to combine these into an iAssembly. To do this, take the following steps:

1. Open `ScrewDriverAssy.iam`. The assembly is fully constrained, but the parts within it are standard parts, not iParts.

2. Right-click on `Handle.ipt` in the **Model Browser**, then select **Open**.

3. The iPart for the handle now needs to be created. The parameters required to change the length of the handle as desired have already been applied. So, select **Create iPart** from the **Author** panel.

4. Select the **Length [150mm]** parameter from the **Parameters** list in the **iPart Author** window.

5. In the **iPart Author** window, select the key symbol on the **Length** parameter.

6. Right-click on the first line of the table, then click **Insert Member**. Repeat this to create a total of three members in this iPart.

7. Rename the members and part numbers, and then change the length to the values shown in *Figure 3.46*:

	💾 Member	Part Number	🔍 Length
1	Handle-Short	Handle-S	150 mm
2	Handle-Medium	Handle-M	180 mm
3	Handle-Long	Handle-L	200 mm

?	Options...	Verify

Figure 3.46: iPart Author window active with required Member, Part Number, and Length values

8. Select **Verify**, then **OK**.

9. The iPart members have been created in the file. Expand the **iPart Table** node in the **Model Browser** and double-click on each member to update the **Handle** part. Observe how the length changes in *Figure 3.47*:

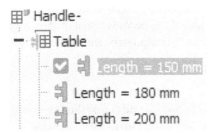

Figure 3.47: iPart table created with the three length variations

10. Double-click and activate the **Length = 150 mm** iPart member so that it is active.

11. Hold *Shift* and select all members in the **Model Browser**. Then, right-click and select **Generate Files**, as shown in *Figure 3.48*:

Figure 3.48: Generating the files from the iPart members

Save the file if prompted.

12. Close the Handle.ipt file.

13. Open the ScrewDriverAssy.iam file.

14. A local update is required on the model. Select the **Update** button to activate the update, as shown in *Figure 3.49*:

Figure 3.49: Local update button location

The **Handle** iPart is now showing in the assembly, but if you attempt to double-click on the various members, the part will not update. The component needs to be replaced with the new version of itself in the assembly.

15. Right-click **Handle** in the **Model Browser**, then select **Component | Replace**.

16. Browse for Handle.ipt and select **Open**.

17. Inventor now prompts you to select the desired iPart value to be imported into the assembly. Leave this as 150 mm. We have yet to define the **Shaft** component and convert the assembly into an iAssembly. Select **OK**.

18. Upon completion, an error will flag. This error states that one of the insert Mates is now unresolved and needs attention to proceed. You can see this from the warning triangle in the **Model Browser**.

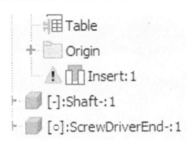

Figure 3.50: Unresolved mate constraint warning

We now need to solve this error. Right-click on the **Insert:1** mate, then select **Edit**.

19. We can now reapply the mate to resolve the issue. Pull the **Shaft** component away from the handle by clicking and dragging it; this makes editing the mate easier.

20. Select the **Insert Mate** and apply it as shown in *Figure 3.51*. Select **OK**, and the mate is resolved.

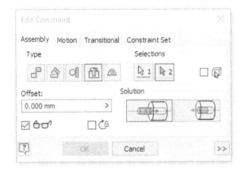

Figure 3.51: Insert Mate to be redefined

21. This process now needs to be completed for the second iPart of the iAssembly: `Shaft.ipt`.

22. Right-click on **Shaft** from the **Model Browser** in the `ScrewDriverAssy.iam` file. Then, click **Open**.

23. Select **Create iPart**. Select **Length [165mm]** as the parameter.

24. Assign the key to **Length**. Insert two more rows/members on the table.

25. Change the values of the table as per *Figure 3.52*:

🖫	Member	Part Number	🔍 Length
1	Shaft-Short	Shaft-Short	165 mm
2	Shaft-Medium	Shaft-Medium	180mm
3	Shaft-Long	Shaft-Long	200 mm

Figure 3.52: Changes required to the Member, Part Number, and Length
parameters for the iPart configurations of Shaft.ipt

26. Select **Verify**. Select **OK** to finish.

27. Hold *Shift* and select all three members of the Shaft.ipt iPart. Then, right-click and select **Generate Files**, as per *Figure 3.53*:

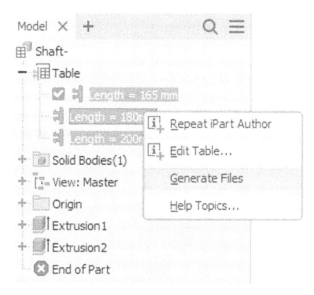

Figure 3.53: Generation of files from the Shaft iPart

28. Save and close Shaft.ipt. Then, open ScrewDriverAssy.iam.

29. Perform a local update on the file. In the **Model Browser**, right-click **Shaft**, then select **Component | Replace**.

30. Browse in the Chapter 3 folder for Shaft.ipt and select **Open**. Leave the default value as 165 mm and select **OK**.

31. Accept the unresolved mate constraints.

32. Open the **Shaft** node in the **Model Browser** and note the unresolved mates, as shown in *Figure 3.54*. We will need to rectify an error in the mates once more.

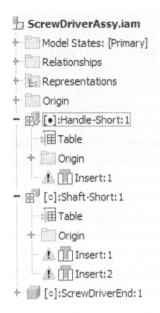

Figure 3.54: Unresolved mates as a result of a Component | Replace operation with Shaft.ipt

33. Right-click on **Insert:1**, then click **Edit**.

34. Pull the two components away from each other with a mouse drag and reapply the **Insert Mate**, as per *Figure 3.55*:

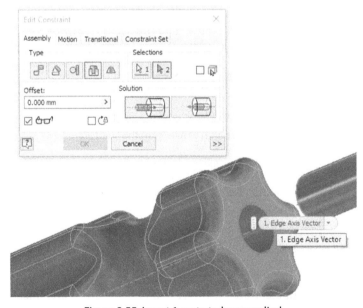

Figure 3.55: Insert:1 mate to be reapplied

35. Repeat the process for **Insert:2**, as per *Figure 3.56*:

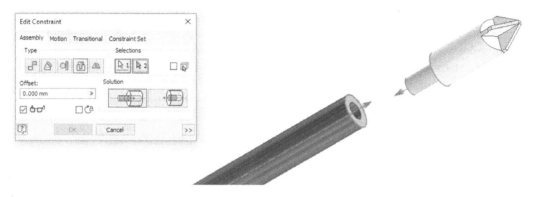

Figure 3.56: Insert:2 mate to be reapplied

36. The mates have now been resolved, and the iParts are completed and imported into the assembly. We now need to convert the assembly to an iAssembly and link the two iParts together. Select **Create iAssembly** from the **Manage** tab.

37. Select **Handle-:1:Table Replace** from the right side of the menu and select the >> button to move it across to the left.

38. Repeat this for **Shaft-:1:Table Replace**. The result of this should be that both **Handle-:1:Table Replace** and **Shaft-:1:Table Replace** have been moved to the right-hand side of the window, as per *Figure 3.57*:

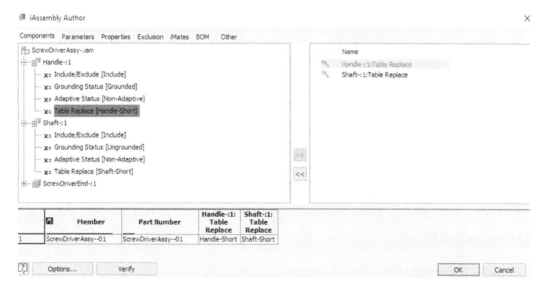

Figure 3.57: iAssembly author of ScrewDriverAssy, shown with components selected

39. Assign keys to both components by clicking on the gray key next to the name; this will turn it blue.

40. Right-click the first member of the table and insert two new rows/members.

41. Update the members by clicking on the text in the table, as per *Figure 3.58*. You will notice a dropdown is available from the iParts on some columns. This is what we defined in the earlier stages. Many more configurations can be made, but for this example, we will keep just three: **Short**, **Medium**, and **Long**.

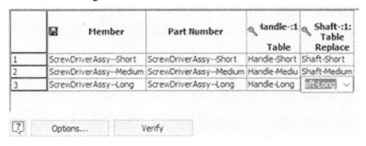

Figure 3.58: iAssembly Member, Part Number, and iPart Table values required

42. Select **Verify**, then **OK** to complete.

43. The iAssembly has now been completed. Note in the **Model Browser** of the assembly that the iAssembly configuration has appeared. Open the node, as shown in *Figure 3.59*, and double-click each iAssembly member to update the model. The screwdriver should update to a **Short**, **Medium**, or **Long** version, with both iParts working together. Once saved, this iAssembly can be configured and placed within a new assembly, or independent files can be generated from it.

Figure 3.59: Completed iAssembly options that change the assembly configuration

You have successfully created an iAssembly of a screwdriver with three length variations that comprises two iParts. This means that assembly configurations of this product can now be edited, updated, and generated efficiently, as existing data is being copied and reused.

Model credits

The following is the model credit for this chapter:

Steering Wheel Lower Lid (by Furkan Güler): `https://grabcad.com/library/steering-wheel-lower-lid-1`

4

Freeform, Surface Modeling, and Analysis

Surfaces in Inventor are non-solid, zero-thickness features that define contoured shapes and geometry. They are often used to depict complex forms where traditional solid modeling tools would be limited in communicating the design intent. Surfaces can be used in context with solid modeling techniques or independently. Surfaces in Inventor can also be used as references to aid in the creation of other geometry in the model. In most cases, surfaces in Inventor are used to create surface models that are converted to solid models, define complex curves, cuts, and extrusions, or create a skin over imported wireframe models. The Freeform modeling tools within Inventor allow you to create complex organic shaped forms with conventional modeling techniques.

Both surface modeling and freeform geometry modeling environments and tools will be covered in this chapter. We will examine the limitations of the conventional lofts and sweeps within Inventor. It is important to understand the use of these conventional tools so that you will be better placed to decide when to employ a freeform or surface modeling strategy. Then, we will explore surface modeling and freeform geometry modeling environments and tools.

In this chapter, you will learn about the following:

- Creating area lofts
- Creating path and guide rail sweeps
- Creating solid geometry from basic extruded and lofted surfaces
- Creating 3D sketches for surfacing and applying the Trim, Combine, Thicken, and Ruled Surface commands
- Learning how to trim, patch, and convert surfaces to solids with Sculpt
- Replacing a solid face with a surface
- Using the Copy Object command to create a matching contoured part to another part

- Importing surfaces into Inventor
- Validating surface designs: Zebra Analysis, Surface Analysis, and Curvature tools
- Understanding the basic features and techniques of freeform modeling
- Freeform modeling within the context of existing geometry

Technical requirements

To complete this chapter, you will need access to the practice files in the Chapter 4 folder within the Autodesk Inventor 2023 Cookbook folder.

Creating area lofts

An area loft enables you to create geometry by blending multiple profiles; they are similar to rail and centerline lofts, except with an area loft, you can define the area of each section at specific points along its path or centerline.

In this recipe, you will use an **Area Loft** operation to create the base geometry of a glass. The sketches and planes have already been created for you.

Getting ready

Open Glass.ipt file from the Chapter 4 folder.

How to do it...

In this example, we will use **Area Loft** to create the geometry of a glass. This will involve using the standard Loft command and selecting the Area Loft options. To do this, see the following:

1. With the Glass.ipt file open in Inventor, navigate to the **3D Model** tab and select the **Loft** command.

2. The **Loft** command window now opens. From here you can create a range of different lofts but in this example, we will use the **Area Loft** function to create the geometry of this glass from the simple sketches provided. So, select the **Area Loft** option in the **Loft** command window, as shown in *Figure 4.1*:

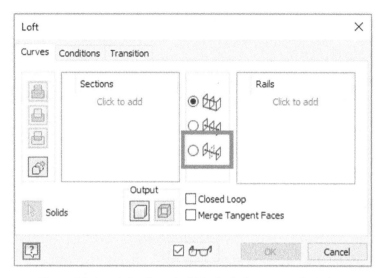

Figure 4.1: Area loft selected within the Loft command window

3. Select **Click to add** in the **Sections** box and proceed to select the two circle sketches. A preview will now be shown of the base loft as per *Figure 4.2*:

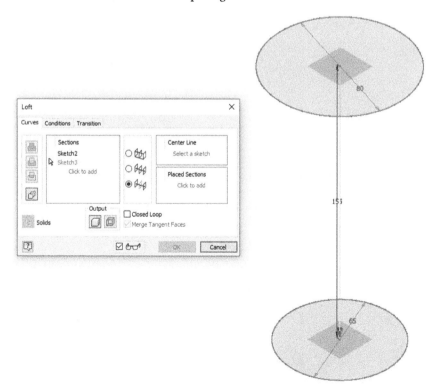

Figure 4.2: Preview of the area loft shown

At this point, the **Loft** command is still creating a normal loft. To start to use the area functionality, we now need to select and define the centerline, and specific points to incorporate different areas or profiles of the loft.

4.   Select **Center Line** in the window and then select the centerline between the two circles in the graphics window. The preview will update and show a basic loft as per *Figure 4.3*:

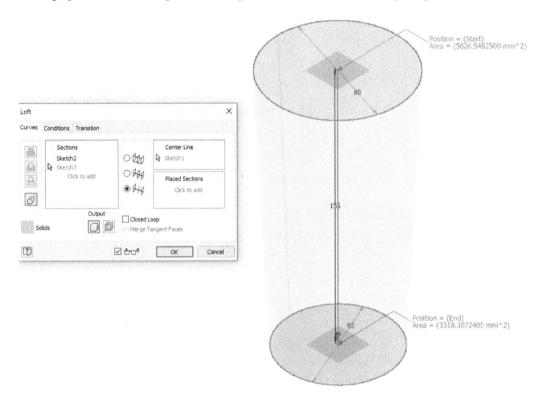

Figure 4.3: Preview of the area loft shown

5. Now that the profiles and centerlines have been defined, we can specify points along the centerline to create different areas for the area loft. Select **Placed Sections** and select a point along the centerline as per *Figure 4.4*:

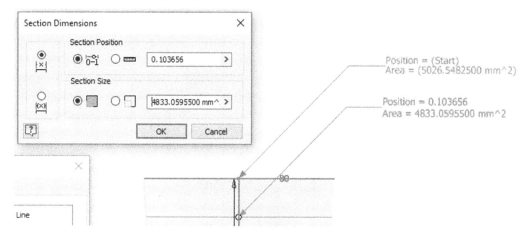

Figure 4.4: The placement of the first placed section

Inventor has now opened another window for us to define the position of the point along the centerline and the section size. The values that need to be entered are as follows:

- Position: 0.15

- Section size: 6000 mm^2

Then, select **OK**.

As you enter these values, the shape of the area loft will change.

6. Now, repeat the process in *step 5*, adding the positions and section sizes as detailed in *Figure 4.5*:

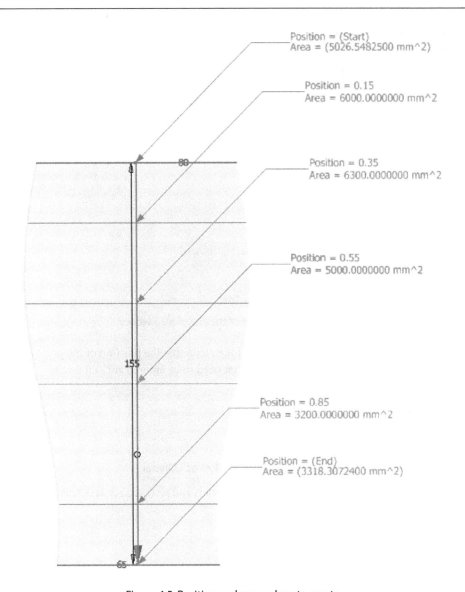

Position = (Start)
Area = (5026.5482500 mm^2)

Position = 0.15
Area = 6000.0000000 mm^2

Position = 0.35
Area = 6300.0000000 mm^2

Position = 0.55
Area = 5000.0000000 mm^2

Position = 0.85
Area = 3200.0000000 mm^2

Position = (End)
Area = (3318.3072400 mm^2)

Figure 4.5: Position and area values to create

7. Once all **Position** and **Area** values have been defined, select **OK** to complete the area loft. With one simple operation of using the area loft and a simple sketch, you have created the basic geometry of a glass.

This is a very effective and easy way to create complex curvature, without using surface modeling or freeform modeling. The operation can also be made as a cut, intersection, or outputted as a surface if required.

8. Now, select the **Shell** command from the **3D Model** tab.

9. Select the top face of the glass to remove in the shell operation and define the thickness in the **Shell** command window as 2mm. This is shown in *Figure 4.6*. Select **OK** to complete the operation:

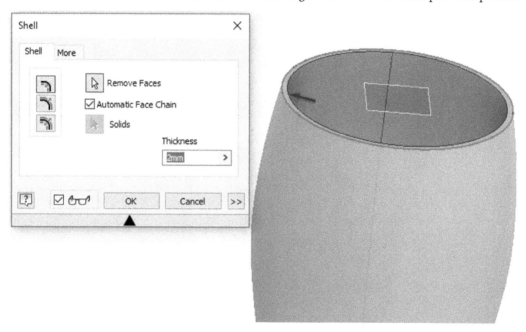

Figure 4.6: Shell thickness and face to remove from model

10. Select the **Fillet** command and apply a 2mm fillet on the two edges, shown in *Figure 4.7*:

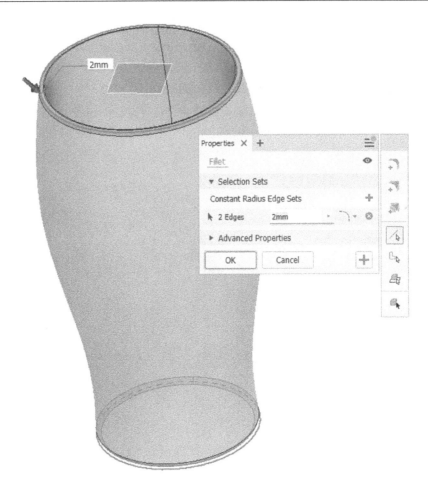

Figure 4.7: 2 mm fillets added to the glass

11. Finally, right-click on **Glass** in the **Model Browser**. Select **iProperties** | **Physical** | **Material** and then select **Glass** from the drop-down menu. Select **Apply**, followed by **OK**. All Inventor files contain iProperties, which are used to manage files, and automatically update linked information such as BOMs, drawing parts lists, and title blocks. iProperties also contain information about the part, such as mass and material.

You have now used the Area Loft command to generate the complex geometry of a glass.

Creating path and guide rail sweeps

Sweeps can be created as cuts, joins, intersections or to create new solids. They are used to sweep a profile along a defined path. Either the addition of a **guide rail** or a surface can be used to further control the output.

In this recipe, you will use the **Path & Guide Rail** function of the **Sweep** command to generate and complete an ergonomic handle of a kettle and apply a swept groove into the body of the kettle. The outcome of this is shown in *Figure 4.8*:

Figure 4.8: The completed Kettle.ipt model

Getting ready

Open Kettle.ipt from the Chapter 4 folder.

How to do it...

To start with, we will examine the model and then start to utilize the existing sketches provided, to create the cuts. To do this, let's see the following:

1. With the Kettle.ipt file open in Inventor, select **Sketch24** from the **Model Browser**. This highlights the sketch. A spline about the centerline of the curved handle will be shown in the graphics window.

    Sketch 24 is what has been created to act as the path for the sweep operation. Navigate to the **Create** tab and select **Sweep.**

2. In the **Sweep** window, do the following:

    • Select **Profile** and proceed to pick the circle profile of the curve.

    • Select **Path** and then set **Spline** to be the **Curve/Path** option.

    • Select **Follow Path** as the **Orientation** option.

    • Select **Intersect** as the **Boolean** option.

    These options are shown in *Figure 4.9*:

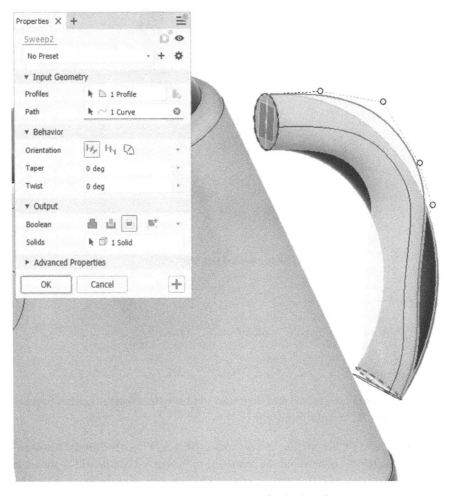

Figure 4.9: Sweep preview for the handle

3. Select **OK** to complete the sweep. The sweep is generated and the handle is intersected with the sweep to create a more ergonomic design.

4. The ends of the handle need to be defined and joined or blended to the profile of the kettle housing. Select **Start 2D Sketch** and create a new sketch on the flat circular face shown in *Figure 4.10*. For the sketch, select **Project Geometry** and then select the face:

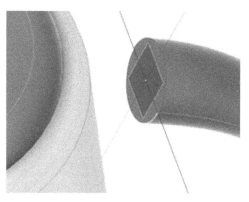

Figure 4.10: Face to create a new sketch and project geometry

5. Select **Extrude** and then select the newly created projected sketch as the profile. Set **Distance** in the menu to the **To** option and select the surface of the kettle. The preview should resemble *Figure 4.11* with the extrusion blending into the body of the kettle:

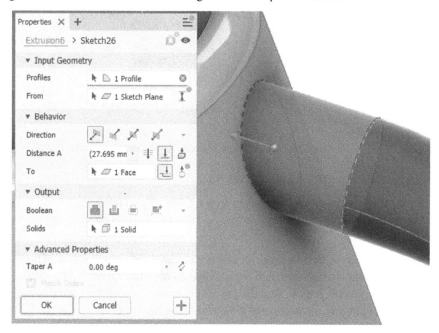

Figure 4.11: Extrusion to create that completes the top of the handle

6. Select **OK** to complete.

7.   Repeat the projected sketch and extrusion. Remember to define the extrusion as **To** for the bottom of the handle; the result of this is shown in *Figure 4.12*:

Figure 4.12: The completed handle on the kettle

You have now completed the handle of the kettle.

We will now create a cut sweep along a path with a guide rail and reference surface to create an aesthetic detail on the kettle housing. The base sketches and planes have already been defined for you in the assembly.

A surface that intersects the model has already been created, following a spline of how we want the detail to be defined. This has then been projected onto the curved surface using the 3D **Intersection Curve** sketch tool. Near the top of the handle, a circle has been drawn to define the profile of the cut on a plane created at the end of the intersection.

8. Select **Sweep** and navigate to the top of the handle. Locate *5mm Diameter Circle* and select this as the profile for the sweep operation. This is shown in *Figure 4.13*:

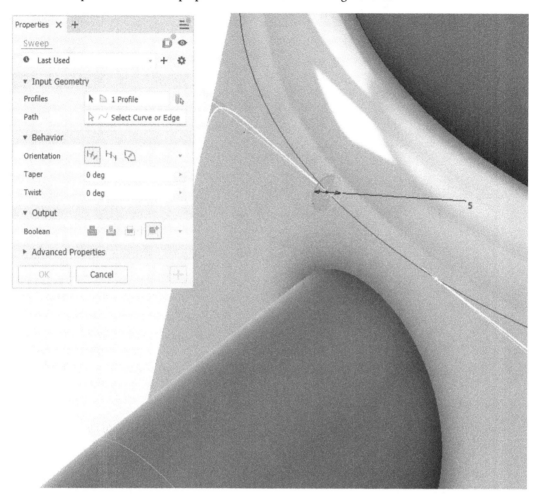

Figure 4.13: Profile to select for the second sweep operation

9. Select **Curve** in the **Sweep** menu, and then select the projected intersection curve on the housing. It is displayed in *yellow* by default. *Figure 4.14* shows this selected:

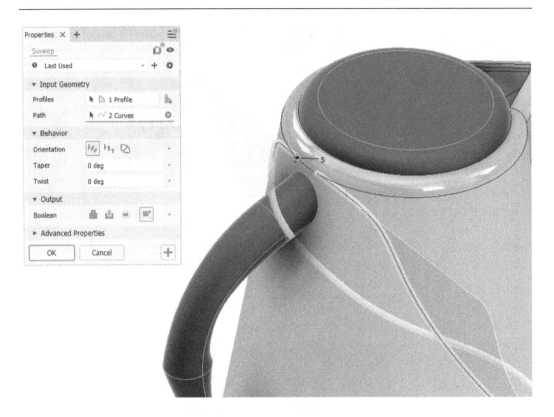

Figure 4.14: Intersection curve selected to form the path of the sweep

10. For **Orientation**, select **Guide**.

11. Select the cylindrical face of the kettle as the given **Reference Surface**, illustrated in *Figure 4.15*:

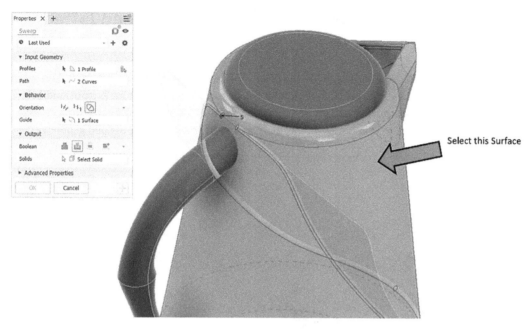

Figure 4.15: Surface to select

12. **Boolean** should default to **Cut**; if it's not, select **Cut.**

13. Select **Solid** in the window and then navigate to the **Model Browser** and select **Solid1** from the **Solid Bodies** folder, as per *Figure 4.16*:

Figure 4.16: Selection of Solid1 for the sweep operation

14. Select **OK** to complete the operation.

You have completed the kettle, by using advanced sweep functionality, including sweeping along a path, sweeping along a path with a guide rail, and sweeping along a path and guide surface.

Creating solid geometry from basic extruded and lofted surfaces

With this recipe, you will complete the modeling of an EV charger using basic surface tools. You will learn how to create a basic extruded surface for use as a termination point for an extrude and how to create and apply a lofted surface, which transitions between two edges on different surfaces. The resultant surfaces will then be combined with a stitch operation and converted to solid geometry.

The completed model of the EV charger you will create is shown in *Figure 4.17*:

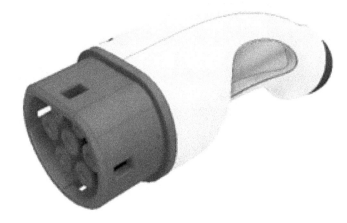

Figure 4.17: Completed EV charger model

Getting ready

You will need to access the EV Charger.ipt file in the EV Charger folder within the Chapter 4 folder.

How to do it...

The EV charger model is partially created and you do not have to build this from scratch. We will be applying surface modeling techniques to complete the model. To do this, let's see the following:

1. Open EV Charger.ipt. The unconsumed sketches required have been left visible and are shown in *black* and *yellow* on the model.

2. The first objective is to create the curved geometry in the center of the EV charger where there is currently a large opening, shown in *Figure 4.18*. We will extrude **Sketch 2** as a surface and select the **Extrude** command:

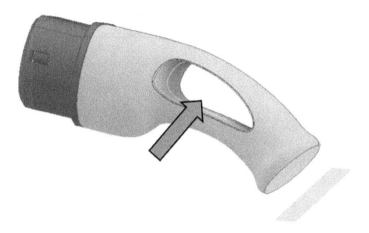

Figure 4.18: The EV charger model with the opening we will complete using surface modeling

3. In the **Extrude** command, navigate to the top right of the window and select the **Surface Mode ON** button as per *Figure 4.19*. This changes the **Extrude** command to extrude a surface instead of a solid geometry:

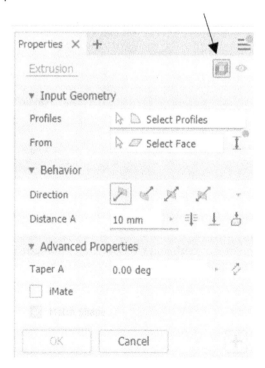

Figure 4.19: Surface mode on selected in the Extrusion window

4. In the **Profiles** section of the window, deselect **Profile1** if this is selected by default; do this by clicking the **x** symbol next to **Profile1**. Then, select **Profile** and pick the sketched spline as shown in *Figure 4.20:*

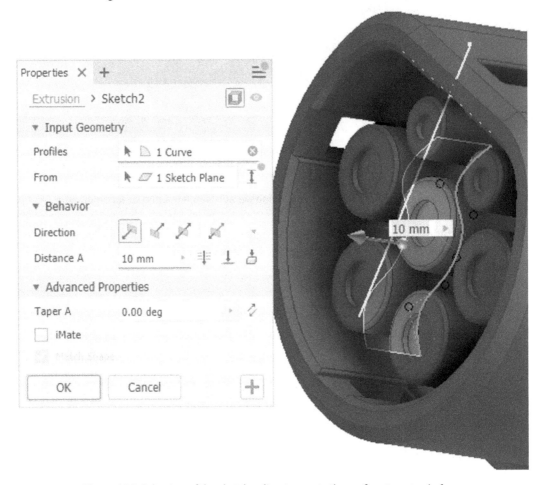

Figure 4.20: Selection of the sketch spline to create the surface to extrude from

5. In the **Behavior** section of the command window, under **Direction**, flip the direction of the extrusion by selecting the **Flipped** icon, as shown in *Figure 4.21:*

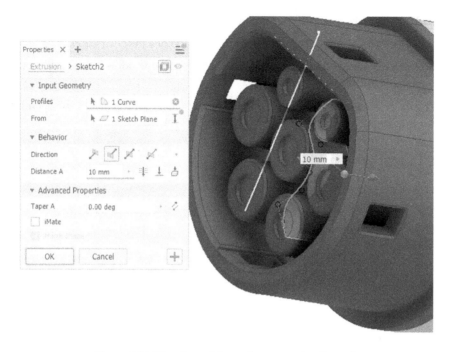

Figure 4.21: Direction of the surface extrusion flipped

6. Extend the **Distance** value by clicking and dragging the orange display arrow. Extend this to around 150 mm. The preview should display as per *Figure 4.22* once complete:

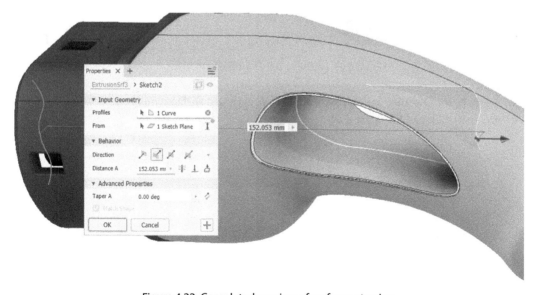

Figure 4.22: Completed preview of surface extrusion

7. Select **OK** to complete the **Surface Extrude** operation.

8. Select the **Extrude** command again to start the creation of a new extrusion.

9. Select **Surface Mode OFF** in the top right of the command window if this is still active from the previous extrusion.

10. For **Profile**, select **Sketch 1,** which is the projected sketch in the center of the opening of the EV charger, as shown in *Figure 4.23*:

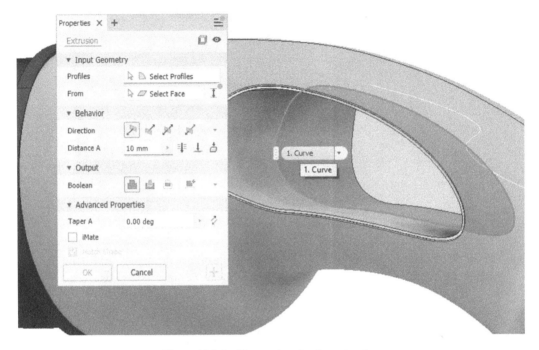

Figure 4.23: Profile to select for the extrusion

11. Select **To** for **Distance** and select the face of the extruded surface you previously created. The preview will look like *Figure 4.24*:

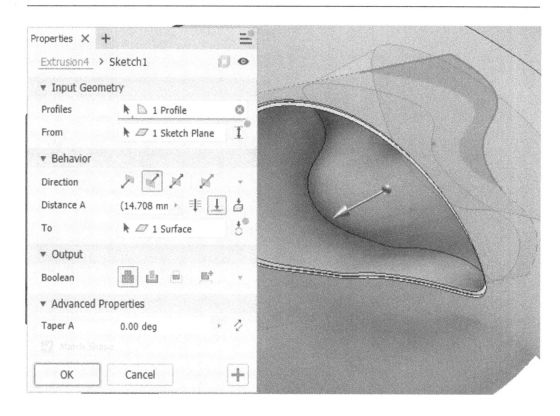

Figure 4.24: Preview of the extrusion after the profile and the surface have been created

12. Select **OK**.

This completes the operation. You have now successfully used a basic extruded surface for use as a termination point to create the desired geometry.

This geometry needs to be replicated on the other side of the model.

13. Navigate to the other side of the opening in the graphics window and select **Mirror** from the **Pattern** commands.

14. Select the solid extrusion created by *step 12* as the feature to mirror.

15. Select **Mirror Plane** in the **Mirror** window and then select the surface shown in *Figure 4.25*:

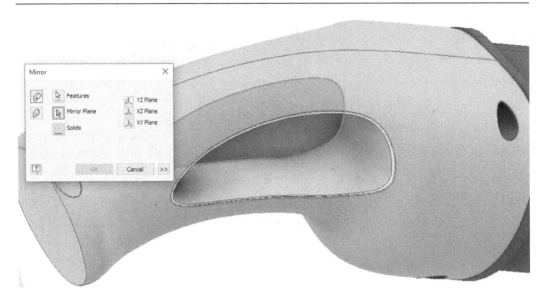

Figure 4.25: Showing the mirror plane to select

16. Select **OK** to complete.

17. Navigate to the **Model Browser**, right-click **ExtrusionSf#**, and then select **Visibility**. This removes the visibility of the surface extrusion, as it is no longer required.

18. Use **Start 2D Sketch** on **XY Plane**.

19. Select **Project Geometry** and then select **Work Plane 1**.

20. Create the sketch as shown in *Figure 4.26*:

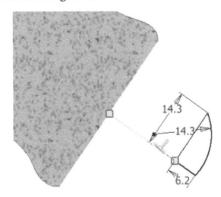

Figure 4.26: Sketch to be created on the XY plane

21. Select **Finish Sketch** and then select **Revolve**.

22. Select **Surface Mode ON**.

23. Select the recently created sketch from *step 20* as the profile.

24. Select the line that is 6.2 mm long as the axis. The preview generated is shown in *Figure 4.27*:

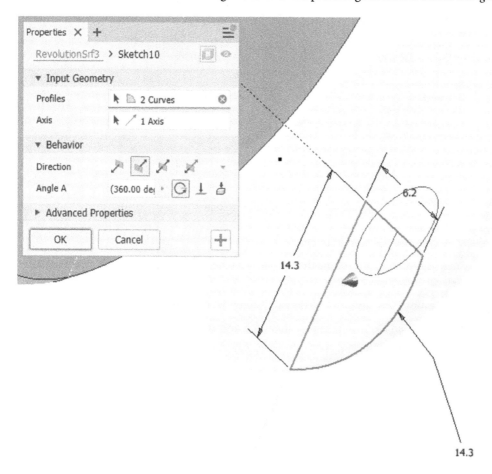

Figure 4.27: The preview of the extruded revolving surface

25. Select **OK** to complete the revolve as a surface.

26. Navigate to the **Model Browser**, right-click **Work Plane 1**, and then select **Visibility**.

27. Select **Loft** from the **Create** tab. Select **Output** as **Surface**.

28. Click **Sections** and then **Click to Add**. Select the sketch lines shown in *Figure 4.28*:

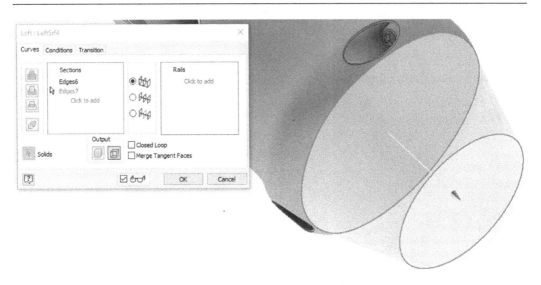

Figure 4.28: Sketch sections to select

29. Select **OK** to complete the lofted surface.

30. We now need to stitch and combine the two surfaces, so that they form a quilt or composite surface. Select **Stitch** from the **Surface** tools.

31. Change **Maximum Tolerance** to 0 . 2 mm and select the two surfaces to stitch together.

32. Select **Apply** and then **Done** to complete the stitch operation.

33. Now, we can convert the quilted surface into solid geometry. Select the **Thicken/Offset** command from the **Modify** panel.

34. Select the two faces as shown in *Figure 4.29*. Change **Thickness Distance** to 4 mm.

35. Select **OK**:

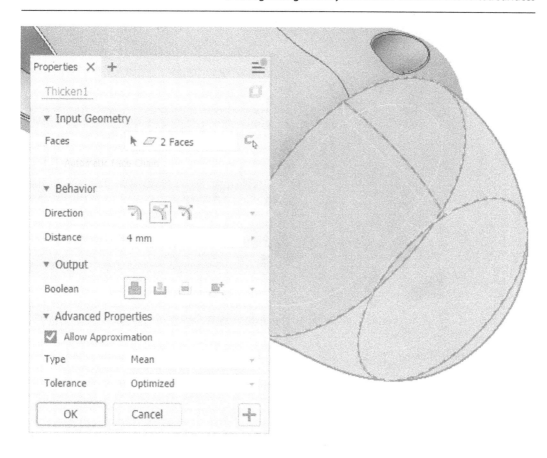

Figure 4.29: Thicken/Offset command with faces selected and thickness distance defined

36. Use **Start 2D Sketch** on **Work Plane 1**.

37. Then, create the following sketch on **Work Plane 1** as shown in *Figure 4.30*. The internal circle needs to be 15 mm in diameter:

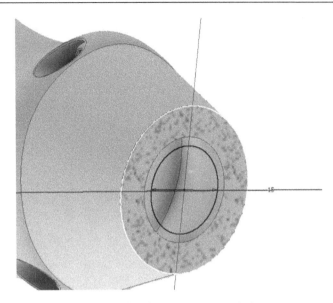

Figure 4.30: Sketch to create on Work Plane 1

38. Select **Finish Sketch**.

39. Select **Extrude**.

40. Select the sketch previously created as the profile, change **Direction** to **Symmetric**, change **Distance** to 50 mm, and change **Boolean** to **Cut**. The preview is shown in *Figure 4.31*:

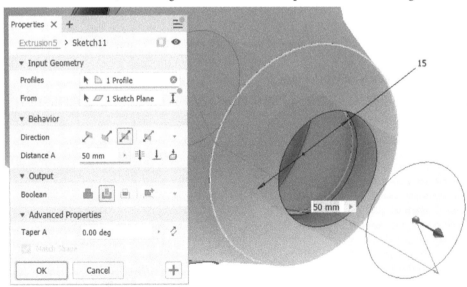

Figure 4.31: Extrusion settings to apply

41. Select **OK** to complete:

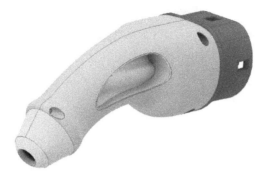

Figure 4.32: Completed EV charger

You have successfully completed the EV charger using basic surface tools such as **Extruded Surface**, **Lofted Surface**, and **Stitch**.

Creating 3D Sketches for Surfacing and applying the Trim, Combine, Thicken, and Ruled Surface commands

In this recipe, you will use a range of 3D surface tools to create an exhaust manifold. Initially, 3D sketches will be created to act as guides for the surfaces to follow. You will then perform surface sweeps and refine these with the **Trim**, **Combine**, **Thicken**, and **Ruled Surface** commands.

Getting ready

To begin this recipe, you will need to open the Exhaust_Manifold START.ipt file from the Chapter 4 folder.

How to do it

To begin, we will have to create 3D sketches to form the basis of the 3D surface sweeps. We will then use these to perform surface sweeps, apply a thickness, and edit as desired to create the manifold. So, to do this, see the following:

1. Look at the existing model of Exhaust_Manifold START.ipt in the graphics window.

   You can see that already a base unconsumed sketch of a circle has been created, and the ends of the manifold have already been detailed. The first task is to create the swept geometry of the manifold, using assorted surface tools available. To create the swept surfaces of the manifold, we will need to construct a 3D sketch to act as a guide curve.

   Under **Start 2D Sketch**, open the dropdown and click **3D Sketch**, as shown in *Figure 4.33*:

Figure 4.33: Location of the 3D Sketch tools

2. With the *3D Sketch tools active*, the ribbon will change and show the 3D sketch commands. Select **Line**.

3. Hover the cursor over the circle in the manifold as shown in *Figure 4.34* and, once the green midpoint is visible, left-click to start the sketch:

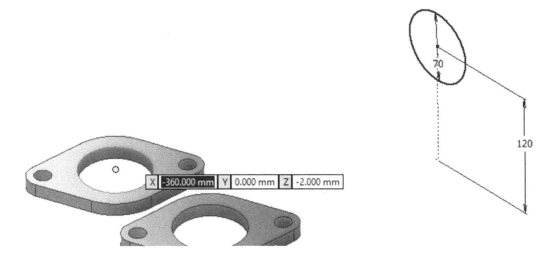

Figure 4.34: Circle on manifold selected and midpoint active

4. Create a 3D sketch line in the **Z**-axis. Type 112mm and hit *Enter*. The 3D sketch line is created as per *Figure 4.35*:

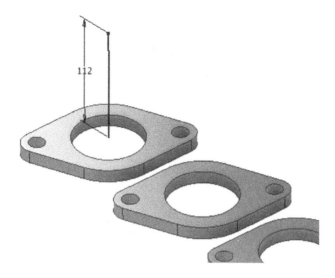

Figure 4.35: 112 mm L 3D Sketch line to create

5. Select **Line** again from the **3D Sketch** tools and create the line shown in *Figure 4.36*. The length must be 65mm and in the direction of **Y**.

6. Create another 3D sketch line that joins the two previous lines together, as per *Figure 4.36*. Select the previous endpoint that was created in *step 5* to complete this:

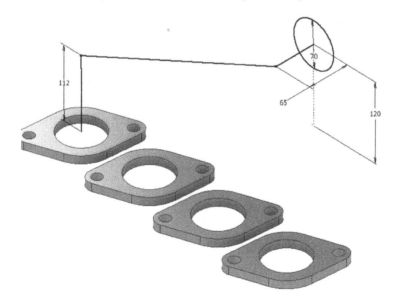

Figure 4.36: Additional 3D sketch lines to create

7. Select **Bend** from the **3D Sketch** tools.

8. Type 8 0mm for the **Bend Radius** option.

9. Select both the 112 mm line and the last interconnecting line created.

10. Change the value in the **Bend** window to 6 5mm.

11. Select the 65 mm L line and the middle connecting line to apply the 65 mm bend to the 3D sketch. The result is shown in *Figure 4.37*:

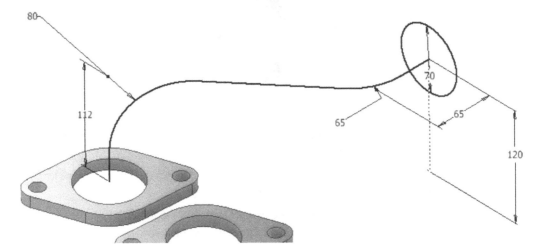

Figure 4.37: Bends to be created on the 3D sketch lines

12. Select **Finish Sketch**.

13. This defined 3D sketch line will form the basis for the surface sweep that will create the first of the exhausts for the manifold. Select the **Sweep** command from the **3D Model** tab.

14. Ensure that the option for **Surface Mode** is turned on for the **Sweep** command.

15. Select the 70 mm circle as the profile and the 3D sketch spline as the centerline for the sweep or (path). The preview will resemble *Figure 4.38*:

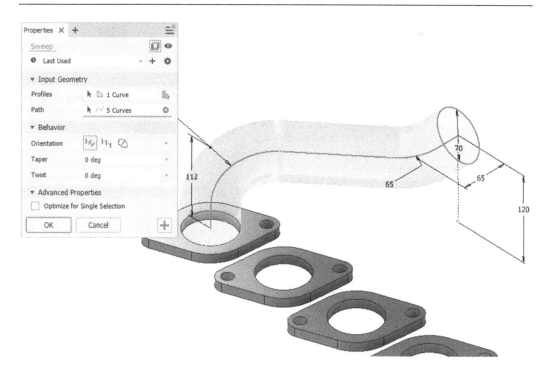

Figure 4.38: Surface Sweep options required to create the first exhaust of the manifold

16. Select **OK** to complete. You have now created a surface that uses a 3D sketch as a reference. The basis of the first exhaust is complete.

17. The surface has been created but may not be very visible. Right-click on **SweepSrf1** in the **Model Browser** and then select **Translucent**. This makes the surfaces much easier to see on a white background.

18. We must now define the path for the second sweep for the second exhaust. The process is very similar to the first exhaust, except there are slightly different dimensional values in this case. Select **Start 3D Sketch.**

19. Create a 3D sketch line from the second manifold that is 112mm in length and in the direction of the **Z** axis. This is shown in *Figure 4.39*.

20. Create a 150mm 3D sketch line from the 70 mm diameter circle in the **Y** axis, shown in *Figure 4.39*:

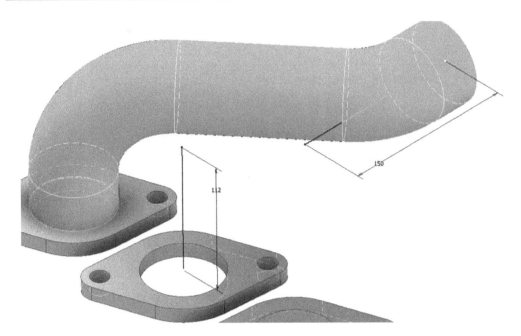

Figure 4.39: 3D sketch lines to create for the second exhaust

21. Create a final 3D sketch line that joins the two lines previously created.

22. Select the **Bend** command. Apply a 50mm bend to the corner point of the sketch path, as shown in *Figure 4.40*:

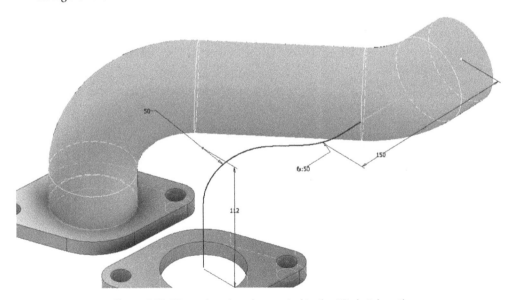

Figure 4.40: 50 mm bends to be created in the 3D sketch path

23. Use **Start 2D Sketch** on **Work Plane 1**.

24. Select **Project Geometry** from the **3D Sketch** tab and select the edge of the surface to project the surface geometry, as shown in *Figure 4.41*:

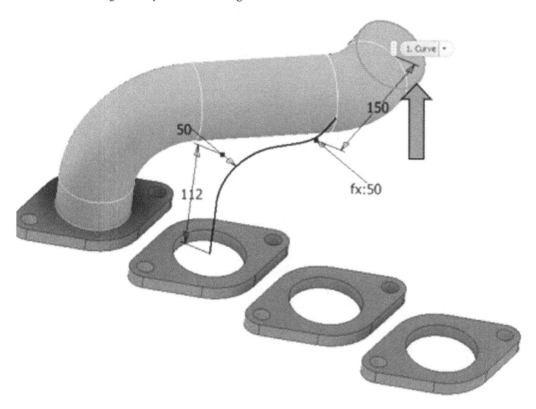

Figure 4.41: Project Geometry used to project the surface of the surface sweep

25. Select **Finish Sketch**. Then, select **Sweep**.

26. With **Surface Mode ON** selected, set the circle as the profile, and the unconsumed 3D sketch as the path, shown in *Figure 4.42*.

27. Select **OK** to complete the surface sweep:

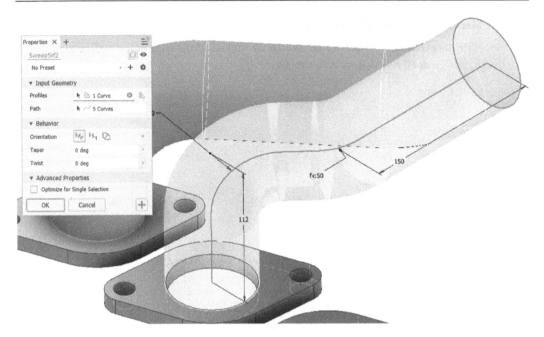

Figure 4.42: Second surface sweep operation to complete

28. We will now perform a half-section view to examine the sweep path created. Select the **View** tab.

29. In the **Visibility** panel on the far left, select **Half Section View**, and then select the XY plane from the **Model Browser**.

30. Drag the section view by left-clicking and holding the orange arrow, then moving the mouse upward. Do this until the section view resembles *Figure 4.43*:

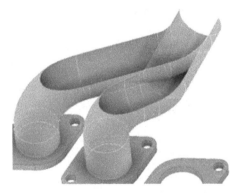

Figure 4.43: Creating the section view until the model resembles this

Additional unwanted surface geometry has been created internally because of doing two surface sweeps that intersect. This needs to be cleared out with a **Trim Surface** operation.

31. Select the **3D Model** tab, navigate to the right of the ribbon, and in the **Surface** tools, select **Trim**.

32. You are now able to specify the **Cutting Tool** settings and the piece of geometry to use the **Remove** operation on. Select the first sweep created as the cutting tool and select the internal unwanted geometry to remove it. The preview is shown in *Figure 4.44*:

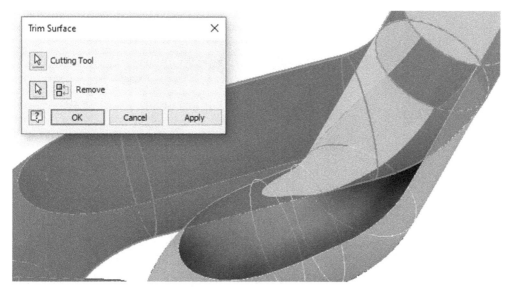

Figure 4.44: Trim Surface cutting tool and removal surfaces shown

33. Select **Apply** to complete the trim. The internal geometry is now removed.

34. Repeat this **Trim Surface** operation once more for the other internal geometry, until the surfaces resemble those in *Figure 4.45*:

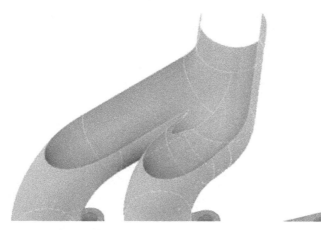

Figure 4.45: The result of two Trim Surface operations to remove the internal unwanted surfaces

35. Use **Start 2D Sketch** on **Work Plane 1**.

36. Sketch the geometry shown in *Figure 4.46* on **Work Plane 1**. Use **Project Geometry** and, when complete, select **Finish Sketch**:

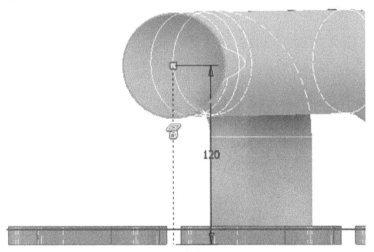

Figure 4.46: Geometry to create on Work Plane 1

37. Select **Start 3D Sketch** and create the line as shown in *Figure 4.47*, from the 120mm sketch line endpoint. Then, select **Finish Sketch**:

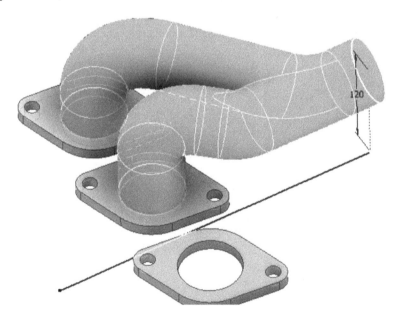

Figure 4.47: 3D sketch line to create from the 120 mm sketch line

38. Using these sketch lines, we will now create a three-point plane to act as a mirror plane, which will enable us to copy the two existing surfaces onto the remaining manifold plates. In the **3D Model** tab, select **Plane**, and then the three-point plane.

39. Select the three points on the sketch lines, as shown in *Figure 4.48*, to complete the plane:

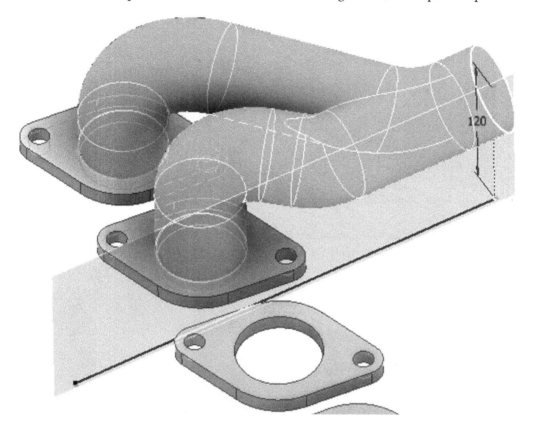

Figure 4.48: Three-point plane to be created from the sketch lines

40. With the **3D Model** tab active, navigate to the right of the ribbon and select **Mirror**.

41. Select the two surface sweeps as the features to mirror and the newly created **Work Plane 2** as the **Mirror Plane**. Select **OK** to complete.

    Right-click the new mirrored surface features in the **Model Browser** and select **Translucency**, to make the surfaces more visible. The result is shown in *Figure 4.49*:

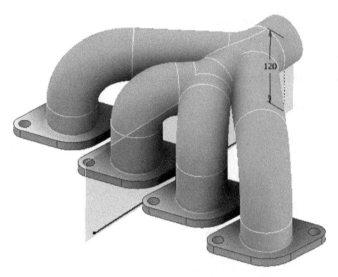

Figure 4.49: Mirror operation of the existing swept surfaces completed

42. The basis of the exhausts for the manifold has now been completed. During the **Mirror** operation, some unwanted internal geometry has been created, which we will now remove. Create a half-section view of the model as per *Figure 4.50*:

Figure 4.50: Half-section view of the manifold showing unwanted internal surfaces

43. Select the **3D Model** tab and then select the **Trim Surface** command:

  • Select the **Cutting Tool** and **Remove** references shown in *Figure 4.51* as the first **Trim Surface** operation and select **Apply**:

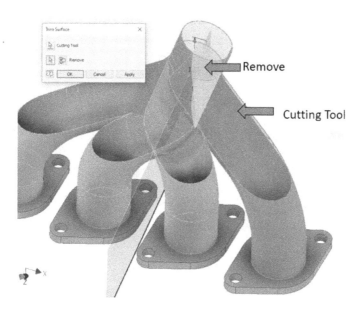

Figure 4.51: Trim Surface operation 1

- For the next **Trim Surface** operation, select the **Cutting Tool** and **Remove** references as per *Figure 4.52*. Then, select **OK**:

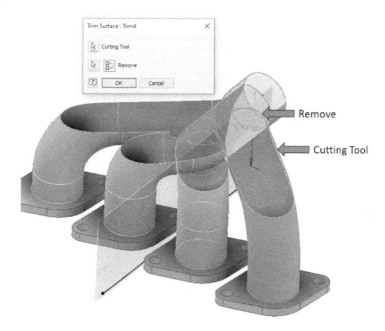

Figure 4.52: Trim Surface operation 2

- Select the **Cutting Tool** and **Remove** references as per *Figure 4.53*. Then, select **OK**:

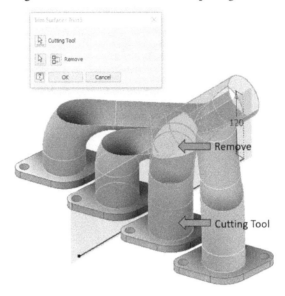

Figure 4.53: Trim Surface operation 3

- Select the **Cutting Tool** and **Remove** references as per *Figure 4.54*. Then, select **OK**:

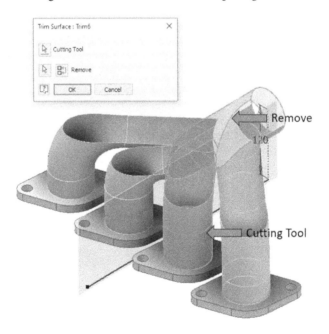

Figure 4.54: Trim Surface operation 4

- Select the **Cutting Tool** and **Remove** references as per *Figure 4.55*. Then, select **OK**:

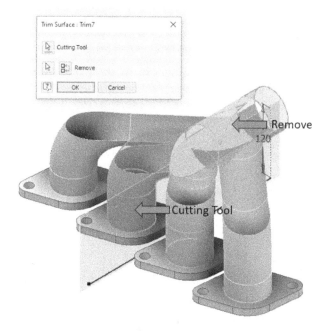

Figure 4.55: Trim operation 5

Figure 4.56 shows the completed result after all the successive **Trim Surface** operations. All unwanted internal surface geometry has been removed:

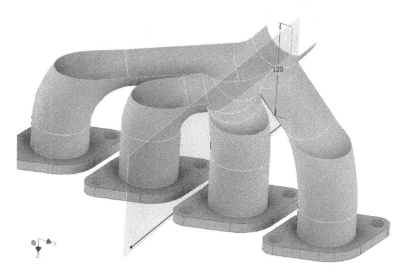

Figure 4.56: Completed surface of the manifold

44. The successive trim operations have removed all the internal surfaces so that the manifold is correct. Now, select the **View** tab, followed by **End Section View**.

45. Next, the multiple surfaces must be combined using a **Stitch** command. Select the **3D Model** tab and navigate to the **Surface Modelling** tools. Select **Stitch**.

46. With the **Stitch** command active, select all four exhaust surfaces and select **Apply**. The surfaces are stitched together as a **composite surface**. In doing this, the translucency will become active again, as this is a new surface now. *Figure 4.57* shows this:

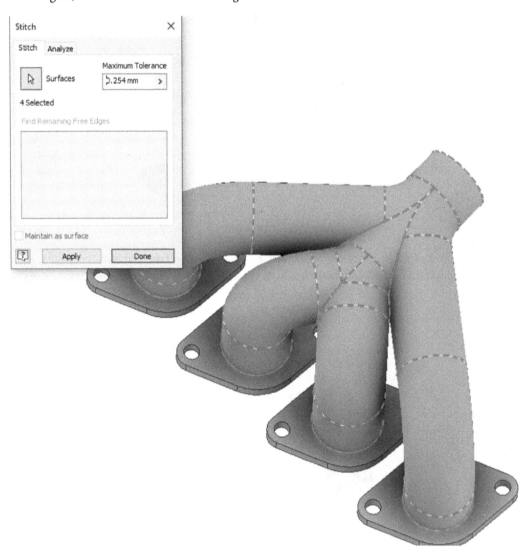

Figure 4.57: Surfaces stitched

47. Right-click on **Stitch Surface 1** in the **Model Browser**. Select **Translucent** to make this more visible.

48. Now that a surface is fully defined, we can begin to use the **Thicken** command to convert the surface to a solid body. To do this, in the **3D Model** tab, under the **Modify** panel, select **Thicken/Offset**.

    This command is used to thicken or offset geometry and works for surfaces too. We will use this to add a specific thickness of material onto the surfaces we have just created. With the **Thicken** command active, it prompts you to select surfaces or faces to add material. We could select all the surfaces in one go and select **OK**, but this is likely to fail due to the complexity of the surfaces. Instead, we will achieve this in several operations and then combine the solids.

49. In the **Thicken/Offset** menu, under **Direction**, select **Inside** and change **Thickness** to 1 mm.

50. Select the faces shown in *Figure 4.58*. For the **Boolean** option, select **New Solid**. Then, select **OK**. The 1 mm thickness is added to the surfaces:

Figure 4.58: Thicken options required and surfaces to select

51. Apply another 1 mm **Thicken** with **Inside Direction** selected, followed by **New Solid** on the faces, as shown in *Figure 4.59*.

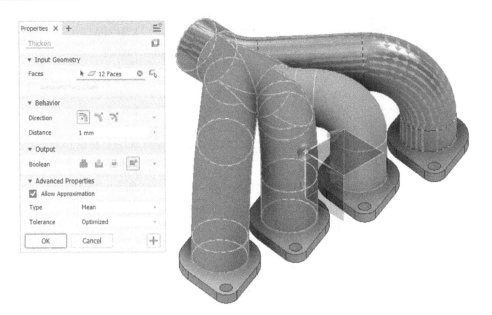

Figure 4.59: Second Thicken operation to complete and faces to select

52. Create a new **Thicken** operation with the same values as *step 51* on the remaining faces as per *Figure 4.60*:

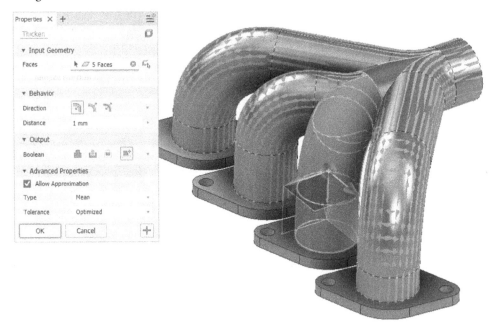

Figure 4.60: The last Thicken operation to complete and faces to select

53. The preview of the model will display the material and the surface at the same time. This can be confusing, so now that the **Thicken** operation is complete, we can hide the visibility of the underlying surface. In the **Model Browser**, right-click on **Stitch Surface1** and select **Visibility**.

54. To make modifications to the model easier, we will now combine the several solid bodies of the model into one. Navigate to the **Model Browser** and note the **Solid Bodies** folder. It is populated with seven separate solid bodies.

55. With the **3D Model** tab active, navigate to the **Modify** panel and select **Combine**.

56. Select **Basebody** and then select all the solid bodies in the graphics window, as shown in *Figure 4.61*:

Figure 4.61: The Combine operation and settings to apply

57. Select **Join** and then **OK**. The solids have now been combined into a single solid. The **Solid Bodies** folder in the **Model Browser** should now say (**1**).

58. Select **Fillet** and then select the edges shown in *Figure 4.62*. Add a **Fillet** value of 15 mm and select **OK**. If the **Combine** operation had not been carried out, the fillet would not have been computed this way, as Inventor would have still recognized this model as separate solid bodies:

Figure 4.62: 15 mm fillet to apply to the model

59. We now need to design the next plate on the outlet of the exhaust manifold. To do this, we will use the **Ruled Surface** command.

    The **Ruled Surface** feature creates surfaces that extend a specified distance and direction from a selected edge. They are effective to use in casting or molding designs, as a draft can easily be added or to create smooth extensions to complex surfaces.

    As we will be working with overlapping surfaces, it is easier to change the background from white to something else temporarily to make seeing the surfaces much easier.

    Go to **File | Options | Colors Tab**. Change **In-canvas Color Scheme** from **Presentation** to **Forest** (this can be changed back later). Then, select **Apply**.

60. From the **Surface Modelling** tools, select **Ruled Surface**.

61. There are several types of ruled surfaces that can be created – for this, we will select the **Normal** option. This will create a surface normal for the selected edge.

    Select the outer edge of the exhaust manifold, as shown in *Figure 4.63*.

62. Set **Distance** to 40 mm (you may need to flip the direction if the preview does not display). Then, select **Apply** to create:

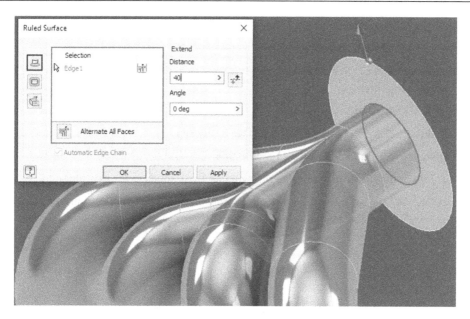

Figure 4.63: Ruled surface applied to the opening of the exhaust manifold

63. With the **Ruled Surface** command still active, select the edge of the ruled surface created in *step 62*. This is shown in *Figure 4.64*.

64. Change **Angle** to -20 degrees and **Distance** to 30 mm:

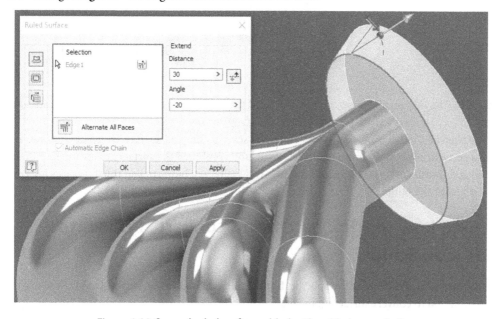

Figure 4.64: Second ruled surface added with a -20-degree draft

65. Select **Apply** to complete.

66. Then, select the **Stitch** command.

67. Select the two RULED SURFACES created as the surfaces and select **Apply**, followed by **Done** to complete the stitch.

68. Select the **Thicken/Offset** command.

69. Select the stitched ruled surface as the surface to thicken.

70. Set **Distance** to 3 mm and **Direction** to **Outside**.

71. Select **Join** as the **Boolean** operation; this will join these solids to the existing solid.

72. Select **OK** and hide the internal surface once complete. The **Thicken** operation should result in *Figure 4.65*. You can also add a 10 mm fillet to the outer edge and two 20 mm holes to complete it:

Figure 4.65: The Thicken operation complete on ruled surfaces with an optional fillet and holes added

Using 3D sketches, you have created centerlines for several complex surface sweep operations. You have also successfully trimmed, stitched, thickened, and combined the surfaces to create the exhaust manifold and transform this into a single solid body. Finally, you have applied ruled surfaces to create the end plate for the outlet.

Learning how to trim, patch, and convert surfaces to solids with Sculpt

In this recipe, you will learn how to apply the **Trim** and **Patch Surface** commands and how to use the **Sculpt** command to convert surfaces into solids.

Getting ready

Open `Surface_Part.ipt` from the `Chapter 4` folder.

How to do it...

With `Surface_Part.ipt` open, we will start to trim the existing surfaces to the desired profiles, followed by extending a profile and patching the gaps in the surface. We will then convert this to a solid model. To get started, see the following:

1. Select the **Trim** command from the **Surface** tools. We will use this to trim one surface from another to create the desired profile.

2. Select **ExtrusionSrf2** as the cutting tool and **Extrusion Srf1** as the section to remove, as shown in *Figure 4.66*:

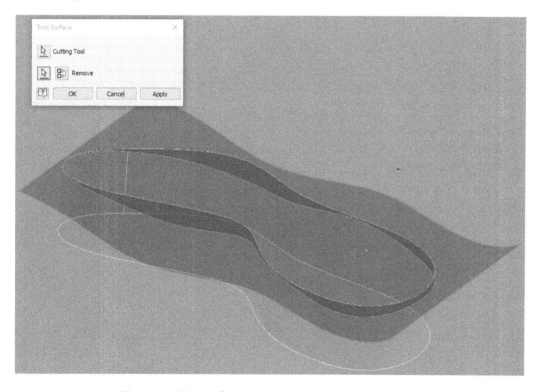

Figure 4.66: Trim Surface options to select on Surface_Part.ipt

3. Select **Apply** to complete.

4. With the command still active, select the following surfaces shown in *Figure 4.67* to trim them from the model:

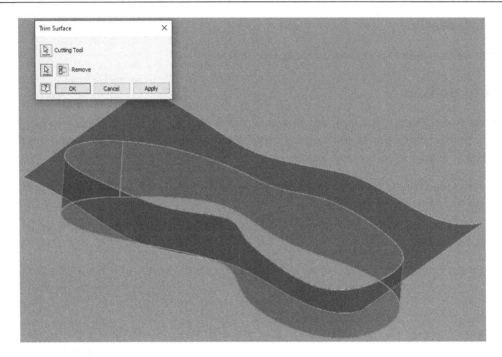

Figure 4.67: Second Trim operation to apply

5. **Extend** can be used to extend surfaces the desired amount. Select **Extend** from the **Surface** commands.

6. Select the flat edge as shown in *Figure 4.68*, change the extension to 4 mm, and select **OK**:

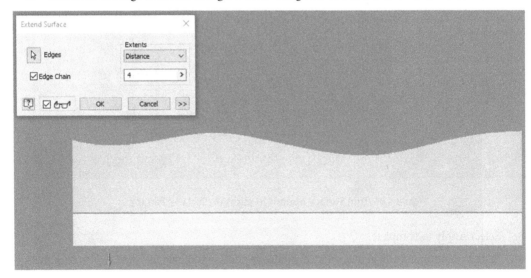

Figure 4.68: The edge to apply an extended surface

7. One end of the surface is open-ended. In this example, we need to close this boundary. One of the most important and useful surface commands that we can use to accomplish this is the **Patch** command. **Patch** allows you to close a gap between surfaces and tangency and G2 curvature can also be applied if required. The addition of guide rails can be used further to control the surface's continuity. So, select the **Patch** command from the **Surface** tools.

G2 curvature

If the radius of curvature is the same value at the common endpoint, the curvature is classed as continuous (G2).

---

8. Select the edge shown in *Figure 4.69*, and in the edge selection area, activate the drop-down menu and select **Tangent Condition**. This applies a tangency to the boundary patch.

9. Modify the **Weight** setting of the tangency from .5 to .2 and then select **OK**:

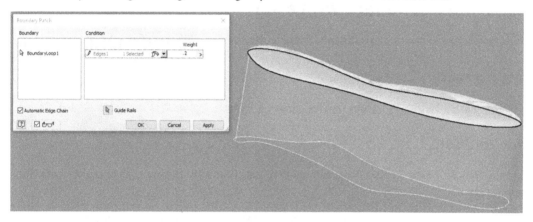

Figure 4.69: Edge to select for the patch

10. Select **Stitch** and combine the surfaces together to create a composite surface.

11. Now, we need to convert the surfaces into a more usable 3D solid or solid body. To do this, we can use the **Sculpt** command. Select **Sculpt** from the **Surface** commands.

12. Select the surfaces to sculpt. Ensure **New Solid** is active in the window and select **OK**. This will convert your stitched surface into a solid body:

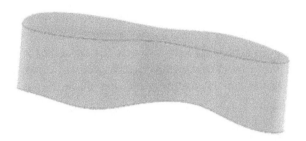

Figure 4.70: The surface converted to a solid body model using Sculpt

You have successfully applied **Surface Trim**, **Extend**, and **Patch** to create the desired outcome of this surface and converted the surface into a solid body.

Replacing a solid face with a surface

In this recipe, you will take an existing model and replace the face with that of an intersecting surface.

Getting ready

Open `Replace Face.ipt` from the `Chapter 4` folder.

How to do it...

To begin, we will use the **Replace Face** command to select the solid face we wish to replace with the chosen surface:

1. A swept surface profile has been swept along a 3D block. The surface intersects with and passes through the block. Using **Replace Face**, we can transform the flat surface of the block to reflect the surface sweep that intersects it. Select **Replace Face** from the **Surface Modeling** tools.

2. Select the existing face as the flat top of the block.

3. Select the swept surface as the new face, as per *Figure 4.71*:

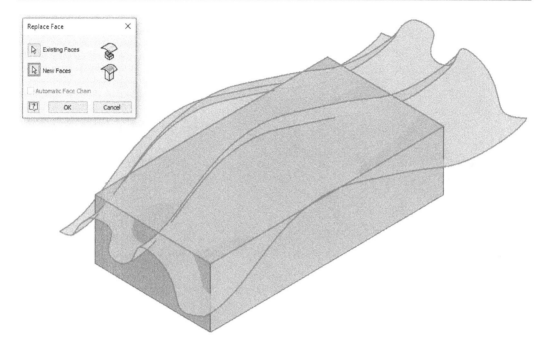

Figure 4.71: Existing and new faces to select with the Replace Face tool

4.   Select **OK**. The flat face has been replaced with the complex swept surface, as shown in *Figure 4.72*.

5.   Right-click on **SweptSrf2** in the **Model Browser** and select **Visibility** to hide the surface:

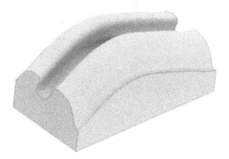

Figure 4.72: The completed surface of Replace Face.ipt

6.   The original sketch that was used to create the swept surface adapts to the changes we have made in the **Replace Face** command. We will now observe this, expand **SweepSrf2** in the **Model Browser**, right-click on **Sketch 3**, and then **Edit Sketch.**

7.   With the mouse, pull one of the spline points so that the sketch resembles *Figure 4.73*. Select **Finish Sketch**. The 3D model updates as per your changes to the spline:

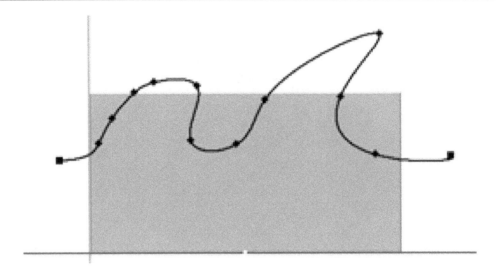

Figure 4.73: The change to make to one of the spline control points in Sketch 3

You have used **Replace Face** to replace an existing solid body face with that of a complex surface.

Using the Copy Object command to create a matching contoured part to another part

In this recipe, you will use **Copy Object** to copy the existing faces on a part file as a surface and then use **Thicken/Offset** to create a contoured part. We will then make changes to the original part using **Direct Edit** and the changes will be reflected in the overall assembly automatically

Getting ready

Open Copy_Object.iam from the Chapter 4 folder.

How to do it...

We will begin looking at the existing model and then creating an in-place component within the assembly. So:

1. With Copy_Object.iam open, select the **Assemble** tab and then click **Create** from the ribbon. Here, we will start the creation of an in-place component within the context of the assembly.

2. Rename the new component Casing.

3.   Browse the **Template** drop-down list and select `Metric\Standard (mm).ipt`.

4.   Leave the file location and BOM structure as their defaults. Then, select **OK**:

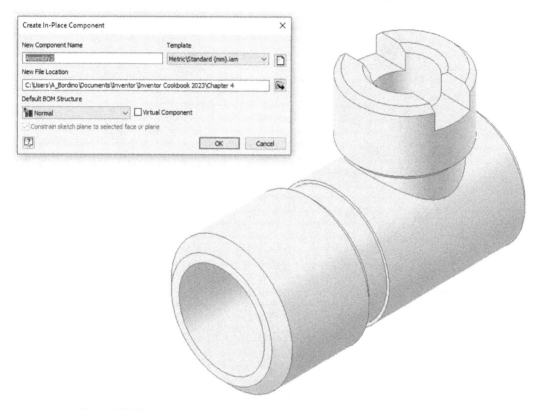

Figure 4.74: Creating an in-place component within the context of the assembly

5.   You are now prompted to select a plane to start the creation of the part. Select the YZ plane from the **Model Browser**.

6.   The **Edit Part** environment is now open for the new `Casing.ipt` part we are creating. Open the **Modify** drop-down menu and select **Copy Object**.

7.   Then, click **Bodies** and select the **Copy Object** part from the graphics window, as shown in *Figure 4.75*:

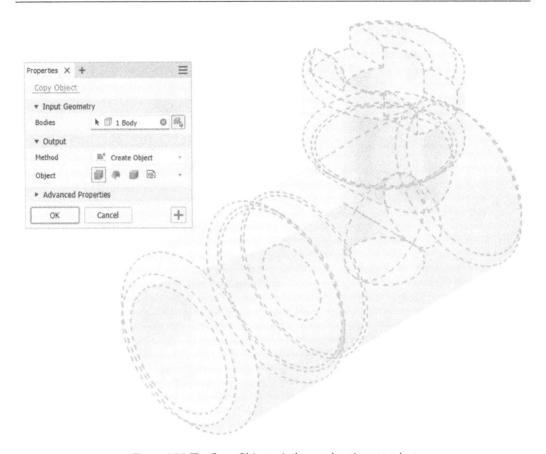

Figure 4.75: The Copy Object window and options to select

8. Ensure that the **Method** option is **Create Object** and the **Object** type is defined as **Surface**.

9. Select **OK**. The **Copy Object** command has now created a surface skin of the model.

    We can now adapt this and use a **Surface Delete** tool to remove unwanted surfaces.

10. Right-click on **Surface 1** in the **Model Browser** and disable **Translucency**.

11. In the **Modify** tab, select **Delete Face**. This is used to delete unwanted surfaces and faces.

12. To delete a face, with the command active, select a face or surface on the model in the graphics window and select **OK**. Note more than one face can be removed in one operation.

13. Delete all faces until the surface resembles *Figure 4.76*:

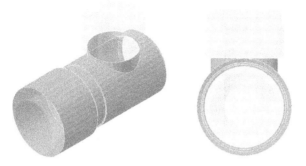

Figure 4.76: Deleting all surfaces until your model resembles this

We now have defined the extent of the contoured casing with surfaces.

14. Now, we will add solid geometry to this. Select **Thicken/Offset** from the **Modify** tab to add solid geometry to the surfaces.

15. Select the eight faces shown in *Figure 4.77* to apply **Thicken**. Set **Direction** to **Outside** and **Distance** to .5mm:

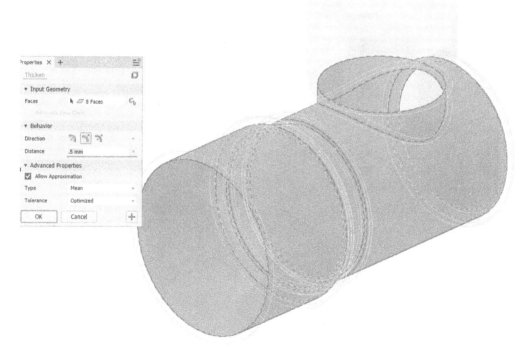

Figure 4.77: Surface faces to apply the Thicken/Offset command to and the values to apply

16. Select **OK** to complete.

17. Right-click on **Casing:1** from the **Model Browser**. Select **iProperties** and then open the **Physical** tab. Change the **Material** option to **Rubber** and select **Apply**.

18. Close the **iProperties** menu and select **Return** in the ribbon to return to the assembly.

19. We have now created a contoured part from the original using surfaces. As the new part is adaptive, if changes are made to the original, they will automatically translate to the casing created. Right-click on `Copy_Object.ipt` from the **Model Browser** and select **Edit**.

20. Select **Direct Edit** from the **Modify** tab.

21. Select **Scale** as the operation and then select the solid of `Copy_Object.ipt`, shown in *Figure 4.78*:

22. For the scale value, change this from `1.0` to `1.5` and ensure **Uniform** is selected. As the new scale value is entered, it will be reflected in the graphics window in real time:

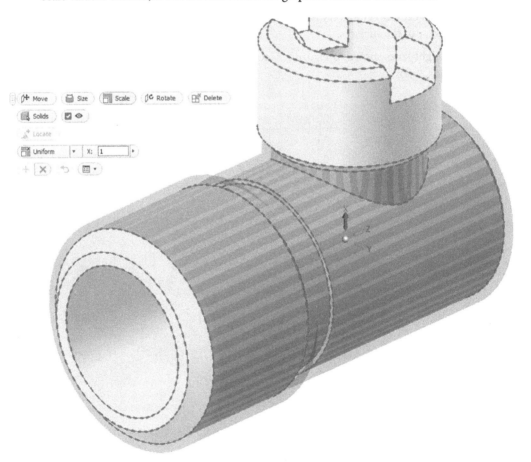

Figure 4.78: Direct Edit in progress on Copy_Object.ipt

23. Select the green + icon to complete the scale.

24. Select **Return** from the ribbon to return to the assembly file.

25. The contoured `Casing.ipt` we created previously has automatically updated in scale to suit `Copy_Object.ipt`, shown in *Figure 4.79*:

Figure 4.79: Returning to the assembly activates the adaptive casing
we created to update in scale to suit the original pars

In this recipe you have used **Copy Object** to select the desired faces on an existing model, convert to surfaces, edit, and delete unwanted surfaces with **Delete Face**, apply solid material to the surfaces, and create an adaptive contoured casing, that has updated from the result of a scale operation using **Direct Edit**.

Importing surfaces into Inventor

It is sometimes necessary to import surfaces or CAD files from non-native CAD formats into Inventor. Inventor is very capable of importing and using non-native CAD data and in this section, we will examine how Inventor can import surfaces from non-native CAD files and what modifications are possible.

It is not uncommon for imported surfaces to translate with some errors. Fortunately, Inventor has an array of **Surface Repair** tools that can be used to identify, locate, and correct surface bodies.

In this recipe, you will import a `.iges` file as a surface. The same process is used to import any non-native CAD data into Inventor.

Getting ready

For this recipe, you will need a new Inventor part file open.

How to do it...

To begin, we will need the new Inventor part file open; we will then start the process of importing in the `.iges` file. To do this, see the following:

1. Select **File | Open | Import CAD Files**.

2. In the dropdown for **File Type**, select **All Files**.

3. Find and select the `Import.igs` file. Then, click **Open**.

4. The import options are now open (*Figure 4.80*) and you can choose how you would like to import the file:

    • Deselect **Solid** and select **Surface**.

    • You can also define the units at this point – however, we will leave this as the default.

    • Under **Surfaces**, the dropdown allows you to select from **Individual**, **Composite**, or **Stitch**. In this example, we want to import the entire model as a surface, so **Composite** will work best:

Figure 4.80: Import options available on the IGES file to be imported into Inventor

5. Then, select **OK** to complete the import. The `.iges` file has now successfully been imported into Inventor as a composite surface.

6. Inventor will notify you as the user if errors are present in the model. To fix these, you must navigate to the **Repair Bodies** command in the surface commands area of the ribbon. This opens a new environment for the **Surface Repair** tools. In this case, the surface has been imported without issue, as a green tick is visible next to the composite name in the **Model Browser**.

    As this is an error-free surface with healthy geometry, we can work on the model without issue.

7. Now, close the model.

You have now imported a non-native `.iges` file into Inventor as a composite surface.

There's more

In some cases, it may be necessary to perform repairs on an imported surface model. Issues that may arise could be missing faces, intersecting geometry, or missing aspects of the model that were not translated across in the model.

For the full list of compatible files that Inventor 2023 can import, please see `https://help.autodesk.com/view/INVNTOR/2022/ENU/?guid=GUID-AF41FA87-7588-4698-9C41-756A01EBE7F4`.

The main objective when repairing an imported surface with errors is to create a watertight surface, with closed boundaries, so that a solid geometry can be generated from it.

To repair an imported surface, the typical workflow is as follows:

*   Select **Repair Bodies** to open the **Repair** environment
*   Select **Find Errors** to identify where the errors are
*   Use **Stitch** and close the surfaces together
*   Conduct an error check
*   Use **Stitch** and close further if necessary

The **Repair** environment can be accessed by selecting **Repair Bodies** in the **Surface Modeling** area of the ribbon. The ribbon will then transform as *Figure 4.81* shows. The **Repair** functions are designed to be used left to right and you can find the **Find Errors** command on the far left:

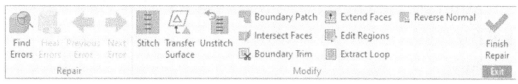

Figure 4.81: The Inventor Surface Repair ribbon

Upon activating **Find Errors**, you can then choose **Select All Surfaces** to select all the surfaces in the model. It is advised that in most cases, this is what should be done to ensure all the imported surfaces are analyzed.

The error check performed will then analyze all the imported surfaces and report back any errors to you via the **Model Browser**. Surfaces with errors will show with either an exclamation mark in a circle or a warning triangle. Geometry that has passed the check will show with a green tick. *Figure 4.82* shows the errors visible in the **Model Browser** and graphics window:

Figure 4.82: Warning symbols on a model

Inventor will group similar errors into folders within the **Model Browser**. You can also use the **Next** and **Previous** commands in the ribbon to cycle through the errors identified.

Once an error has been identified, it can be manually fixed through stitching and other functions or the **Heal Errors** command can be used. This tool only becomes visible once an error has been identified.

When the relevant bodies have been selected, the **Analyze Selected Bodies** function within **Heal Errors** can be used to resolve issues within the specified tolerance range. If the error cannot be healed automatically, the tools in the **Modify** panel must be used to manually correct the geometry.

Once the errors have been identified, use the **Stitch** command to close the gaps and ensure you are left with a complete surface. The **Patch** and **Extend** tools can also be used to achieve this.

Validating surface designs with Zebra Analysis, Surface Analysis, and Curvature Analysis

Inventor has a range of analysis tools that focus on surfaces and surface quality. This is important, as often as a designer, you need to ensure and validate the surface design before manufacture and ensure that the surfaces blend or transition appropriately without any sharp changes in curvature or smoothness. This can be particularly important for molded components and products.

In this recipe, you will make use of three of the common surface analysis tools: **Zebra Analysis**, **Surface Analysis**, and **Curvature Analysis**.

Getting ready

For this recipe, open `Bike_Saddle.ipt` in the `Chapter 4` folder.

How to do it...

First, we will examine the part, then apply the relevant surface analysis tool and examine the results. To do this, see the following:

1. To begin, we will examine the **Surface Inspection** tools. Select the **Inspect** tab, go to the **Analysis** panel, and then select **Surface**.

2. From the **Surface Analysis** drop-down menu, select the type of curvature calculation you want to perform. There is a range of options that you can choose from and each represents different aspects of curvature. They are as follows:

    *   **Gaussian**: Displays a gradient representing the product of the curvature – the curvature in the U direction of the surface multiplied by the curvature in the V direction of the surface. Use this for looking at shapes such as coils or sweeps.

    *   **Mean Curvature**: Displays a gradient representing the mean curvature of the U and V surface curvature values. Used for cylindrical or cone features, typically suited to lofted profiles.

    *   **Max Curvature**: Displays a gradient representing the greater of U or V surface curvature values. Ideal for analyzing areas of low and high curvature.

    For this example, select **Max Curvature**.

3. We can now specify the ratio gradient range that we want to look at in the model.

    The options to pick from are as follows:

    *   **Minimum Curvature Ratio**: Sets the value that maps the start (blue end) of the color gradient

    *   **Maximum Curvature Ratio**: Sets the value that maps the end (green end) of the color gradient

    *   **Auto Range**: Sets the minimum and maximum curvature ratio that the analysis display uses

    For our analysis, select **Maximum Curvature Ratio**.

4. You can also specify the gradient values and set these. In this recipe, we need to select **Auto Range**.

5. There is a slide in which you can alter the display quality, which determines the resolution or quality of the gradient or color bands. As we are not using the gradient setting, leave this as the default.

6. We now need to select the faces or surfaces that we want to analyze. By default, under **Selection**, this checkbox is checked, meaning the whole model will be analyzed. Select the check box to deselect **All**.

7. Select the two faces shown in *Figure 4.83*. Ensure that your curvature settings also match those shown in the screenshot:

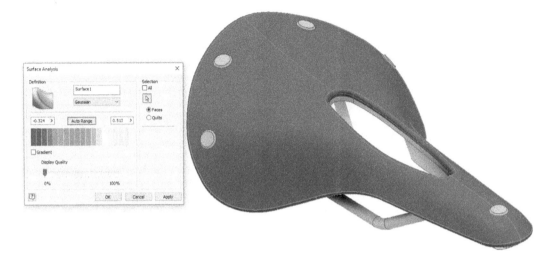

Figure 4.83: Curvature surface analysis settings and faces to apply to Bike_Saddle.ipt

8. Select **OK**. The curvature surface is displayed on the faces as per *Figure 4.84*:

Figure 4.84: The Maximum Curvature gradient displayed on the model

Now that this has been applied, you will notice in the **Model Browser** that an additional folder called **Analysis Surface** has been created automatically. This stores all the surface analysis studies so they can be turned on or off and edited within the model.

9. Next, we will look at the **Zebra Analysis** tool. This can also be used to validate the surface continuity smoothness and gradient. From the **Inspect** tab, select **Zebra**.

> **Note**
>
> To understand and interpret **Zebra Analysis** surface continuity, see `https://knowledge.autodesk.com/support/autocad/learn-explore/caas/CloudHelp/cloudhelp/2016/ENU/AutoCAD-Core/files/GUID-82E5989F-C943-49A6-A6C6-834B81CC203B-htm.html`.

10. We can now define the thickness and orientation of the analysis. Leave all settings as default in this case, except for **Thickness**, which you should reduce slightly as per *Figure 4.85*.

11. Deselect **All** in **Selection**. Select the two faces you previously analyzed for curvature and then select **Apply**:

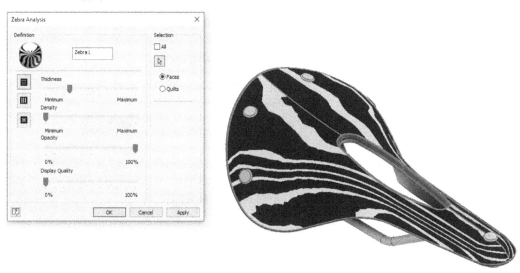

Figure 4.85: Zebra Analysis applied to Bike_Saddle

12. With **Zebra Analysis** active, orbit around the model to examine how the curves blend.

    Note that the **Analysis** folder now has both the surface and zebra studies within it. You can double-click each instance in the **Model Browser** to flip between **Zebra** and **Curvature**. Edits can also be made by right-clicking on the study in the **Model Browser**.

13. Now, we will look at and examine **Curvature**. The **Curvature** tool provides a visual analysis of the curvature and overall smoothness of model faces, surfaces, sketch curves, and edges. Select **Curvature** from the **Inspect** tab.

14. With **Curvature Analysis**, you can select and alter the height and density of the combs that will be placed on the chosen surfaces. Change these so they reflect the settings in *Figure 4.86*:

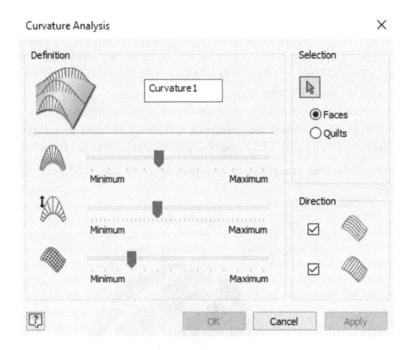

Figure 4.86: Curvature Analysis settings to apply on the surfaces

15. Select the two faces of `Bike_Saddle.ipt` previously selected for the other two **Surface Analysis** studies. Select **OK** to complete.

You can now apply several **Surface Analysis** tools to examine, validate, and check designs.

Understanding the basic features and techniques of freeform modeling

The Freeform modeling workflow and tools within Autodesk Inventor allow you to easily create complex visual forms and models without the constraints and precision required in solid modeling workflows. The **Freeform tools** enable you to push and pull material in an organic way to create true curves and specialized organic forms. Freeform modeling poses G2 continuity across all created surfaces.

The freeform workflow can also be combined with the uses of surface modeling and solid body modeling. Learning all the freeform modeling techniques takes time, persistence, and dedication. In this section of the chapter, we will focus on the basics to get you started with freeform modeling in Inventor.

In this first Freeform recipe, you will be introduced to the Freeform environment and create a basic Freeform box shape – you will then manipulate this to learn how some of the basic tools and commands work in this environment.

Getting ready

For this recipe, you will need to open a New Standard (mm.ipt part file.

How to do it...

To get started with the Freeform environment, see the following:

1. Firstly, we need to define how we want to work in the Freeform environment. In this instance, we will start with one of the preset freeform shapes and edit it to suit.

    Within the **3D Model** tab, under **Create Freeform**, select **Box** (note that in the dropdown, there are other options we can choose from, but this is what we are choosing for now).

2. You will then be prompted to further define the box shape. We can specify the length, width, height, and number of faces. The more faces that you apply, the denser the freeform mesh will appear on the part, and the more control you can apply to the form. With freeform modeling, it is often the case that you will also want to apply a degree of symmetry to the model and this can also be achieved in this window.

    For our example, apply the values shown in *Figure 4.87* for **Length**, **Height**, and **Width**. Then, select **Width Symmetry**.

3. Select the XY plane to generate the base form of the box as defined, also shown in *Figure 4.87*. When the origin is selected, the freeform box is created:

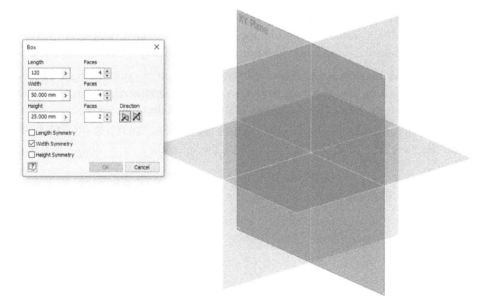

Figure 4.87: Freeform box options to select

4. Select **OK**. The freeform box as we previously defined in the freeform box creation window is now in the graphics window and we can begin to edit it. Note that the **Width Symmetry** rule we applied will remain throughout. Your graphics window should now display the following shown in *Figure 4.88*. Note also that the ribbon has now transformed to reveal the **Freeform Modeling** tools:

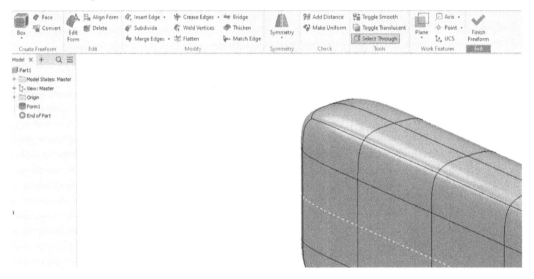

Figure 4.88: Freeform modeling environment active

5. Navigate to the ribbon, then click **Toggle Smooth** from the **Tools** panel. This enables you to switch the model from a smooth freeform display to a block display style. This is a visual tool that is used to improve performance. In this instance, we will keep the display **Smooth**.

6. There are several freeform model tools, and at any time, you can exit **Freeform Modeling** and resume surface or solid modeling workflows. The main command used in manipulating freeform shapes is the **Edit Form** command. In the ribbon, select **Edit Form**.

7. We are now able to specify entities to pick. On the **Filter** options, **All** is selected, which means we can select any face, edge, or point on the form; this offers the most flexibility, so leave this as default.

In the **Transform** section, we can define whether we want to move the existing faces or scale them. You can also define how the faces are selected with **Selection Options**, based on individual items.

Apply the settings shown in the window in *Figure 4.89*:

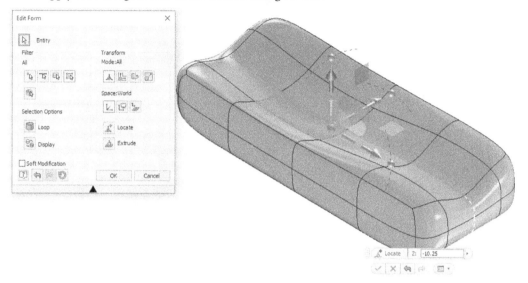

Figure 4.89: Vertexes to select and pull downward to manipulate forms

8.   Then, hold *Shift* and select the three vertex sections along the symmetry line on the form.

9.   With the graphical arrows displayed, select the up arrow and pull this down around -10.25 mm. The result is also shown in *Figure 4.89*.

Tip

As well as manually dragging the arrows in **Edit Form**, you can also type specific values if required.

---

10.   With the **Edit Form** command still active, select the face shown in *Figure 4.90*. Note how the opposing identical face on the other side of the model is also selected:

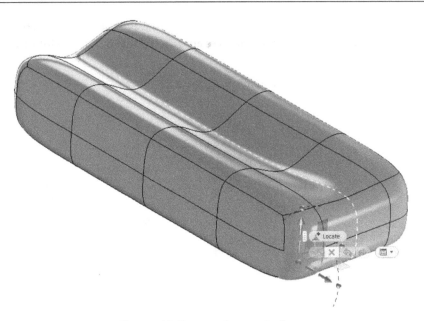

Figure 4.90: Face to select on the form

Pull the face by 30 mm in the direction shown in *Figure 4.91*. The same is applied on the opposite side of the mode. This is because **Width Symmetry** was defined in the **Form Creation** stage:

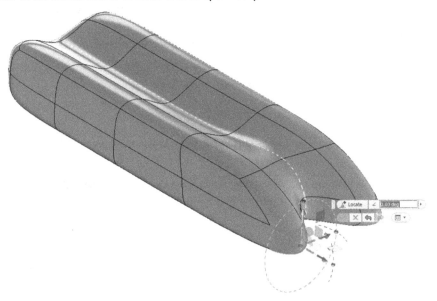

Figure 4.91: Face pulled out 30 mm and replicated through the line of symmetry

11. Select the rotation node and twist it **90** degrees in the direction shown in *Figure 4.92*:

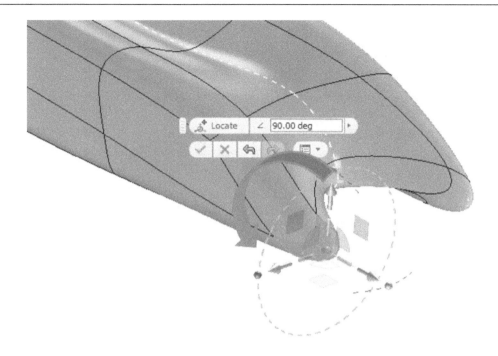

Figure 4.92: Rotation node selected and moved 90 degrees, twisting the previous extension from the form

12. Instead of a face or vertex, this time, select the point shown in *Figure 4.93*:

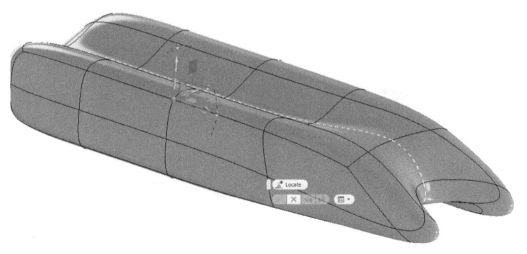

Figure 4.93: Point on the form to be selected

Pull the point inward towards the center of the model by **13** mm as shown in *Figure 4.94*:

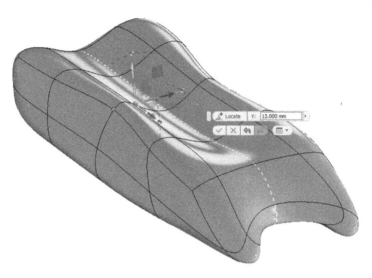

Figure 4.94: Point on the form selected and pulled inward 13 mm

13. Faces in the form can also be further subdivided. This is ideal if there is refined detail that needs to be added or if you require greater control of a face to create the desired geometry.

    Select **Subdivide** from the **Modify** panel.

14. Increase **Faces**, **Width**, and **Length** from 2 to 3.

15. Select the face shown in *Figure 4.95*; the face will then subdivide as specified:

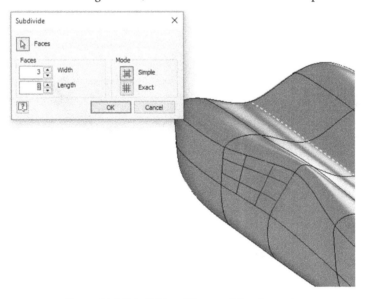

Figure 4.95: Subdivide settings and face to apply

16. Select **OK**.

17. Select the three faces shown in *Figure 4.96* and pull them outward by -11.25 mm:

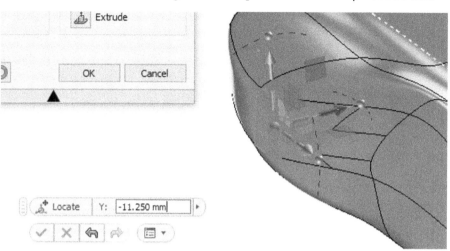

Figure 4.96: Three subdivided faces on the form selected and pulled outward

18. Once an edit has been made to a subdivided face, the edge can be deleted if required. Select **Delete** from the **Edit** panel, select an edge, and then click **OK.** The result is shown in *Figure 4.97*:

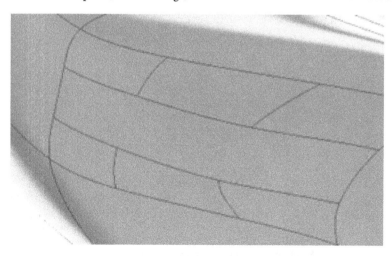

Figure 4.97: Edges deleted from a subdivided face using Delete

19. Select **Finish Freeform**. The model reverts from the freeform environment to the modeling environment (*Figure 4.98*) and the form is transformed into a solid geometry. Standard solid modeling workflows can now be used:

Figure 4.98: Completed form transformed into a solid body model

At any point, you can return to the freeform tools by right-clicking on **Form 1** in the **Model Browser** and selecting **Edit Form**. You have now used the basic Freeform tools to define a shape and convert it into a solid.

Freeform modeling within the context of existing geometry

Freeform modeling is mostly conducted in the context of an assembly or with existing components. In this recipe, you will perform a freeform modeling exercise to create an ergonomic handle for a hand grinder in the context of an assembly. You will then convert the freeform into a surface, make further edits, and then sculpt it into a solid geometry.

The completed model you will create is shown in *Figure 4.99*:

Figure 4.99: Completed angle grinder model

Getting ready

In the Chapter 4 folder, open the Hand Grinder folder, and then open 4.5 in Grinder Assembly.iam.

How to do it...

Once 4.5 in Grinder Assembly.iam is open, you will see that the hand grinder is complete except for the handle. This is the segment of the design you will complete using the **Freeform Modeling** tools. *Figure 4.100* shows the model open:

Figure 4.100: 4.5 in Grinder Assembly.iam

To begin with, we will have to find the component we want to edit within the context of the assembly.

1. In the **Model Browser**, right-click on **Grinder Handle** and select **Edit**. You are now editing the handle in the context of the assembly.

2. Select **Create Freeform** and select **Cylinder**.

3. Select the face of the **Grinder Handle** part, and then select the midpoint to place the **Cylinder** freeform, shown in *Figure 4.101*:

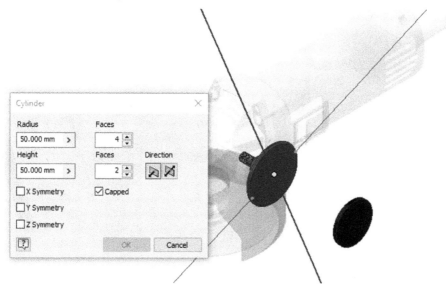

Figure 4.101: Placement of the Cylinder freeform

4.   Left-click to place the freeform cylinder. You can see the default cylinder in *Figure 4.102*:

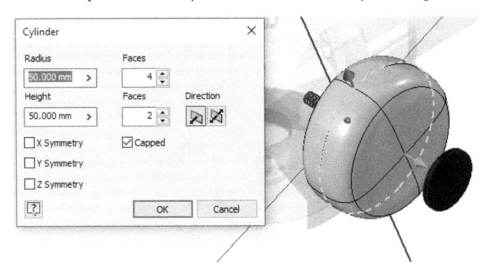

Figure 4.102: Default freeform cylinder placed

5.   We will now refine the cylinder so that it is closer to our desired shape. Define the cylinder properties as per *Figure 4.103* and then click **OK** to complete the cylinder:

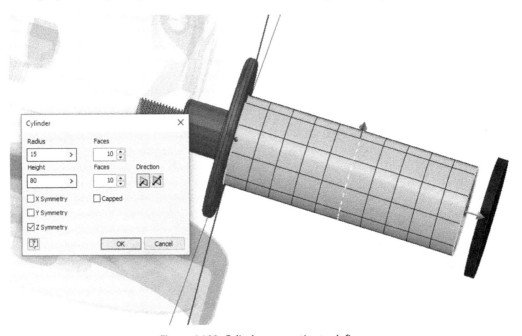

Figure 4.103: Cylinder properties to define

6. Select **Edit Form**. Select the Freeform faces as shown in *Figure 4.104*. The Z symmetry will apply symmetry and any changes to the other side of the model:

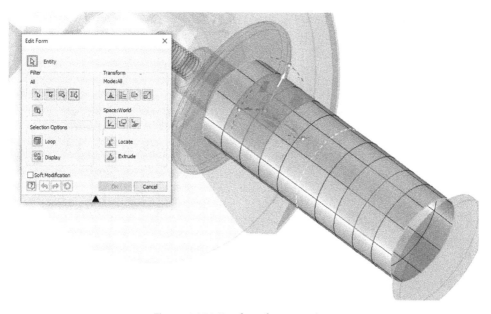

Figure 4.104: Freeform faces to select

7. Pull these faces downward by -4.24 mm as shown in *Figure 4.105*. It is best to navigate to the left view in **View Cube** to do this:

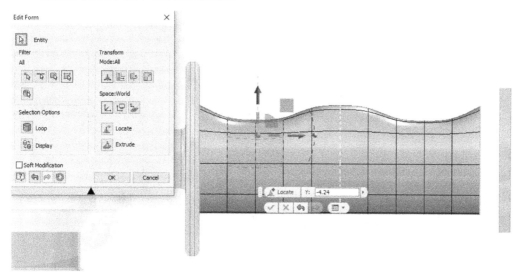

Figure 4.105: Faces pulled downward -4.24 mm

8. Select the green tick to complete the edit form operation.

9. Select **Edit Form** and navigate to the bottom view on the **View Cube**.

10. Then, select the faces shown in *Figure 4.106*:

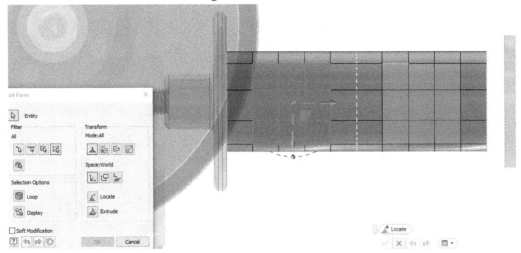

Figure 4.106: Faces selected to perform the next edit

11. Navigate to the left view on **View Cube**.

12. Select the sphere node and pull upward; this will flip the orientation of the arrows so we can pull the feature in the correct direction, as shown in *Figure 4.107*:

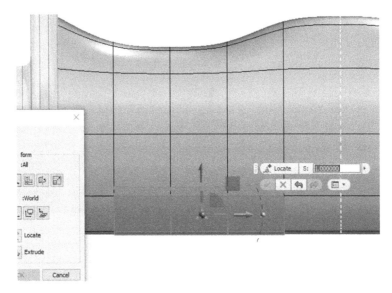

Figure 4.107: Sphere to select to re-orientate the Freeform control arrows

13. Pull the arrow pointing upward up so that the model resembles *Figure 4.108*:

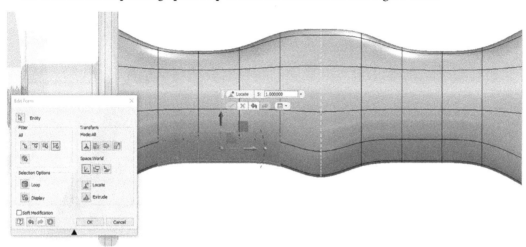

Figure 4.108: Pull the freeform arrow shown to create the shape shown

14. Select **OK** and then select **Finish Freeform** in the ribbon.

Because we uncapped the ends of the original freeform cylinder, the freeform model has transformed into a surface model.

15. We now need to create a joining surface where there is currently a gap. In the **Surface Tools** panel, select **Patch**.

16. Select the two boundaries shown in *Figure 4.109* and select **OK**:

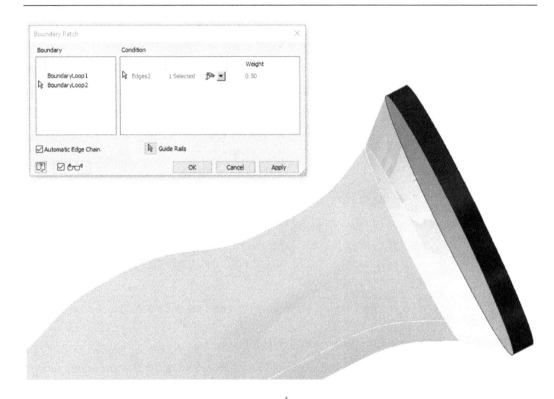

Figure 4.109: Boundaries to select shown and the Patch operation performed

17. Select **OK**. From the **Surface** panel, select **Sculpt**.

18. Select the two surfaces that make up the handle, as shown in *Figure 4.110*:

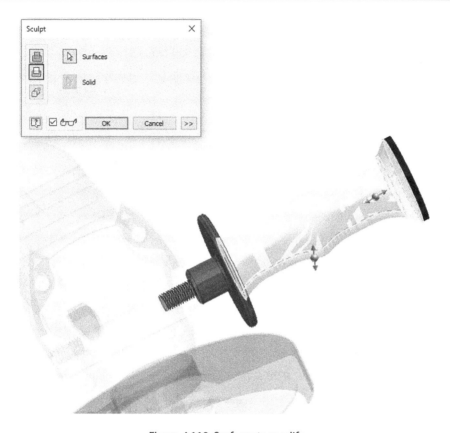

Figure 4.110. Surfaces to modify

19. Select **Add** as the option, followed by **OK**. This transforms the surfaces into a solid geometry.

20. Select **Return** to return the assembly.

You have successfully used the Freeform tools to design and create an ergonomic handle for the hand grinder within the context of an assembly.

Model credits

The model credits for this chapter are as follows:

- **EV charger type 2 plug(female)** (by Brandon) `https://grabcad.com/library/ev-charger-type-2-plug-female-1`

- **4.5 Inch Grinder** (by David Guza) `https://grabcad.com/library/4-5-inch-grinder-1`

- **Brooks C17** (by max morozov) `https://grabcad.com/library/brooks-c17-1`

Advanced CAD Management and Collaboration – Project Files, Templates, and Custom Properties

Setting Inventor up correctly for your organization is important to ensure that all users are using the right files, content, and template files. This enforces and encourages standardization and a code of best practice within the design department. The correct setup and project management save time and improve the efficiency and flow of work. Most designers and engineers spend far too much time trying to locate the correct files for a job. Ensuring that the right templates are active and that Inventor is set up for your organization's workflows is crucial. In this chapter, we will learn about the CAD management aspects of Inventor, how you can set up templates, and the collaboration tools at your disposal.

Inventor can work with a dedicated **Product Data Management** (**PDM**) solution – **Autodesk Vault**. This chapter will not cover workflows, setup, or project management involving Autodesk Vault, as this is a separate product. If you wish to learn more about Vault, contact your local authorized Autodesk partner, who will be able to best advise on this.

In this chapter, we will learn the following:

- Creating and managing an Inventor `.ipj` project file
- Managing the Content Center
- Style and Standard Editor – creating and editing templates
- Collaborating with shared views
- Collaborating and sharing designs with Pack and Go

Technical requirements

To complete this chapter, you will need access to the practice files in the `Chapter 5` folder within the `Inventor Cookbook 2023` folder.

Creating and managing an Inventor .ipj project file

Before embarking on a practical recipe for creating and managing an Inventor project file, it is important to understand the terminology and nomenclature and how the project file system works within Inventor.

Inventor uses **project files** to organize and collate all the design data associated with a job or project that you are working on. Projects can consist of parts and assemblies but also standard components, and libraries of the standard components used, such as fasteners.

Inventor projects act as a central repository and location for your design files, so Inventor knows the location of all parts and assemblies that are associated with a particular design job.

There is no limit to the amount of project files that you can have to manage your work within Inventor. There are normally two schools of thought about managing projects in an organization:

- A single project with an internal folder structure within it for each job/project (which is also arguably the best setup for projects if you use Autodesk Vault)
- Multiple individual project files for separate jobs

There is no right or wrong setup, and this is entirely dependent on how your organization wants to work, and your design process. The *Inventor 2023 Cookbook* project is a single project, with all design files stored within a folder structure.

An advantage of using a single project is that you can open multiple design files from multiple jobs and projects at once. If you had parts or assemblies stored in different projects that were required for a job, you would not be able to open them at the same time, as Inventor only allows the user to have one project active at any given time. If the files are stored in another project, then you must close all current data for the current project and then activate the required project to access the files.

By default, in Inventor 2023, three projects are installed with the Inventor software: default, sample and tutorial files. Migrating the files to another project once a design is completed is difficult and complex. It pays dividends to spend time ensuring your setup is right the first time!

Inventor projects are stored as project files (`.ipj` files) to store the paths to folders where your design data is located.

A `Project.ipj` file is a text file in `.xml` format. The project wizard creates this automatically when a project is created. The file specifies the paths to folders that contain the files in the project. These stored paths assure that links between files work properly. When you open a file in a project, Inventor uses the search paths in the order they appear so that it can find the file and any referenced files. For example, if an assembly is opened, Inventor can find all the related parts and assemblies within the project that relate to this. If parts are stored in a different project or separate `.ipj` file, Inventor will flag errors, stating that it cannot find the path. You will then have to manually find the specific file.

In this recipe, you will create a new Inventor project file, complete the setup and configuration of this, and then test your project file by adding some parts to it.

Getting ready

You will not need any practice files for this recipe. Just navigate to the Inventor **Home** screen to begin.

How to do it...

We will first begin on the Inventor **Home** screen:

1. Select the vertical ellipses button next to the active project, and select **Settings**, as shown in *Figure 5.1*:

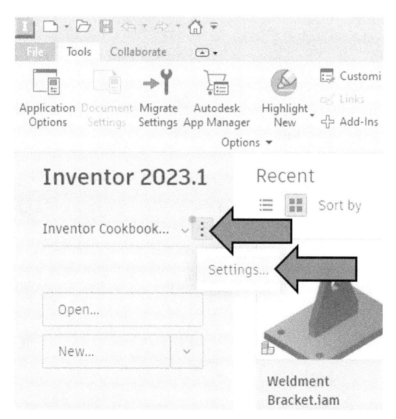

Figure 5.1: The location of the projects area of Inventor

The project wizard will now open, which displays your current active project and a list of other available projects. Note that the list of projects shown in *Figure 5.2* may be different on your machine.

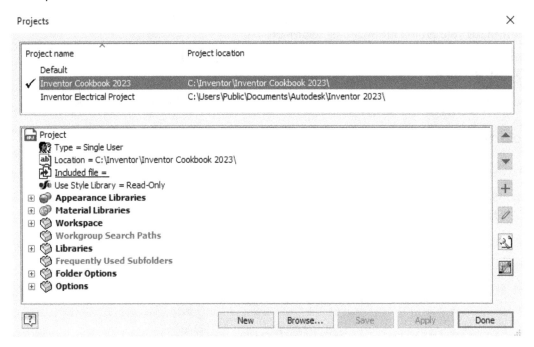

Figure 5.2: The Projects window active

2. Select **New** in the **Projects** window.

3. Ensure that **New Single User Project** is selected, and then select **Next**. **New Single User Project** does not mean that the project is limited to its use by one individual user; it just means that this is an Inventor project that is not using Autodesk Vault.

    The **New Vault Project** would be selected if we were using the dedicated PDM product, Autodesk Vault, to manage our CAD data, which includes extra setup, settings, and options that must be defined to ensure we link a project to a Vault. In this case, we are not using Autodesk Vault, so **New Single User Project** will be fine:

Cookbook 2023   C:\Users\A_Bordino\Documents\Inventor\Inventor Cookbook 2023\
Electrical Project   C:\Users\Public\Documents\Autodesk\Inventor 2022\

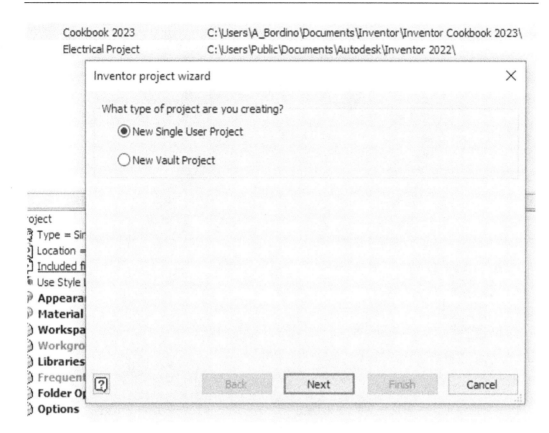

Figure 5.3: The project wizard window

4. The next step is to name our project. Projects typically are named after the organization or the client the project is for, although you are free to name your projects whatever you want. In this example, type the name of the project as Test Project.

5. The next important aspect to define is the project location or **workspace**. This is the top level of the project that will contain all other subfolders and files associated with the project. All files created will be saved underneath this folder. You can browse a location on a drive to find a place to store this. In this example, leave it as default.

    Select the three dots next to the **Project (Workspace) Folder** option and browse the new location of your project file. Select the C:\ drive and create a new folder called Test Project. *Figure 5.4* shows the result of completing this. Your project location should now be C:\Test Project:

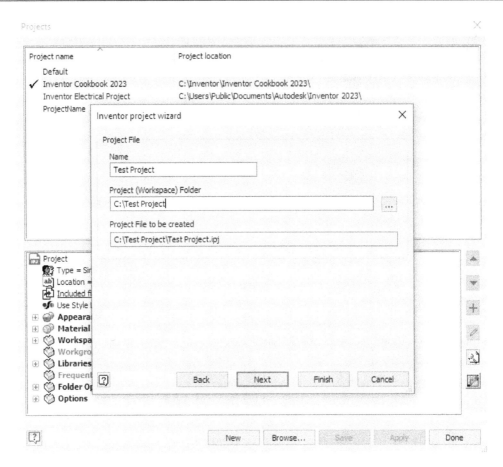

Figure 5.4: Creation of the new project workspace on the C:\ drive

Collaborative project files

If working in a collaborative environment, the project file should be stored on a network drive so that all other users can collaborate and access project data. With Autodesk Vault, you have much more control in terms of user permissions and who can access what files and at what revision or life cycle state. Out-of-the-box Inventor Professional does not contain these features.

6. Select **Next**. The following window allows you to set up and share standard library parts across projects. For example, you may have another folder location on a network drive that contains several standard parts required in your projects. In this case, it is not necessary.

To proceed, hit **Finish** to complete the creation of the new project, **Test Project**, which will now become the *active project* in Inventor. This is shown by the check mark that is displayed next to the project name. Remember that only one project can be active at any one time in Inventor.

7.   Navigate to the C:\ drive and open the Test project folder. Note that a Test Project. ipj file has now been created:

Figure 5.5: Test Project.ipj created in the Test Project folder or workspace

8.   Navigate back to the Inventor **Home** screen.

9.   Ensure that **Test Project** is still the active project in Inventor, and the check mark should be displayed next to the name in the **Projects** area of the **Home** screen. If this is not the case, double-click **Test Project** to activate this:

Figure 5.6: Test Project active

Select **Open** from the **Home** screen. Automatically, the **Test Project** workspace should open and show there are no files. We will now test whether the project is set up correctly by creating some parts and saving them in the workspace:

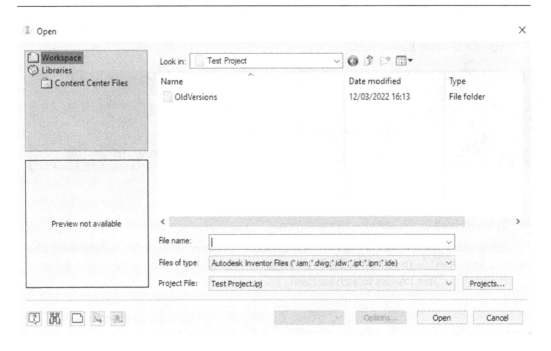

Figure 5.7: Test Project empty

10. Select **New**, and then select `Standard.ipt` from the templates, followed by **Create**.

11. Inventor will now open a new `.ipt` file, without creating any geometry. Now hit **Save**. Inventor should prompt you to save the Inventor `.ipt` part file in the **Test Project** workspace. Then, select **Save** again.

12. Close the file.

13. Select **Open**; your newly created `.ipt` file should now populate the Inventor **Test Project** workspace that we previously created, in *step 12*. This means the project file is working as intended. If desired, multiple folder structures could be created within **Test Project** to further organize designs and projects.

14. Navigate to the **Home** screen and select **Projects**.

15. Under the **Projects** area of Inventor, you will see there are many other options to configure for the setup of your Inventor project. We have already covered the location of the project previously in *step 10* and *step 11*. Expand materials and appearances by selecting + next to the **Material Libraries** section, in the bottom half of the menu:

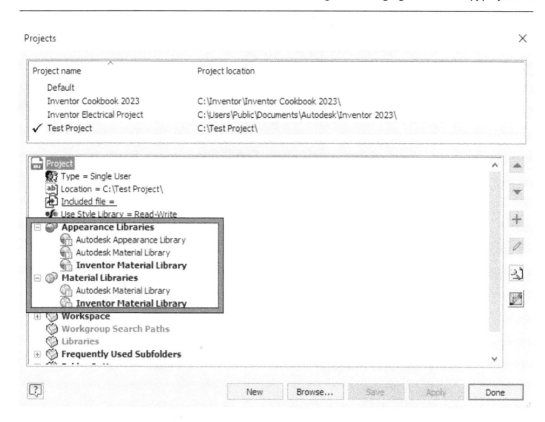

Figure 5.8: Inventor Test Project configurations

Here, you can change and alter what materials and appearances are available and accessible in your Inventor project. For example, you may have a specific material or appearance type for a specific customer or project. This can be loaded here.

The libraries area is where you can add paths for different folders that are not in the workspace. These could be stored on a server or network, outside of your machine. You can create paths to libraries by right-clicking on **Libraries** and selecting **Add Path**. This would allow Inventor to access these library paths and folders within them if required, even though they are not part of your workspace for this project.

Frequently Used Subfolders is for storing links to folders that you use regularly as part of the project. The subfolders act in a similar way to bookmarking a tab within an internet browser. We will now create a **Frequently Used Subfolders** link.

16. To begin, we will create a folder within the **Test Project** workspace. In the Windows browser, navigate to the C:\ drive, and then select the Test Project folder. Right-click and select **Create a New Folder**. Call this folder Project 1.

17. Within the Inventor project, right-click on **Frequently Used Subfolders** and select **Add Path**:

Figure 5.9: Frequently Used Subfolders

18. Select the **Browse** button and then the `Project 1` folder previously created. Type `Project 1` in the text field where **Folder** is set as default, as shown in *Figure 5.10*

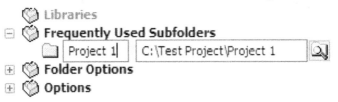

Figure 5.10: Project 1 typed in the folder text field

19. Now, if you select **Open** In Inventor, you will see your subfolder link displayed on the left-hand side, as shown in *Figure 5.11*:

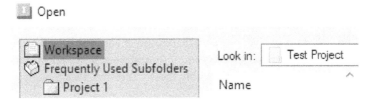

Figure 5.11: The Project 1 subfolder created and displayed

You have now created a new project file, tested it, and set up a frequently used subfolder.

Managing the Content Center

In this recipe, you will learn how to manage the Content Center as part of your Inventor project. You will also learn its configuration and setup.

Getting ready

You will not need any practice files for this recipe. You will need to ensure you have completed the previous recipe, *How to create and manage an Inventor .ipj project file*, as we will be using the Test Project.ipj file.

How to do it...

The **Content Center** is a repository of standard nuts, bolts, washers, fasteners, structural members, and more that come as standard with Inventor 2023. The Content Center allows you to pull in existing standard design and manufacturing components and use them within your designs. It is important that, along with the Styles and Standards library, this is set up correctly as part of an Inventor project file:

1. From the Inventor **Home** screen, ensure that **Test Project** is the active project.

2. Expand + next to **Folder Options**:

Figure 5.12: Folder Options expanded in Test Project

Folder Options is where you can tell Inventor where your **Design Data** and **Templates** folders will be stored and where you want your **Content Center Files** folder to be. By default, they are set to **Default**. If you are working in a collaborative environment, you must not leave **Content Center Files** as **Default**. If this is left as **Default**, when you use the Content Center and place a nut or bolt from this into your design, it will save this into your MyDocuments folder on your machine. This means that others who require the specific content will not be able to access this from the library. We will now set this up as if we are working in the aforementioned collaborative environment.

3. Navigate using the Windows browser to the **Test Project** workspace on the C:\ drive. In reality, this would be a network drive.

4. Create a folder at this level (as shown in *Figure 5.13*) called Content Center:

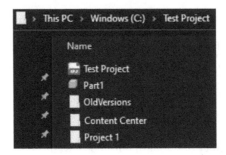

Figure 5.13: Content Center folder created

5. Navigate back to the Inventor project window, right-click Content Center on **Folder Options**, and select **Edit**. Then, click **Browse**, select your newly created Content Center folder, and then select **OK**:

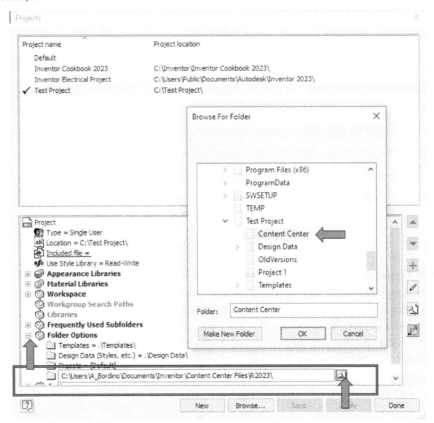

Figure 5.14: The Content Center folder selected in Folder Options

The same can be done for **Templates** so that everyone on a network uses the same templates for parts, assemblies, and drawings. This is important to set up, as without this, there are no centralized styles, templates, or standards in an organization.

We will now configure the Content Center so that we can filter specific Content Center libraries that we want to use. Some users will prefer to have access to the whole Content Center, but other organizations may only want to work with certain standards and, therefore, require the Content Center libraries to be filtered.

6. Select the **Configure Content Center Libraries** button shown in *Figure 5.15*:

Figure 5.15: The Configure Content Center Libraries button

7. You can now pick and choose what Content Center libraries are available within the project. To deselect a standard library, remove the tick from the tick box. In this example, deselect **Inventor DIN** from the list, as per *Figure 5.16*. Then, select **OK**:

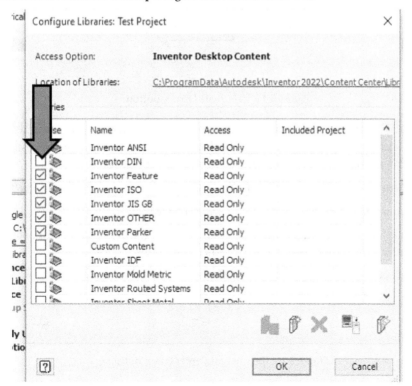

Figure 5.16: Inventor DIN content deselected

8. Select **Save** to complete this section of the project setup for the Content Center.

   The next aspect of the Content Center is how to create your own libraries and publish them to the Content Center.

9. Select the **Configure Content Center Libraries** button. Then, select the **Create Library** button in the bottom right of the menu as shown in *Figure 5.17*:

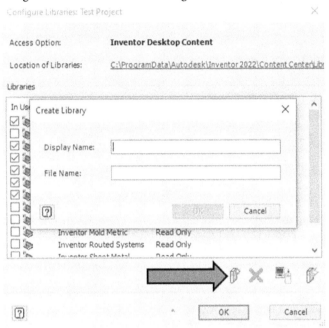

Figure 5.17: The Create Library button

We must now name the library. Call this Test Library and select **OK**. If you are presented with an **Attempt to write a read-only database** error, close Inventor and then start it again using **Run as Administrator** from Windows, and then repeat *step 8*.

10. A read/write test library is now created in the Content Center library for **Test Project**. Select **OK**.

11. Select **Done**, and then **Save**. Select **Yes** to overwrite the project.

12. We will now take an existing Content Center component, edit it, and save it in our new **Test Library** for future use. Create a new Standard Assembly.iam file.

13. In the assembly file now open, select the **Manage** tab, and then **Editor**.

Figure 5.18: The Content Center Editor button

14. Select the *filter icon* and then **ISO** as the standard to view:

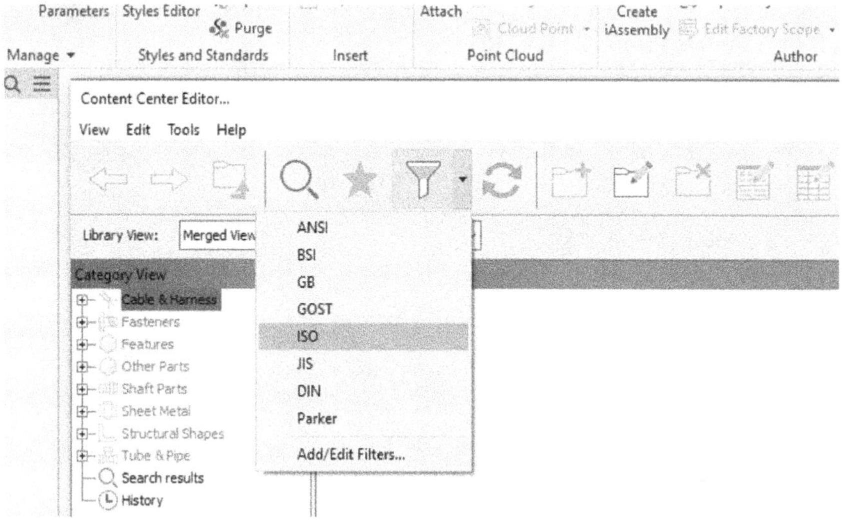

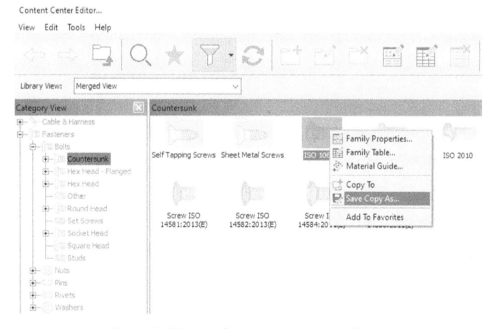

Figure 5.19: The Content Center Editor – filtering the ISO standard content

15. In the Content Center **Category View** pane on the left, select **Fasteners**, then **Bolts**, and then **Countersunk**. Hover over the **ISO 10642** bolt, right-click, and select **Save Copy As…**:

Figure 5.20: Selecting a fastener to create a copy of to edit

16. In the **Save Copy As** window, ensure that the library to copy to is **Test Library** and that **Independent family** is checked. Change Copy of in the **Family Name** and **Family Description** fields to TP (for **Test Project**). Select **Review**:

Figure 5.21: Changes to be made in the Save Copy As window

17. In the next window, replace Copy of in the top two cells with TP:

Expression:	TP|"ISO 10642" & " - " & "M" & {NND} & " x " & {NLG}	
Family Name	**File Name**	**Part Number**
▶ TP ISO 10642	TP "ISO 10642" & " - " & "M" & {NND} & " x " & {NLG}	TP "ISO 10642" & " - " & "M" & {NND} & " x " & {NLG}
	TP ISO 10642 - M3 x 8	TP ISO 10642 - M3 x 8
	TP ISO 10642 - M3 x 10	TP ISO 10642 - M3 x 10
	TP ISO 10642 - M3 x 12	TP ISO 10642 - M3 x 12

Figure 5.22: Changes made to the File Name and Part Number cells

18. Select **OK**, and **OK** again.

19. In the **Library View** dropdown, pick the **Test Library** option. The copied **TP ISO 10642** part should now be visible in this library. We now need to edit this Content Center part, as shown in *Figure 5.23*:

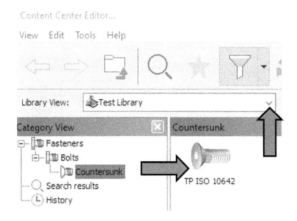

Figure 5.23: The new Content Center library active, with the copied ISO 10642 part included

20. Right-click on **TP ISO 10642** and select **Family Properties**. Change **Standards Organization** from **ISO** to TP. Select **Apply**, followed by **OK**.

21. Right-click the drop-down arrow next to the filter icon and select **Add/Edit Filters…**:

Figure 5.24: Add/Edit Filters… selected from the filter icon

22. Select **Add** and create a new filter called TP. Deselect all ticked boxes in the middle and right sections of the window, except for the **TP** option, as shown in *Figure 5.25*. Select **Apply** and **OK** to complete:

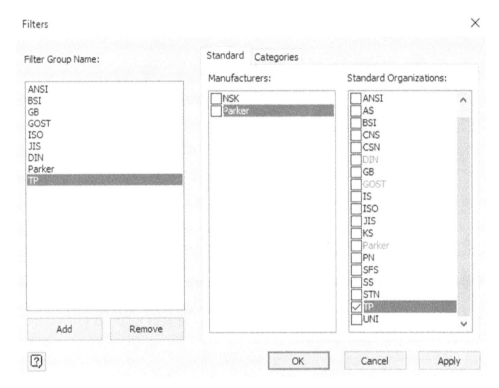

Figure 5.25: TP filter created and TP selected as the standard to view in this filter

23. Navigate back to **TP ISO 10642**. Right-click on it and select **Family Table....** We will now configure the new option within our custom content fastener:

Figure 5.26: Family Table… selected

24. This opens the various options for this specific piece of content. In this case, we will make a new option or variant in the M3 class. We will create an option for an M3 bolt that is 3 X 8.5, which currently is not available.

Select **Row 1**. Right-click and select **Copy**.

25. Right-click and select **Add Row**. Then, right-click on the newly added row and select **Paste**.

26. Navigate to **Bolt Length** on the newly created row and change the length from **8** mm to 8 . 5 mm:

Family Table: TP ISO 10642

RowStatus	File Name	Material	Size Designation	Part Number	Nominal Diameter [mm]	Thread description	Class	Thread Type	Bolt type	Nominal Thread Diam [mm]	Bolt Length [mm]	Thread Pitch [mm]	
1	TP ISO 10642 - M...Steel		M3x8	TP ISO 10642 - M...3		M3x0.5	6g	ISO Metric profile	M3	3	8	0.5	6
2	TP ISO 10642 - M...Steel		M3x8.5	TP ISO 10642 - M...3		M3x0.5	6g	ISO Metric profile	M3	3	8.5	0.5	6
3	TP ISO 10642 - M...Steel		M3x10	TP ISO 10642 - M...3		M3x0.5	6g	ISO Metric profile	M3	3	10	0.5	6
4	TP ISO 10642 - M...Steel		M3x12	TP ISO 10642 - M...3		M3x0.5	6g	ISO Metric profile	M3	3	12	0.5	6
5	TP ISO 10642 - M...Steel		M3x16	TP ISO 10642 - M...3		M3x0.5	6g	ISO Metric profile	M3	3	16	0.5	6
6	TP ISO 10642 - M...Steel		M3x20	TP ISO 10642 - M...3		M3x0.5	6g	ISO Metric profile	M3	3	20	0.5	6
7	TP ISO 10642 - M...Steel		M3x25	TP ISO 10642 - M...3		M3x0.5	6g	ISO Metric profile	M3	3	25	0.5	6
8	TP ISO 10642 - M...Steel		M3x30	TP ISO 10642 - M...3		M3x0.5	6g	ISO Metric profile	M3	3	30	0.5	6
9	TP ISO 10642 - M...Steel		M4x8	TP ISO 10642 - M...4		M4x0.7	6g	ISO Metric profile	M4	4	8	0.7	8

Figure 5.27: Bolt length changed to 8.5 mm

27. Select **OK** to complete. This now publishes your new Content Center part to your custom Content Center library.

28. Close the Content Center window and navigate back to your assembly file. Select the **Assemble** tab, then **Place**, followed by **Place from Content Center**:

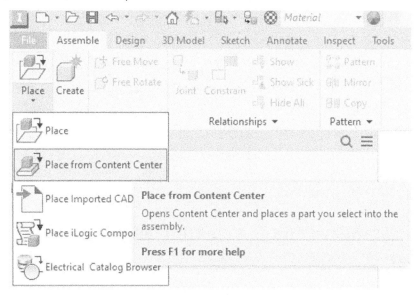

Figure 5.28: Place from Content Center

29. Select the **TP ISO 10642** fastener and select **OK**. On the import options, you will see **8.5** mm is now available to pick as a configuration for an **M3** bolt type. Select **8.5** mm under **Bolt Length**. Then, select **OK** to place this from the TP Content Center:

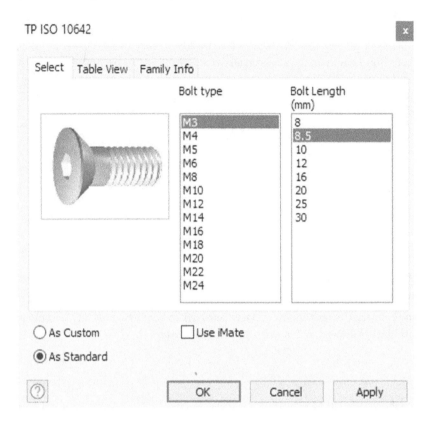

Figure 5.29: M3 x 8.5mm TP ISO 10642 selected

30. The custom Content Center item is imported into your assembly model.

You have successfully configured a custom Content Center library and created and published new custom content to it, for use in future designs. If you navigate to the Content Center folder in the C:\ drive, you will see that the M3x8.5.ipt file is created inside this folder.

Style and Standard Editor – creating and editing templates

In this recipe, you will learn how to use the **Style and Standard Editor** to edit a drawing template and configure style libraries so that a template is accessible to others in your organization.

A **style library** contains the definition of individual style types. When you apply a style to an object in a document, the attributes of the style are retrieved from the style library. Inventor gives you the ability to copy and then edit existing style libraries to suit your individual or organizational requirements. This could be, for example, the way that a drawing template is configured for a company or a material list that needs to be accessible to all team members.

By default, all styles in the style library associated with the active drafting standard are available to format objects in documents. Your style libraries should be managed by your CAD administrator or engineering manager, as it should not be something that all users have read/write access to. This is because, without these securities, it is harder to enforce specific standards and templates.

You can use a single global library so that all designers use the same styles, or you can specify a library for a specific set of design files or each project.

The main reasons why you may want to use the Style and Standard Editor are as follows:

- You work with large assemblies and need to share style information between multiple documents
- Collaboration in a workgroup/company or project that needs a common source of styles throughout
- The enforcement of internal CAD standards and best practices within an organization

In this recipe, we will create a new Inventor style library for an Inventor drawing template and enforce this across a project. There are countless changes and customizations that can be applied to a drawing template. However, in this recipe, we will make some simple changes. A comprehensive list of what is possible to change from within the **Style Editor** can be found here: `https://knowledge.autodesk.com/support/inventor/learn-explore/caas/CloudHelp/cloudhelp/2019/ENU/Inventor-Help/files/GUID-33EE57FB-9343-4647-83F0-B737F80B7E9C-htm.html`.

Getting ready

You will not need any practice files for this recipe. You will need to ensure you have completed the *How to create and manage an inventor .ipj project file* recipe, as we will be using the `Test Project.ipj` file.

How to do it...

1. From the Inventor **Home** screen, ensure that **Test Project** is the active project.
2. Select **New**, and then **New** again. Expand **en-US**, select **Metric Folder**, and scroll down. Pick `ISO.dwg` as the template. Select **Create**.
3. Select the **Manage** tab from the ribbon bar, and then select the **Styles Editor** command.

    The Styles Editor will now open. Here, you can make changes to settings that are displayed in your drawing template standard. This could be dimension text, leader type, border thickness, and more – the list of customizations is extensive. This is all controlled from the **Style and Standards Editor** menu:

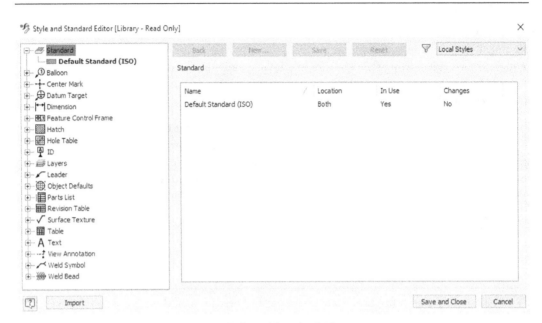

Figure 5.30: Style and Standard Editor open

If changes were made from this menu now, the styles and standards would be updated for this local document, but they would not be transferred to the global style and standard libraries. This would mean that every time a new drawing is created, these settings would have to be manually applied again. Fortunately, there is a way to prevent this and enforce our new style customizations across all users from our global style manager.

4. Close the Inventor drawing previously opened in *step 2* and navigate back to the Inventor **Home** screen.

5. Open the project settings with `Test Project.ipj` active.

6. Expand **Folder Options** with +:

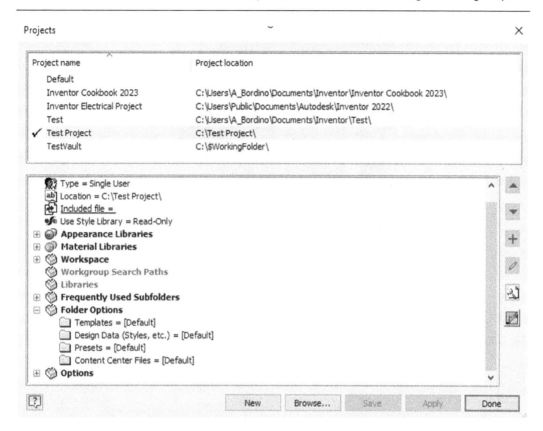

Figure 5.31: Folder Options expanded in Projects

7. **Folder Options** is still in the **Default** setting. This means it is using the standard default global style manager settings. We will now create a copy of the Design Data folder and then place this into our project workspace.

    It is important to make a copy of the Design Data folder, as if we change the master default settings for Inventor, *these cannot be undone!*

    In the Windows browser, navigate to **Windows (C:) | Users | Public | Public Documents | Autodesk Inventor 2023**.

8. Select the Design Data and Templates folders, right-click, and select **Copy**.

9. Navigate to the project workspace location at **Windows (C:) | Test Project**.

10. Right-click anywhere and paste the files into the Test Project folder:

Figure 5.32: Design Data and Templates copied into the Test Project workspace

11. Navigate back to the Inventor **Home** screen and open the project settings for Test Project. ipj.

12. Expand **Folder Options** if these are not expanded. Right-click on **Templates** and select **Edit**. Browse and select the Templates folder we just copied into the workspace.

13. Right-click on **Design Data** and select **Edit**. Browse and select the Design Data folder we just copied into the workspace.

14. Right-click on **Use Style Library = Read Only** and select **Read-Write**. This now enables us to sync local style changes to our new global style library within the project workspace.

15. Select **Done** and then **Yes**.

16. Select **Open** in Inventor, and then select **Templates**. Browse through **en-US** and **Metric**, and select ISO.dwg. This is the base template we will use for this example. You should always use the default template that is most aligned with your organization's standards from the beginning. Select **Open**.

17. We can now start to make all the changes to the template that we want to be global – for example, the drawing border and title block dimensions. In this example, we will make changes to the border color, **Dimension Style**, and edit the BOM columns on the drawing sheet. These are features we want to be global and always present on our drawing template.

    Select the **Manage** tab, followed by **Styles Editor**.

18. Navigate to the **Layers** area of **Style and Standard Editor**, expand **Layers**, and select **Border (ISO)**:

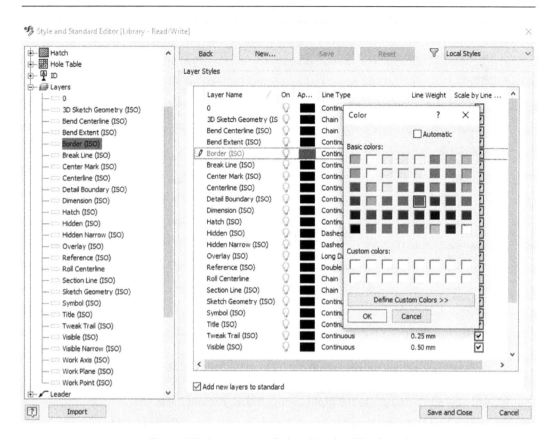

Figure 5.33: Layers expanded and Border ISO selected

19. In the **Layer Styles** area, under **Appearance**, select the *black square* next to **Border (ISO)**. In the color swatch, select *dark blue*. Select **OK**. Then, select **Save & Close**. This will change the border color to blue.

20. We will now change **Dimension Style**. Select the **Manage** tab, followed by **Styles Editor**.

21. Expand the **Dimension** options and select **Default (ISO)**.

22. Change **Decimal Maker** from **Comma** to **Period**.

23. Select **Display | Color** and change it to red.

24. Select **Text**, and then change the size from **3.50 mm** to **4.50 mm**. Select **Save**:

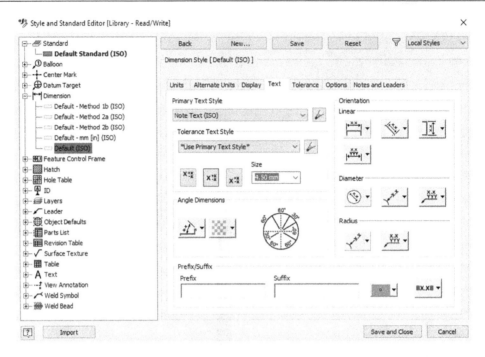

Figure 5.34: Editing the Dimension display from the Style and Standard Editor

25. We will now make changes to the BOM configuration. Expand **Parts List**.

26. Select **Material List (ISO)**, and then select **Column Chooser**. Browse to **MASS** and select **Add ->**. Then, select **OK**:

Figure 5.35: BOM Column Chooser

27. Select **Save**, followed by **Close**.

28. After updating our changes to the border, dimensions, and BOM, we must now confirm this in the global style library. On the **Manage** tab, next to the **Styles Editor** command, select **Save**, as shown in *Figure 5.36*:

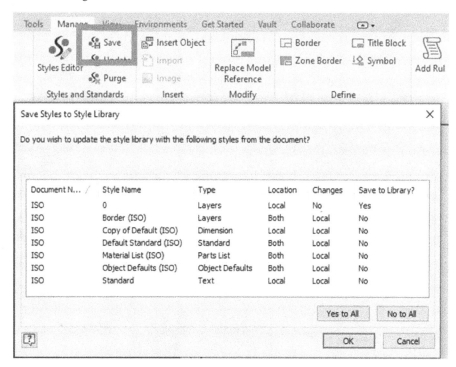

Figure 5.36: The Save icon in the Styles and Standards area

29. Accept any migration that Inventor prompts, and then select **Yes to All** in the next menu. This informs you that your local style has the following differences from the global style and standard library. Then, select **OK** and **Yes**.

30. Close the drawing. You do not need to save any changes, as we have already saved these to the global styles and standards.

31. Select **New**, and browse the ISO.dwg template. Create a new drawing from this template. Select **OK** on any message that appears.

32. Go to **Manage**, and hit **Save** in the **Styles and Standards** area.

33. Select **File | Save Copy as Template**, and name your template TP Template. Select **Save**.

34. Close the drawing.

35. Now, select **New Drawing** in Inventor. **TP Template** will be available to use with all styles and standards we applied and created in this recipe, as shown in *Figure 5.37*:

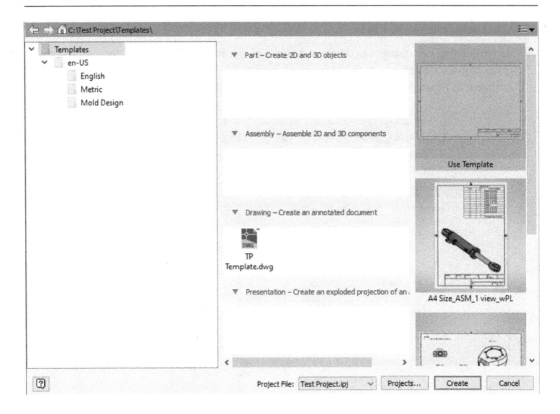

Figure 5.37: TP Template available

You have created and configured a new template for use in Inventor, using the style and standard libraries.

Collaborating with shared views

In this recipe, you will learn how to create a shared view of your Inventor model so that you can send this to collaborators on a project.

Shared views are used to create a read-only, visual representation of your model or design online, via the cloud. Inventor will generate a link that will allow anyone with access to view and comment on the shared view without the need to install an Autodesk product. From within Inventor, you can manage and act on the feedback supplied.

This is especially useful for communicating with internal and external stakeholders in a project, who are outside of the design department or do not have access to Autodesk CAD products such as Inventor.

Getting ready

You will need to access the Chapter 5 folder within the Inventor 2023 Cookbook folder. You will also need to ensure that Inventor Cookbook 2023.ipj is set as the active Inventor project. You will also need access to your email, and it would be beneficial to have a mobile device that can access your email account also.

How to do it...

1. Ensure that your Inventor project is set to Inventor Cookbook 2023.ipj.

2. Select **Open**, and navigate to the Chapter 5 folder.

3. Select the V6 Engine folder. Then, select the V6 Engine.iam assembly file. After that, select **Open**.

4. The assembly opens as normal in Inventor. We will now create and upload a read-only version of this into the Autodesk cloud, and grant access via a link we can share in an email. This means the recipient will be able to view and see this model on a PC or mobile/tablet device.

    Navigate to the **Collaborate** tab. Then, select **Shared Views**:

Figure 5.38: The Shared Views command shown in the Collaborate tab

5. Select **Shared Views**. The **Shared Views** tab will open on the right of Inventor. Select **New Shared View**. You are then prompted to name the viewable active file; leave this as default for now. Renaming this does not change the copy on your local machine.

6. Untick **Hide component names** and **Hide part properties**. Then, select **Share**.

7. The shared view of this assembly will now be created, once successfully uploaded. Select **View in Browser**. This is what the recipient will see once a link has been sent and opened of the design. The view changes slightly on a mobile device.

8. **AUTODESK VIEWER** now loads in your web browser. An email is automatically generated to your email address, registered to your Autodesk account, telling you that your shared view is ready. We can now begin to manipulate and view the read-only version of the CAD model.

Within the browser, there are multiple functions we can use:

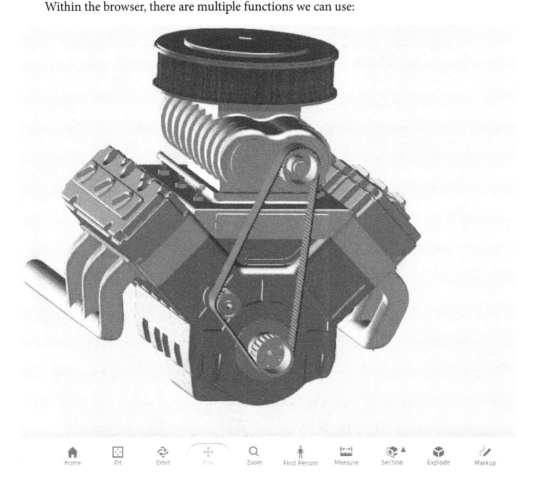

Figure 5.39: AUTODESK VIEWER displaying the shared view in the cloud, created in the browser

Using the commands along the bottom of the browser, we can orbit, pan, zoom, and take measurements.

Next, select **Explode**.

9. Drag the *blue dot* on the slider to explode and reassemble the whole assembly:

Figure 5.40: The Explode function applied to the model

10. Select the **Model** browser in the viewer. From here, you can examine each part that makes up the assembly and hide components. You can also change the transparency of parts:

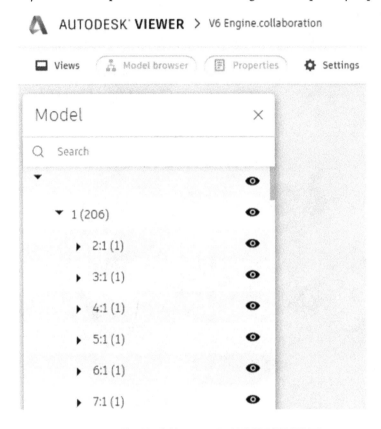

Figure 5.41: The Model browser in AUTODESK VIEWER

11. We will now write a comment back to the designer. In this case, we are simulating what a stakeholder or customer would do if the shareable link had been sent to them. For this, select **Markup** from the bottom toolbar.

12. With **Pencil** selected, click and hold to draw a circle around a component in the exploded view of the V6 engine, as shown in *Figure 5.42*:

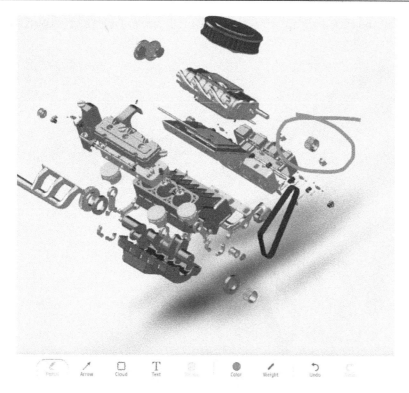

Figure 5.42: Markup applied

13. Select **Arrow** and create an arrow from the circle, by clicking and dragging:

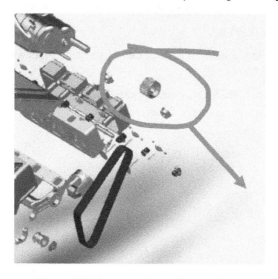

Figure 5.43: An arrow drawn from markup

14. Select **Text**, and click in the space near the head of the arrow. Type TEST COMMENT and click again to complete:

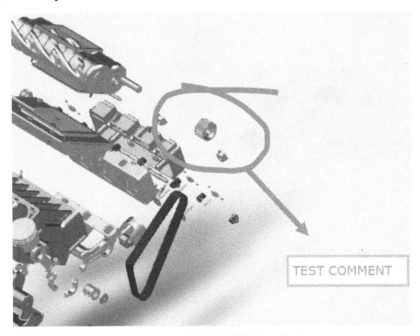

Figure 5.44: TEST COMMENT applied

15. At the top right of **AUTODESK VIEWER**, hit **Save** to complete the markup.

   The history of comments and changes made in the viewer is now shown.

16. Close **AUTODESK VIEWER**.

17. Navigate back to Inventor and the V6 Engine Inventor.iam assembly. Hit the **Refresh** symbol in the **Shared Views** tab.

18. The comment we made in the browser has appeared in the shared view area on the right of the screen, and we can see and act on the design feedback:

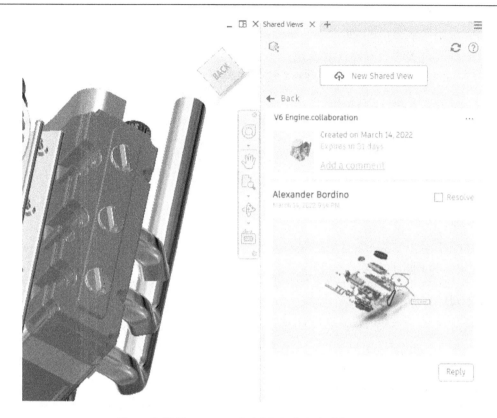

Figure 5.45: The comment visible in Autodesk Inventor

Comments can also be replied to so the recipient will receive this the next time they open the shareable link.

19. Now, we will share the design via email. Select **Copy link** from the **Shared Views** upload:

Figure 5.46: Copy link

20. The link is copied to the clipboard. Open your email, and paste the link into a new email. Send this email to yourself.

21. Open the email on your mobile device, and experiment with the shared views on it.

We have now used shared views to send a read-only viewable file from the cloud, via a link, that can be viewed on a PC or mobile device without Inventor software. We have also learned how to manipulate the model in this interface, and add/receive feedback, markups, and comments.

Collaborating and sharing designs with Pack and Go

In this recipe, you will learn how to use Pack and Go functionality to share a design or project with someone else.

Pack and Go is used to archive a file structure, share or copy a complete set of design files while retaining links to referenced files, or isolate a group of files. It can be used to share data externally as well as internally within your company. Note that recipients must have a valid Autodesk product to open their files once a Pack and Go package is received.

Pack and Go does not create a live link for multiple stakeholders to make edits too but instead sends a copy of data on your local machine as a package. If collaboration on the same files is required within the same organization, then using Autodesk Vault with Inventor is highly recommended for this and is much more efficient.

When Pack and Go is initiated on a file or multiple files, these files are copied to the location specified by the user, and the source files are neither changed nor removed.

Pack and Go will only work with fully resolved files using the current project file.

There are a few ways in which a Pack and Go operation can be initiated:

- From within Autodesk Inventor
- From Microsoft Windows' File Explorer
- Using the Design Assistant outside of Inventor

Getting ready

In this recipe, we will create a Pack and Go of an assembly from within Inventor. You will need access to the Gearbox assembly.iam file located in the Chapter 5 folder, within the Inventor Cookbook 2023 folder.

How to do it...

Within Inventor, we will first open the top-level assembly that we want to share:

1. With Autodesk Inventor open, select **Open** and browse the following location for the practice file: the `Inventor Cookbook 2023` folder | the `Chapter 5` folder | the `Gearbox` folder, and open `Gearbox.iam`.

2. Select **File** | **Save As** | **Pack and Go**:

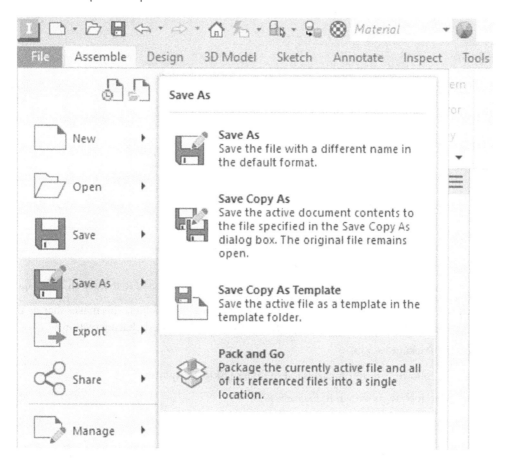

Figure 5.47: Pack and Go selected from within Inventor

3. We now need to specify the destination folder for this Pack and Go. In this case, select the browse button to the right of the destination entry, and browse and select **Desktop**:

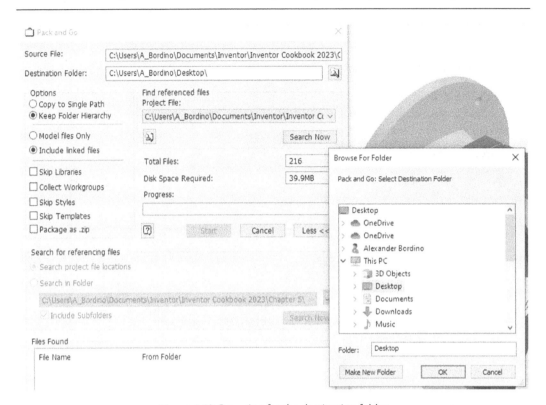

Figure 5.48: Browsing for the destination folder

4. We will leave other options as default to keep the folder hierarchy, and to include all linked files.

    What will be necessary is to package the Pack and Go as a `.zip` folder. This makes sharing the folder and the design much easier, especially over email or a cloud-sharing platform.

    Select the **Package as .zip**.

5. Select the following: **Skip Libraries | Skip Styles | Skip Templates**.

6. Select **Search Now**, as shown in *Figure 5.49*:

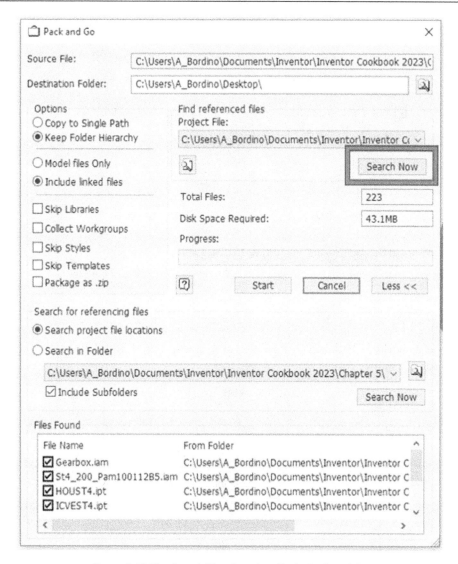

Figure 5.49: The Search Now functionality in Pack and Go

This will ensure that all necessary files are selected for the Pack and Go operation.

7. Once completed, the files found will be listed in the bottom half of the menu. Proceed to select the second **Search Now** option to search for any referenced files.

8. In this instance, no external files are found in relation to this assembly, as shown in *Figure 5.50*. Select **X** to complete:

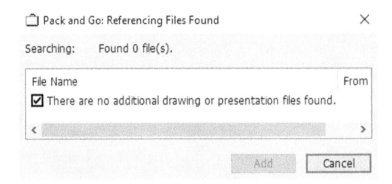

Figure 5.50: No Pack and Go reference files detected

9.   Select **Start** in the menu. This will initiate the Pack and Go.

10.   Once the progress bar has completed, select **X** to close the **Pack and Go** menu.

11.   Browse to your desktop.

12.   Locate the Pack and Go files on your desktop ready to be shared.

You have now completed a Pack and Go operation of an assembly from within Inventor so that files can be shared with another user of it.

Model credits

The model credits for this chapter are as follows:

- **V6 Twin Turbo Engine + supercharger** (by Lê Thảo) `https://grabcad.com/library/v6-twin-turbo-engine-supercharger-1/details?folder_id=9240197`

- **Gearbox-ST4- 200- PAM100112B5** (by Saeed azizi) `https://grabcad.com/library/gearbox-st4-200-pam100112b5-1/details?folder_id=11290519`

Inventor Assembly Fundamentals – Constraints, Joints, and BOMS

The **Inventor Assembly environment** is used to combine multiple part files, subassemblies, and components to create a bottom-down assembly that communicates how all individual parts are combined and interface with each other. The assembly environment can also be used to edit parts and model subsequent parts if required.

The Inventor Assembly environment is instrumental in ensuring tolerance, fit, overall function, and the overall working of a completed product or subassembly. In previous chapters, we used the assembly environment, but in this chapter, we will cover the best practices for assembling parts, managing a **Bill of Materials (BOM)**, and how you can duplicate and replace components within an assembly.

In this chapter, we will learn the following:

- Constraining components in assemblies – the best practices
- Applying joints in assemblies
- Applying and driving Motion constraints
- Duplicating and replacing components in an assembly
- Creating and configuring a BOM
- Detailing and customizing a BOM in a drawing file

Technical requirements

To complete this chapter, you will need access to the practice files in the Chapter 6 folder within the Inventor Cookbook 2023 folder.

Constraining components in assemblies – the best practices

In this recipe, you will use constraints and assemble a brake assembly from multiple Inventor parts and subassemblies. In doing this, we will cover best practices with the Constraint tool.

You may have noticed that within Inventor, there are two distinct methods of combining parts together in an assembly – **Joints** and **Constraints**. Both can be used independently or combined together to achieve the desired design intent.

Both assembly constraints and assembly joints are used to create relationships between components that determine a component's location and the allowable movement.

The **Constrain** and **Assemble** commands are methods of positioning components by gradually eliminating **degrees of freedom** (DOF). This is a legacy feature within Inventor and predates the introduction of assembly joints.

Inventor joints are used to simplify the complexity of component relationships. With a Joint, you not only define the position of a component but also fully define the allowable motion.

In the following recipe, we will just focus on constraints.

Getting ready

To begin this recipe, you will need to create a New Metric (mm) Standard Assembly.iam file and have this open in Inventor.

How to do it...

To begin, ensure that you have a New Metric (mm) Standard.iam file open in Inventor. We will start by placing one component at a time and adding constraints to fully define the assembly. The completed assembly you will create with constraints is shown in *Figure 6.1*:

Figure 6.1: The completed brake assembly

To begin, we will first place the component by following these steps:

1. Within your newly created assembly file, navigate to the **Assemble** tab and select **Place Component**.

2. Navigate to the Chapter 6 folder | the Brake folder, and select DRIVER_ BRAKE_ROTOR. ipt. Select **Open**.

3. The component now opens in your new assembly and is fixed on your cursor. Instead of left-clicking to place the new component, we first need to *ground* it. This is important, as this will fix the first component to the origin of the assembly file. Without this, the component would be undefined, and we would have difficulty assembling the rest of the components to it. A grounding of components is only usually required when the first component of your assembly is placed.

    Right-click anywhere in the **Graphics Window**, and select **Place Grounded at Origin** from the **Graphics Window** options that appear.

4. The first component is now placed, and we can now proceed to add another component. Select **Place** from the **Assemble** tab.

5. Navigate to the Chapter 6 folder | the Brake folder, select BRAKE_HUB.ipt, and then select **Open**.

6. Left-click anywhere to place the BRAKE_HUB file into the assembly. Note that the component is *floating* in the **Graphics Window** and is not fixed. If you click and drag the component, you will see that it moves freely as no DOF have been locked. We will now use the **Assembly** constraint to apply DOF and fix the component in relation to the DRIVER_BRAKE_ROTOR file, as desired.

7. Select **Constrain** from the **Assemble** tab. We are now able to select faces, edges, and planes to constrain the components together. We can also define the Mate type we want to place. Leave the defaults in the **Mate Constrain** window as they are; we will be using standard Mates for now.

    Before assembling a product in Inventor, it is a good idea to know how they go together first. Having a good understanding of the model and parts will enable you to assemble them more efficiently and allow you to use the minimal number of Mates required. Always look for symmetry in the model and shared centerlines or planes with the components!

    With **Constrain** selected, select the bottom face of BRAKE_HUB as the first reference, as shown in *Figure 6.2*:

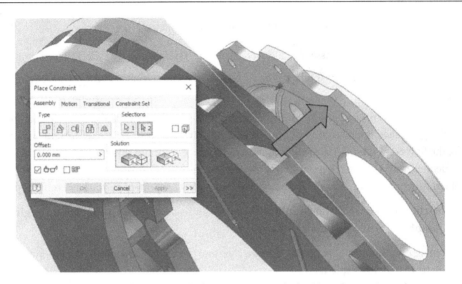

Figure 6.2: First face selected of BRAKE_HUB with the Mate Constrain tool

8. Rotate the model and select the face shown in *Figure 6.3* of DRIVER_BRAKE_ROTOR to act as the second reference for this Mate. This way, we are communicating to Inventor that we want these two faces to be in contact with each other. Then, select **Apply**:

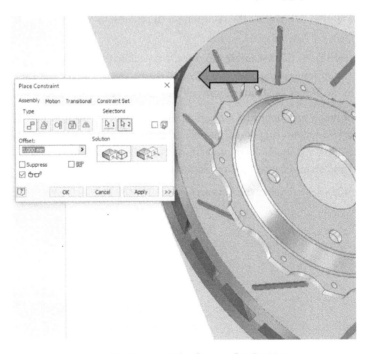

Figure 6.3: The second reference for the Mate

9. The first Mate has been created, although the component is not fully defined as it can still be moved and has DOF. To see what DOF are remaining, you can select the **Degrees of Freedom** button at the top left of the **View** tab within an assembly. Selecting this will reveal to what degrees the components can move.

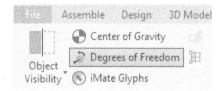

Figure 6.4: Degrees of Freedom

Another option is to grab and move the component and see how it can be manipulated with a mouse drag.

10. We will now define the component further with more mates. In the model browser to the left of the screen, note that the Relationships folder has become populated with **1 Mate**. This is the Mate we created in *Step 8*. At any time, you can right-click on the Mate in the model browser, edit the parameters, and delete or even rename the Mate.

In the **Assemble** tab, select **Constrain**.

11. Navigate to the center of BRAKE_HUB and select the geometry shown in *Figure 6.5*, and then the centerline of this component should be highlighted.

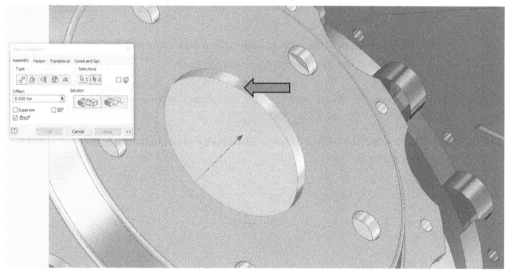

Figure 6.5: The first reference for Mate 2 to select

We will now use the Mate to align the components on a shared centerline.

12. For the second reference of this Mate, rotate to the other side of the model, and select the internal diameter shown in *Figure 6.6*. Select **Apply** to complete the Mate.

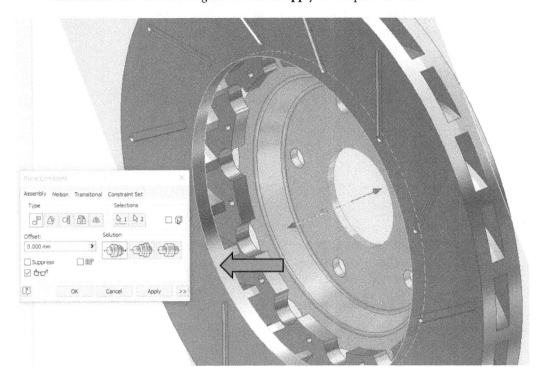

Figure 6.6: Internal diameter to select to complete Mate 2 and
align the components on a common centerline

13. Although the two components are flush together and aligned at the center, BRAKE_HUB needs to be aligned with DRIVER_BRAKE_ROTOR.

    Select the constraint, and then select the internal diameter of one of the small holes on BRAKE_HUB as the first reference.

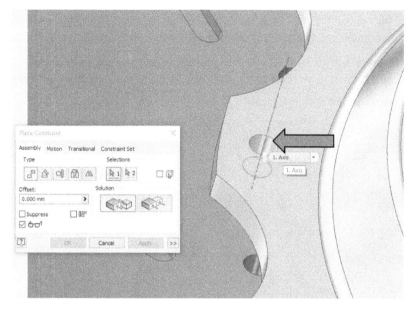

Figure 6.7: The Mate constraint on the internal diameter of small holes in BRAKE_HUB

14. For the second reference, select the corresponding hole on DRIVER_BRAKE_ROTOR, and then select **Apply**.

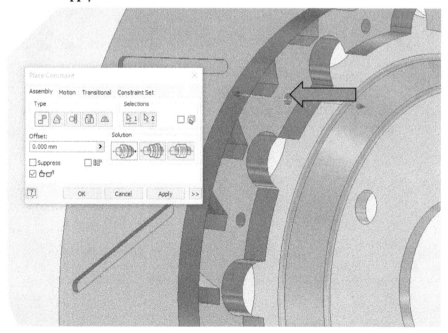

Figure 6.8: The second reference to select for the Mate Constraint

This will fully define the two components so that they are placed correctly and have no more DOF. It is common practice to apply more than one mate to a component to achieve this.

15. Select **Place**, and browse for CALIPER.ipt in the Chapter 6 folder. Select **Open**. Left-click to place the part in your assembly, and then hit *Esc* to exit the placing of components.

16. Select **Constrain**, and select the face of CALIPER, as shown in *Figure 6.9*, as the first reference:

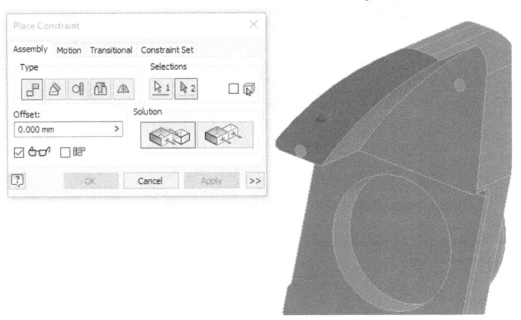

Figure 6.9: The first reference of CALIPER to select

17. Select the work plane of DRIVER_BRAKE_ROTOR as the second reference. As we have an internal central plane, we can align our part to this. Then, select **Apply**:

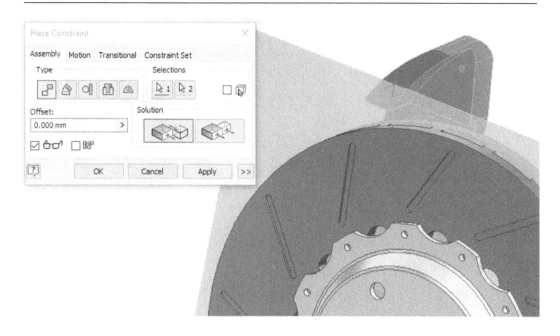

Figure 6.10: The second reference to select of CALIPER

18. Select **Place** and browse for RETAINER_PIN.ipt. Select **Open** and left-click anywhere to place in the assembly.

19. Select **Constrain**. Within the **Model Browser** on the left of the screen, expand the CALIPER part by selecting +. Expand the Origin folder by selecting +. Then, select **YZ Plane** as the first reference:

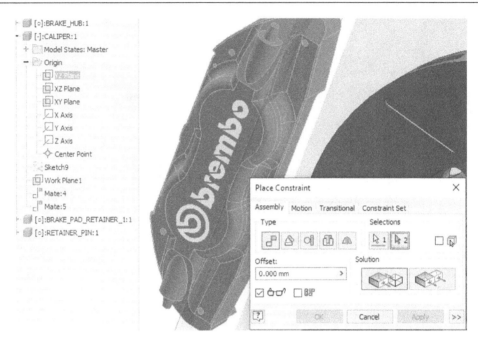

Figure 6.11: YZ Plane of Caliper selected as the first reference

20. Within the model browser on the left of the screen, expand the DRIVER__BRAKE_ROTOR part by selecting +. Expand the Origin Folder by selecting +. Then, select **YZ Plane** as the first reference.

21. In the offset area of the **Place Constraint** menu, type -130mm and select **Apply**:

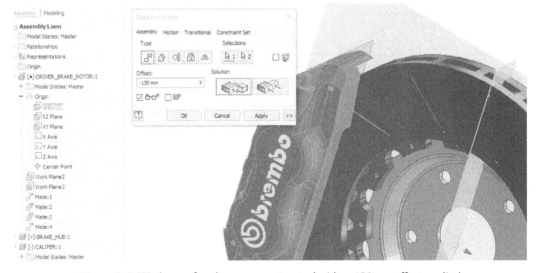

Figure 6.12: YZ planes of each component mated with a -130mm offset applied

There is still one more DOF on the CALIPER. Select **Constrain**, and place a Mate Constraint between **XZ Plane** of CALIPER and **XZ Plane** of DRIVER__BRAKE_ROTOR.

22. Select **Place** and browse for BRAKE_PISTON.ipt. Select **Open** and left-click anywhere to place in the assembly.

23. Select **Constrain**, and then select **Insert Mate Constraint**, followed by the geometry on BRAKE_PISTON, as shown in *Figure 6.13*, as the first reference.

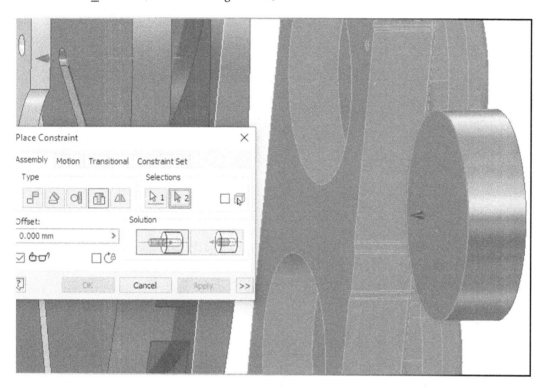

Figure 6.13: The first reference of BRAKE_PISTON to select

24. Select the internal geometry of CALIPER, as shown in *Figure 6.14*, as the second reference, and then select **Apply**:

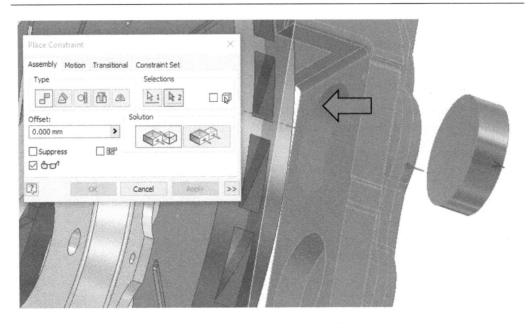

Figure 6.14: The second reference of CALIPER to select

25. Repeat *steps 23 to 24* to add another instance of BRAKE_PISTON and Mate it into position within the assembly. Once this is done, there should be two instances of BRAKE_PISTON on CALIPER, as shown in *Figure 6.15*:

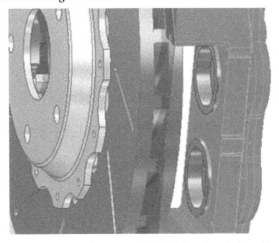

Figure 6.15: Two instances of BRAKE_PISTON defined and placed in the assembly

26. Select **Place** and browse for BRAKE_PAD.ipt. Select **Open** and left-click anywhere to place in the assembly.

27. Select **Constrain** and select the face of BRAKE_PAD shown in *Figure 6.16* as the first reference:

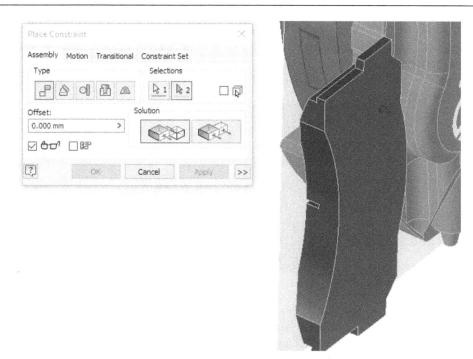

Figure 6.16: The first reference of BRAKE_PAD to select

28. For the second reference, select the top face of BRAKE_PISTON, as shown in *Figure 6.17*:

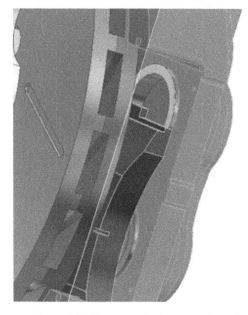

Figure 6.17: The second reference selected

29. Click and drag BRAKE_PAD away from the assembly. The previous mate we created in *step 28* will still be active. Pulling the component away while not fully defined allows us to see all available faces to select for further mates. Another option would be to change the visibility of components, or to adjust the visual style of the model:

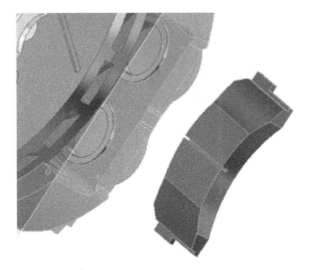

Figure 6.18: BRAKE_PAD manually moved from the assembly

30. Select **Constrain** and proceed to select the geometry shown in *Figure 6.19* as the first reference:

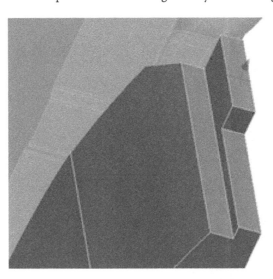

Figure 6.19: The top face of BRAKE_PAD selected as the first reference

31. *Figure 6.20* shows the second reference to select. Select **Apply** to complete:

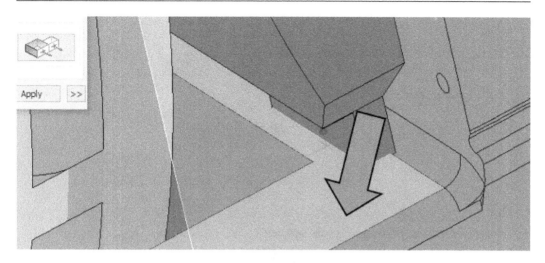

Figure 6.20: The second reference of CALIPER to select

32. Select the geometry shown in *Figure 6.21* to align and eliminate all final DOF on BRAKE_PAD. This is the top face of the pad and the edge of CALIPER, as shown in *Figure 6.21*. Then, select **Apply**:

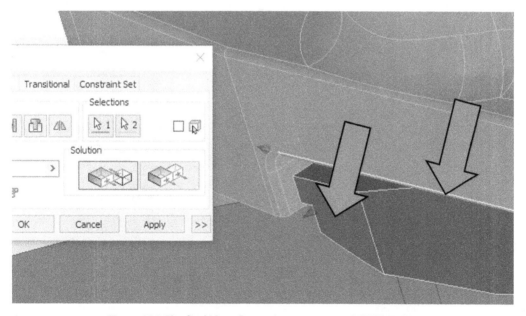

Figure 6.21: The final Mate Constraint to create on RETAINER_PIN

33. Select **Constrain | Insert Constraint**, and then the reference shown in *Figure 6.22* of RETAINER_PIN: 1:

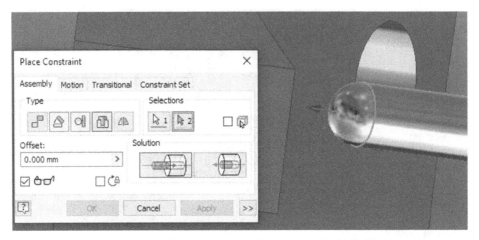

Figure 6.22: The first Insert Mate reference on RETAINER_PIN 1

34. For the second reference, select this geometry on CALIPER. Then, select **Aligned** as the **Solution** option. The result resembles *Figure 6.23*. Select **Apply** to complete.

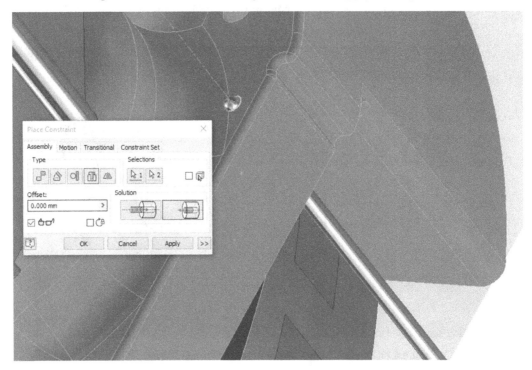

Figure 6.23: The final Insert Mate reference to make on CALIPER
with Aligned selected as the Solution option

35. Repeat *steps 33 to 34* to place and constrain another instance of the RETAINER_PIN in the correct position, as shown in *Figure 6.24*:

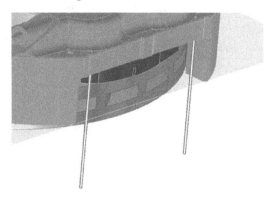

Figure 6.24: The second instance of RETAINER_PIN constrained in position

36. Select **Place** and browse for BRAKE_PAD_RETAINER_1.ipt. Select **Open** and left-click anywhere to place in the assembly.

37. Select **Constrain** in the model browser on the left of the screen, expand + next to DRIVER_BRAKE_ROTOR, and then expand + next to BRAKE_PAD_RETAINER. We will now work to align this component centrally to the model, as is the design intent.

38. Select + and expand the Origin folder of DRIVER_BRAKE_ROTOR. Select **Work Plane 3** as the first reference for this Mate.

39. Select + and expand the Origin folder of BRAKE_PAD_RETAINER_1. Select **YZ Plane** as the second reference of this Mate. Then, select **Apply** to complete.

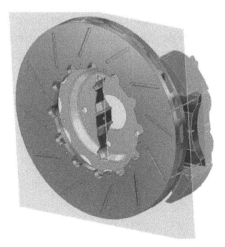

Figure 6.25: The BRAKE_PAD_RETAINER_1. YZ plane mated to WorkPlane3 of DRIVER_BRAKE_ROTOR

40. Select **Mate Constrain**, and create a Mate between BRAKE_PAD_RETAINER_1 and RETAINER_PIN, as shown in *Figure 6.26*. Select **Apply** to complete.

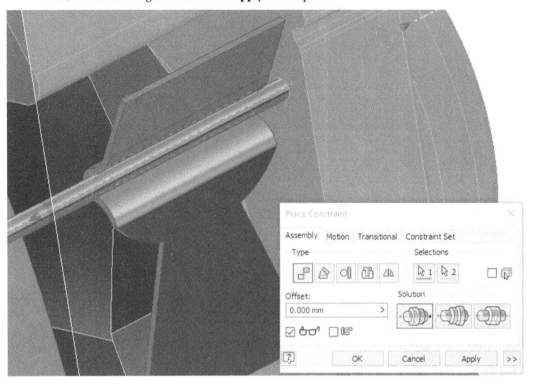

Figure 6.26: The Mate constraint between BRAKE_PAD_RETAINER_1 and RETAINER_PIN

41. Select **Constrain**, and then select **Angle** as the **Type** option. Select the faces shown in *Figure 6.27* as the reference. Then, select **Directed Angle** as the **Solution** option for the **Angle** Mate constraint. Leave the **Angle** value as 0. Select **Apply** to complete:

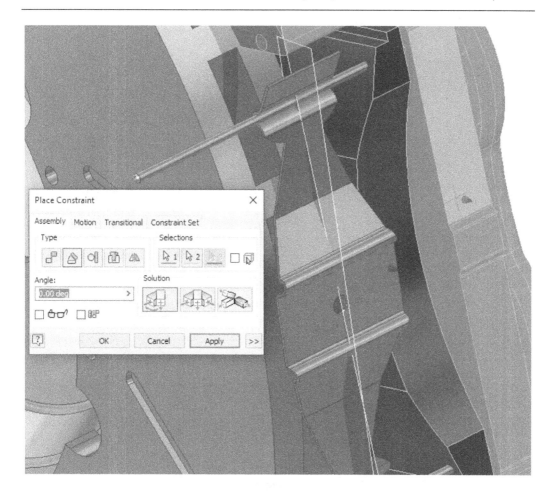

Figure 6.27: The Angle constraint applied to BRAKE_PAD_RETAINER_1 and CALIPER

42. We have now assembled half of CALIPER. Next, rather than placing second instances of all the components again and reapplying the same mates, we will instead look to mirror the components and mates around a central work plane. Always look for symmetry in instances when constraining components, as the use of the **Mirror** command improves efficiencies dramatically.

   Within the **Assemble** tab, navigate to the **Pattern** tools and select **Mirror**.

43. In the **Mirror** command, within the **Graphics Window,** select the components, as shown in *Figure 6.28*. These are the components that will be mirrored.

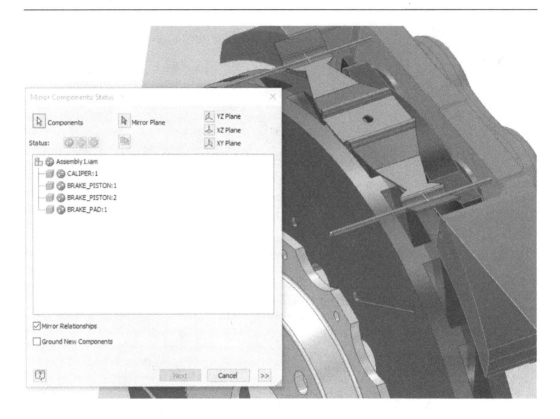

Figure 6.28: Components we need to mirror are selected

44. Select the *green icon* within the **Mirror Components: Status** window, next to `Assembly1.iam`. This will change all the mirror *green icons* of the selected components to a *yellow +* so that they are *reused* instead.

    Mirror, which was the default, would create mirrored copies of the parts with a `MIR` suffix. Selecting **Reuse** makes Inventor reuse the component and create a second instance of this component.

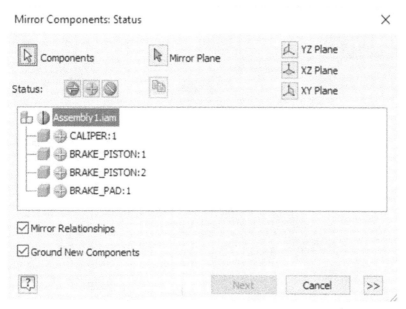

Figure 6.29: The reused components selected

45. Ensure that **Mirror Relationships** and **Ground New Components** are selected. Then, select the *red arrow* next to **Mirror Plane**.

46. Select **Work Plane3** as the mirror plane from DRIVER_BRAKE_ROTOR. A preview is created once this is done.

47. Select **Next**, followed by **OK**, to complete the **Mirror** operation.

48. Select **Place** and browse for BRAKE_PAD_RETAINER_2.ipt. Select **Open** and left-click anywhere to place in the assembly.

49. Select **Constrain** and select the face of BRAKE_PAD_RETAINER_2 as the first reference, as shown in *Figure 6.30*:

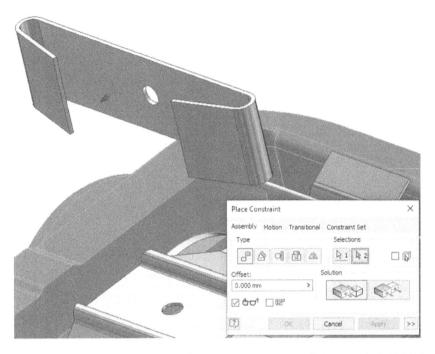

Figure 6.30: The first reference to select for the Mate constraint of BRAKE_PAD_RETAINER_2

50. For the second reference, select the face of BRAKE_PAD_RETAINER_2, as shown in *Figure 6.31*, and then select **Apply**:

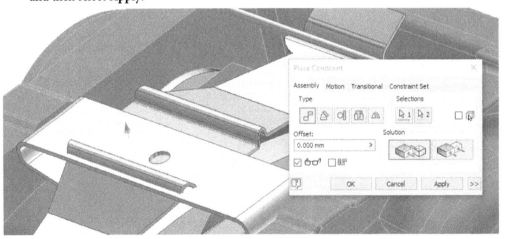

Figure 6.31: The second reference to select for the Mate constraint of BRAKE_PAD_RETAINER_2

51. Create a new Mate constraint between **YZ Plane** of BRAKE_PAD_RETAINER_1 and BRAKE_PAD_RETAINER_2.

52. Create a final Mate between the two central holes to align them:

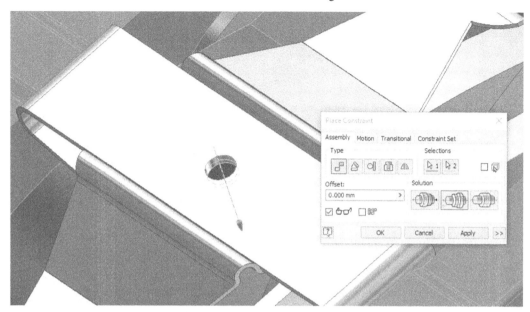

Figure 6.32: Aligned holes of the two components

The assembly of the brake caliper is now complete. You have successfully imported instances of separate parts into an assembly and used various constraints to assemble the completed design. The application of the **Mirror** command has meant we only had to spend time assembling half of the model.

Applying Joints in assemblies

The **Joint** command is an alternative to the **Constrain** command that enables you to define an assembly, with a specified and allowable amount of movement. Joints are selected from a pre-defined list of possible connections and then placed directly onto a model, as constraints are.

The process for assigning a joint to a component is as follows:

1. First, the **Joint** command is selected.

2. The **Joint** type is then selected.

3. References on the component are selected.

4. The limits of movement are defined.

5. The Joint is complete. Once completed, Joints can be flexed or edited to suit.

The types of Joints that you have within Inventor are as follows:

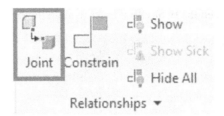

Figure 6.33: The Joint command in the assembly environment

- **Automatic**: This is the default Joint that is selected upon launching the command. This type of Joint enables Inventor to assume the type of Joint connection to create, based on the references picked. If the wrong one is selected, you can edit and change the type if required.

- **Rigid**: This Joint removes all DOF and "fixes" the component in place. In this respect, it is similar to a Mate constraint.

- **Rotational**: This Joint allows for rotation about an axis.

- **Slider**: This Joint allows for translational movement along an axis.

- **Cylindrical**: This Joint allows a component to translate and rotate about a specific axis with two DOF available.

- **Planer**: This Joint allows for movement in a single plane.

- **Ball**: Creates movement of a ball or sphere inside a socket.

In this recipe, we will apply various Joints to assemble several Inventor parts and define the overall assembly.

Getting ready

You will need a new `Inventor.iam Standard (mm)` file open.

How to do it...

In this recipe, we will assemble a simple vice assembly using Joints and some constraints. With the Joints placed, we will be able to set allowable DOF. We will not be applying transitional constraints at this stage to create movement from one component to another, as this will be covered later in the chapter, in the *Applying and driving Motion constraints* recipe. The purpose of this recipe is to introduce some of the Joint constraints.

To begin, we will create a new assembly file. Once created, we will then import ready-made parts and apply the Joints to assemble the vice with allowable DOF. To do this, follow these steps:

1. Create a new `Inventor.iam Standard (mm)` file.

2. From the **Assemble** tab, select **Place**. Browse to the `Chapter 6` folder and open the `Vice` folder.

3. Select all Inventor `.ipt` files in this folder, excluding `Vice Assem .iam`. Then, select **Open.** Left-click to place the components, then press *Esc*. This will bring all the parts in one operation into the assembly, as shown here:

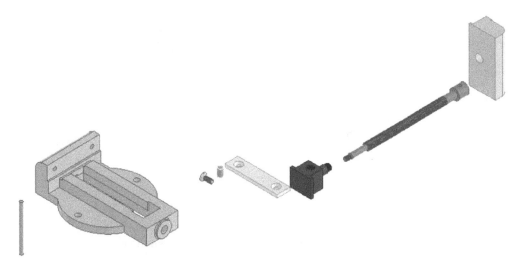

Figure 6.34: Placed Vice components within the new .iam file

4. Move your cursor over `body.ipt` in the **Graphics** window, and then right-click and select **Grounded**. This will ground the main component. With **Constrain**, it is important to do this for the first component.

5. Click and drag `guide.ipt` in the **Graphics** window near the `body.ipt` file you just grounded. This is the first component we will constrain with both Joints and Constraints.

6. Select the **Assemble** tab, select **Joint**, and then select **Cylindrical** from the dropdown. This will allow the component to rotate about the axis defined.

7. For the first reference, select the geometry of `guide.ipt`, as shown in *Figure 6.35*:

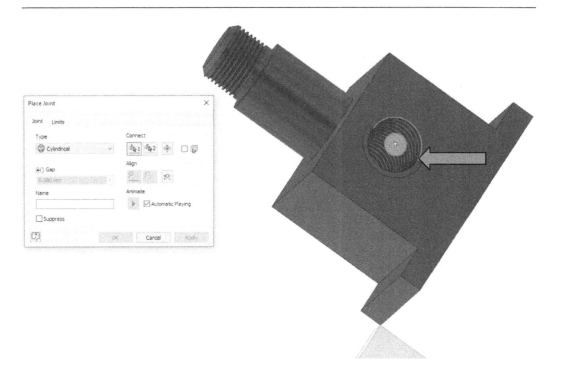

Figure 6.35: The first reference to select

For the second reference, select the geometry of the body, as shown in *Figure 6.36*:

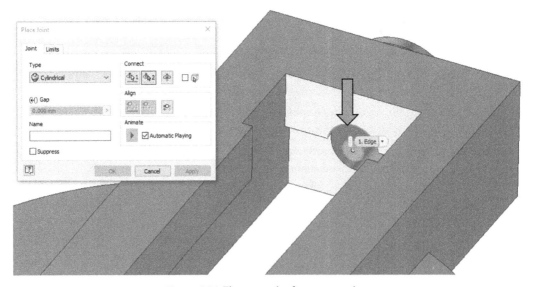

Figure 6.36: The second reference to select

Then, select **OK**.

8. The first joint has been applied, but as you can see, the component is still not properly defined, in relation to other components. You may find that guide.ipt can move inside body.ipt, as a Contact Set has not yet been defined; we will come onto Contact Sets in a few more steps. But for now, we will now define the component further.

9. Select **Constrain** from the **Assemble** tab.

10. Select **Angle Constraint** from the options and then select the face of guide.ipt, as shown in *Figure 6.37*:

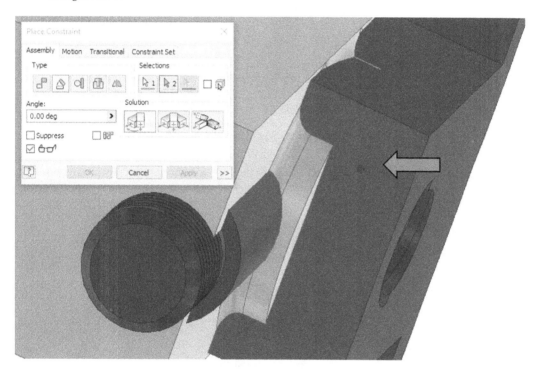

Figure 6.37: The first reference to select for the Angle Constraint

11. For the second reference, select the geometry of guide.ipt, as shown in *Figure 6.38*. Then, select **Directed Angle** from the options and leave the angle as 0 degrees by default. This will orientate the component and further define its allowable DOF. Select **OK** to complete.

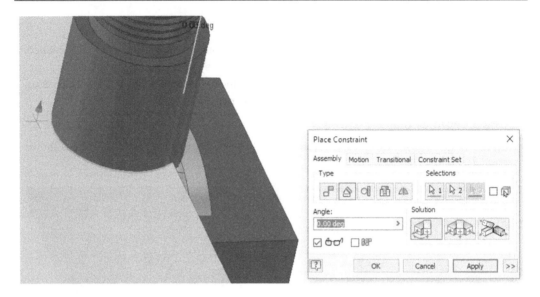

Figure 6.38: The second reference for the Angle Constraint with Directed Angle selected

12. No limits or Contact Sets have yet been applied to the allowable movement of these components. We can observe that none have been applied by clicking and dragging guide.ipt, and you should be able to slide it within body.ipt:

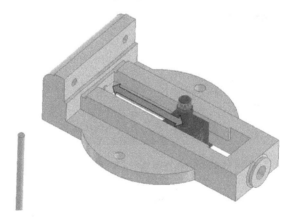

Figure 6.39: Allowable DOF

13. To further define these components, we can establish a Contact Set. This will stop the two components from going through each other when moved and start to establish some real-world physics in the model.

In the model browser on the left of the screen, select guide.ipt and body.ipt. Right-click and select **Contact Set** from the menu.

14. To turn the Contact Set on, navigate to the **Inspect** tab, and select **Activate Contact Solver**. If you click and drag the guide, it will only move bi-directionally within body.ipt and will not collide and merge with the body.ipt geometry. Select **Activate Contact Solver** to turn off the feature.

15. In the **Graphics** window, click and drag sliding jaw.ipt so that it is near body.ipt.

16. In the **Assemble** tab, select **Joint**, and then select **Cylindrical** as the option.

17. Select the first reference, as shown in *Figure 6.40*:

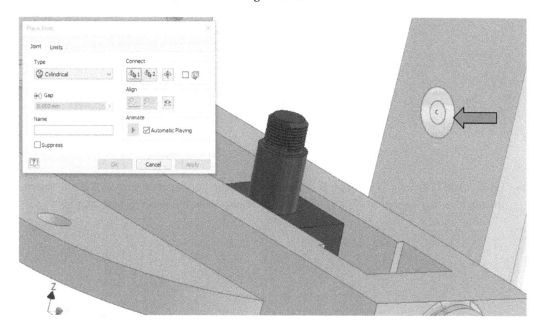

Figure 6.40: The first reference for sliding jaw.ipt

For the second reference, select the geometry shown in *Figure 6.41*:

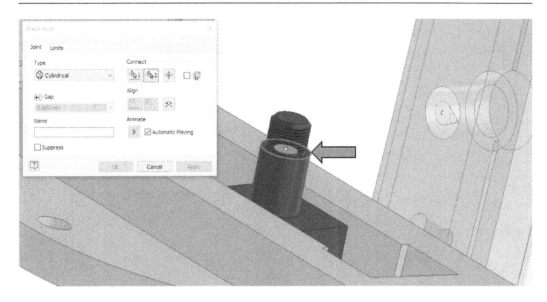

Figure 6.41: The second reference for sliding jaw.ipt

18. Ensure that sliding jaw.ipt is in the correct orientation, as per *Figure 6.42*. If not, select the flip icon highlighted in *Figure 6.42* to flip the component to the correct orientation:

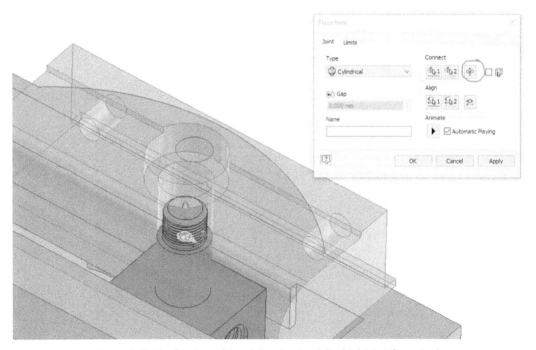

Figure 6.42: The flipping of the sliding jaw.ipt Cylindrial Joint if required

To complete the joint, select **OK**.

19. We now need to apply some **Mate** Constraints to further define `sliding jaw.ipt`. Select **Constrain**, and then **Mate**.

20. Select the bottom face of `sliding jaw.ipt` as the first reference, as shown in *Figure 6.43*:

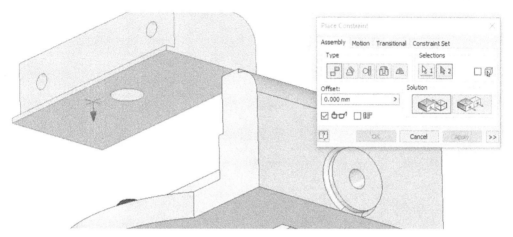

Figure 6.43: The first reference of the Mate constraint for sliding jaw.ipt

For the second reference, select the face of `body.ipt`, as shown in *Figure 6.44*. Then, select **Apply**:

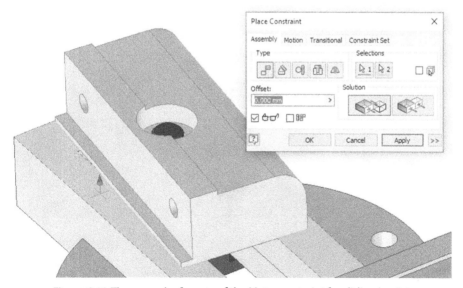

Figure 6.44: The second reference of the Mate constraint for sliding jaw.ipt

21. With the **Place Constraint** menu still active, select **Angle** as the **Constraint** type.

For the first reference, select the face of `sliding jaw.ipt`, as shown here:

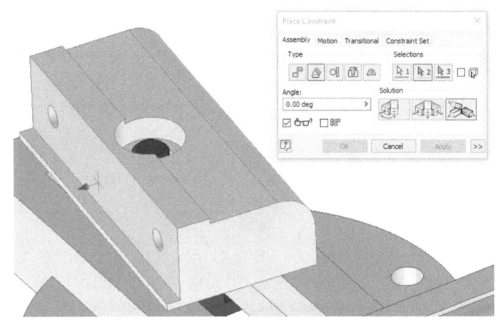

Figure 6.45: The first reference of the Angle constraint for sliding jaw.ipt

For the second reference, select the geometry shown in *Figure 6.46*:

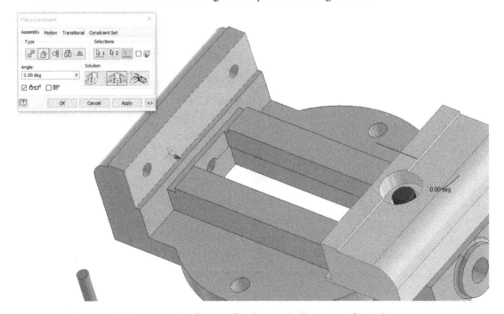

Figure 6.46: The second reference for the Angle Constraint for sliding jaw.ipt

Select **Undirected Angle** as the **Solution** option, and keep the value at 0 degrees as default. Select **OK** to complete.

This results in sliding jaw.ipt and guide.ipt moving as one within the constraints of body.ipt.

22. Click and drag the face plate in the **Graphics** window so that it is near body.ipt.

23. Select **Joint**, and then **Cylindrical** as the type, and then select the first reference as the geometry, as shown in *Figure 6.47*, on face plate.ipt:

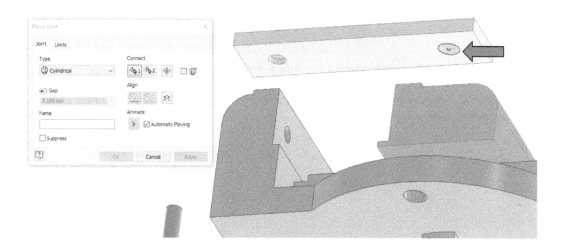

Figure 6.47: The first reference of the Cylindrical Joint for the face plate

For the second reference, select the corresponding hole feature of body.ipt, as shown in *Figure 6.48*:

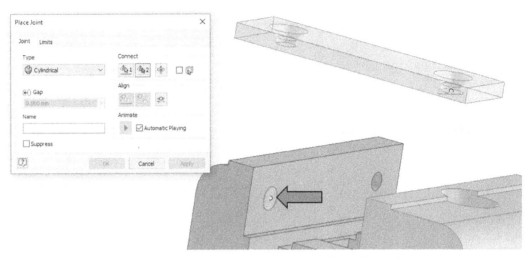

Figure 6.48: The second reference of the Cylindrical Joint for the face plate

If required, select **Flip Component** to ensure that the face plate is positioned in the correct orientation, as shown in *Figure 6.49*:

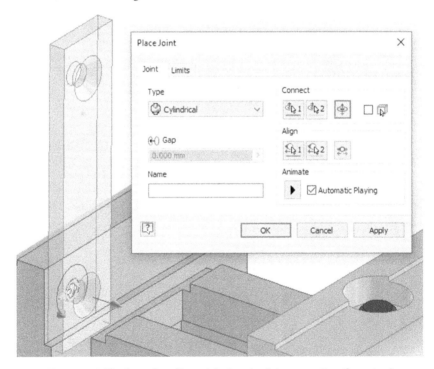

Figure 6.49: The face plate flipped during the Joint operation if required

Select **OK** to complete.

24. Create and apply a **Mate** Constraint between the center lines of the holes of both `face plate.ipt` and `body.ipt`, as shown in *Figure 6.50*. Select **OK** to apply:

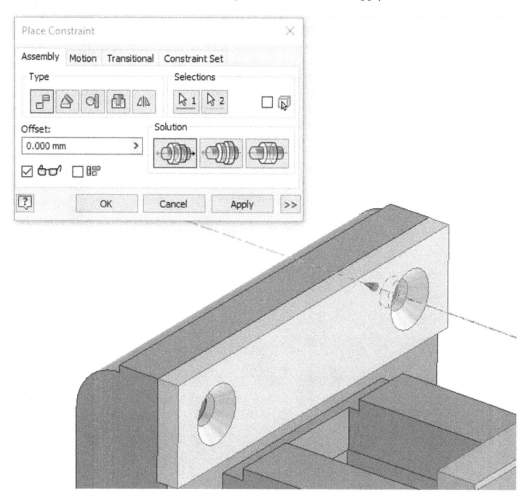

Figure 6.50: The Mate constraint applied to the centerlines of each hole in the face plate and body

25. The face plate still has DOF and can be pulled away. Click and drag the face plate away from the body in the **Graphics** window.

26. Select **Constrain**, then select **Mate**, and apply it to the geometry referenced in *Figure 6.51*:

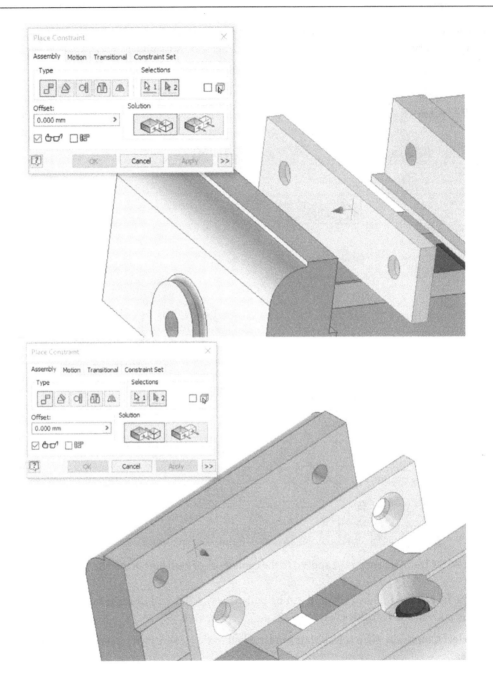

Figure 6.51: The two references selected to perform the Mate constraint

27. In the **Graphics** window, click and drag screw cap.ipt so that it is near face plate.
ipt in the assembly.

28. Select **Joint**, and then select **Rotational** from the options. Proceed to select the edge of `screw cap.ipt` as the first reference, as shown in *Figure 6.52*:

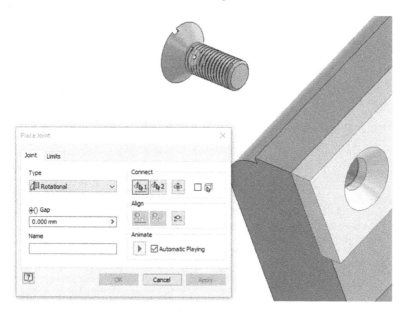

Figure 6.52: The first reference of the Rotational Joint for screw cap.iam

Then, select the following geometry as the second reference for this Joint, as shown in *Figure 6.53*:

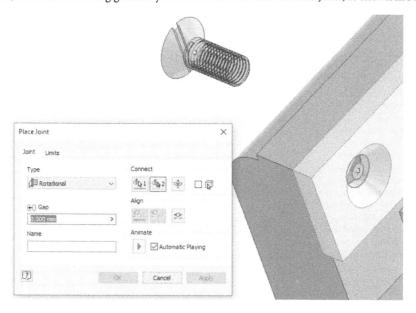

Figure 6.53: The second reference of the Rotational Joint for screw cap.iam

29. You may have to select **Flip component** to ensure that `screw cap.ipt` is in the correct orientation, as shown in *Figure 6.54*:

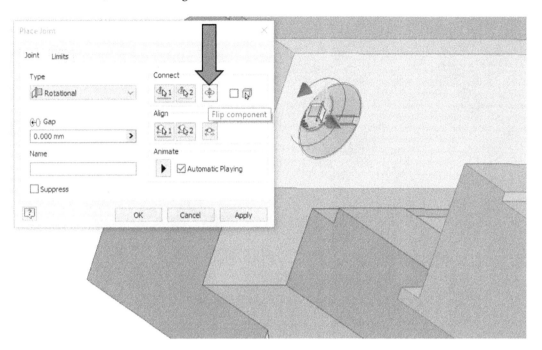

Figure 6.54: The flipped screw cap to move the component to the correct orientation

30. Select **OK** to complete.

31. As an optional extra step, you can also mirror `screw cap.ipt` about the center of the assembly to place another instance of it in the corresponding hole, as shown in *Figure 6.55*:

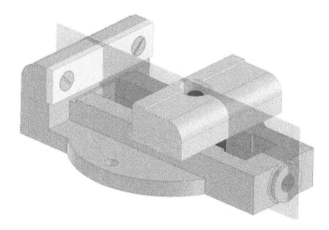

Figure 6.55: Optional step – mirroring screw cap.ipt across a midplane
to create another instance on the corresponding hole

32. Click and drag the `power screw.ipt` toward `body.ipt`.

33. Select **Joint**, and then select **Rotational** as the Joint type.

34. For the first reference, select the geometry of `power screw.ipt`, as shown in *Figure 6.56*:

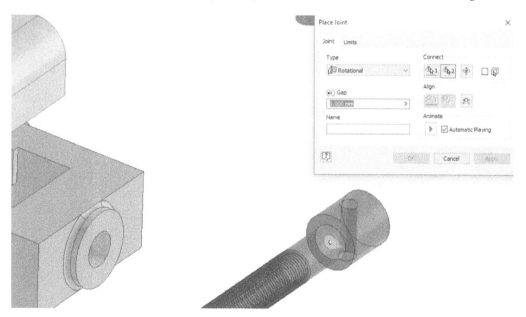

Figure 6.56: The first reference for the Rotational Joint of the power screw

For the second reference, select the geometry shown in *Figure 6.57*:

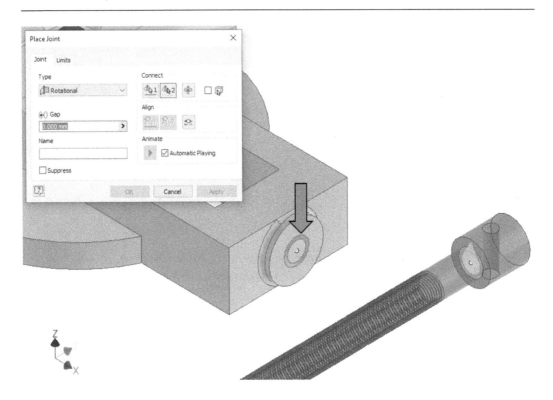

Figure 6.57: The second reference for the Rotational Joint of the power screw

Select **OK** to apply.

Pulling the sliding jaw in the assembly back and forth will not engage the power screw to turn, as a translational motion has not been defined. Therefore, the Joints and Constraints that we have applied are independent of one another. Translational Joints and motion can be applied in Inventor and will be covered in the *Applying and driving Motion constraints* recipe.

35. Select **Constrain**, and then select **Mate**. Place a **Mate** Constraint on the centerline of the cap and arm, as shown in *Figure 6.58*. Select **OK** to complete:

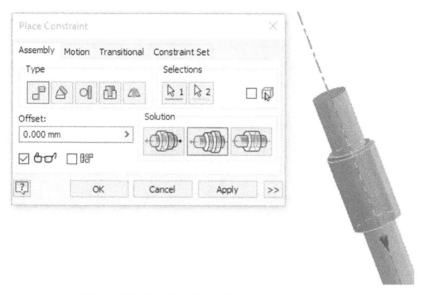

Figure 6.58: The Mate Constraint on the cap and arm

36. Create a second **Mate** Constraint from the top face of the arm and the inside face of the cap, as shown in *Figure 6.59*:

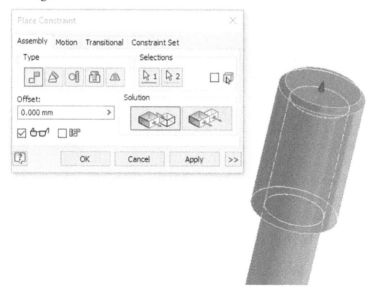

Figure 6.59: The Mate Constraint from the top face of the arm and the inside face of the cap

37. Create another **Mate** Constraint between the center of the arm and the center bore of the power screw, as shown in *Figure 6.60*:

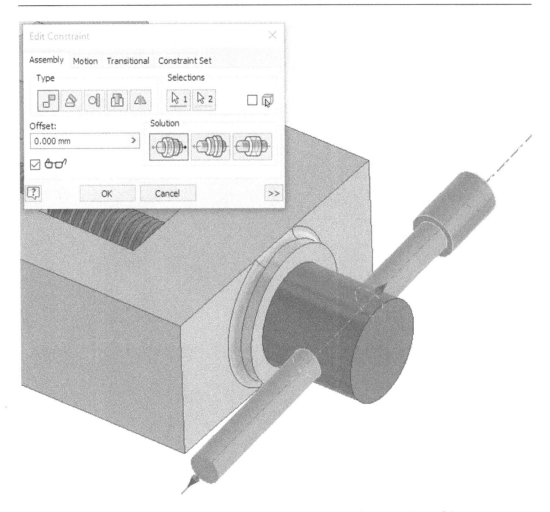

Figure 6.60: The Mate Constraint between the center of the arm and the center bore of the power screw

You have used a combination of Joints and Constraints to assemble several parts into an assembly, where DOF have been defined, ready for contact sets to be established.

Applying and driving Motion constraints

It is sometimes necessary to drive and examine motion between parts within an assembly to assess the end function and performance, and to detect possible collisions of components. With **Motion constraints**, you can apply and drive specified movement between parts.

In this recipe, you will apply Motion constraints to two spur gears, resulting in them being driven from a Motion constraint and meshing correctly.

Getting ready

To begin this recipe, you will need to access the `Chapter 6` folder and open `Gear Assembly.iam`.

How to do it...

To begin, ensure that you have `Gear Assembly.iam` open. The two spur gears have already been created for you using the Spur Gear Design Accelerator, and they have also been locked into position with existing Constraints so that they can only rotate freely about the central axis.

To begin, we will examine the allowable movement and existing relationships in the assembly:

1. Expand the `Relationships` folder by selecting + next to the `Relationships` folder in the model browser. Here, you can see the existing Constraints that have already been applied to the gears to fix them into position at the correct distance.

2. Move your cursor over one of the gears in the **Graphics** window. Then, click, hold, and drag your cursor in a circular motion. You will see that the gear selected will rotate freely in position. The rotating of one gear does not affect another, and the teeth at this point will collide with each other:

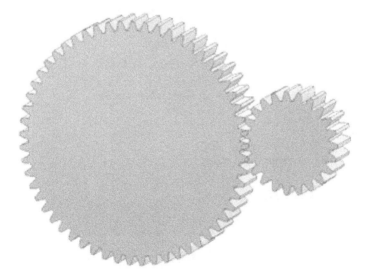

Figure 6.61: The gears rotate freely independent of one another and are not meshing correctly

We will apply Motion constraints to enable them to rotate with each other, and for the gears to mesh correctly.

3. In the **Design** tab, select the **Spur Gear** command from the ribbon.

4. In the **Spur Gears Component Generator** window, highlight the desired gear ratio default value. Right-click and select **Copy**, as shown in *Figure 6.62*.

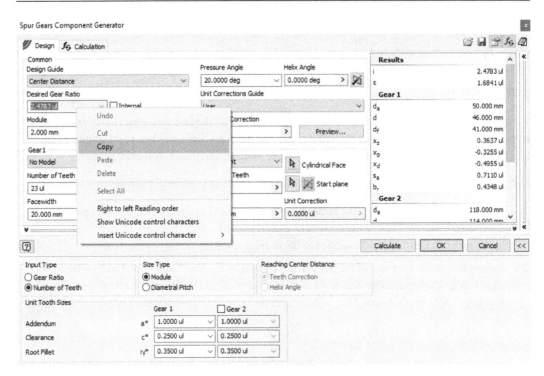

Figure 6.62: The Spur Gears Component Generator window, copying the gear ratio

Hit **x** to close the **Spur Gears Component Generator** window.

5. From the **Assemble** tab, select **Constrain**. Select the **Motion** tab within the **Place Constraint** window.

6. In **Ratio**, paste the ratio from the clipboard, which should be 2.4783 ul:

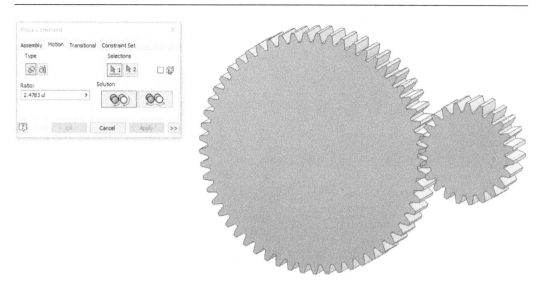

Figure 6.63: The Motion tab selected and a ratio of 2.4783 ul applied

7. Now, we need to select reference geometry for the Motion constraint. Navigate to the area shown in *Figure 6.64*, and select a flat face of the "bottom land" of the second gear:

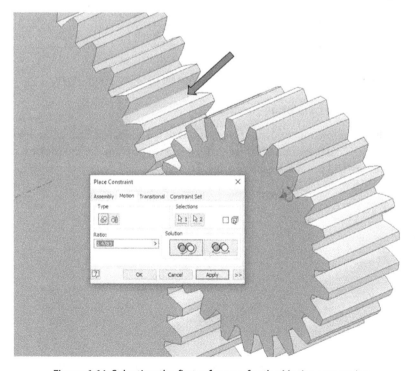

Figure 6.64: Selecting the first reference for the Motion constraint

8.   Select a flat face of the "bottom land" of the second gear for the second reference:

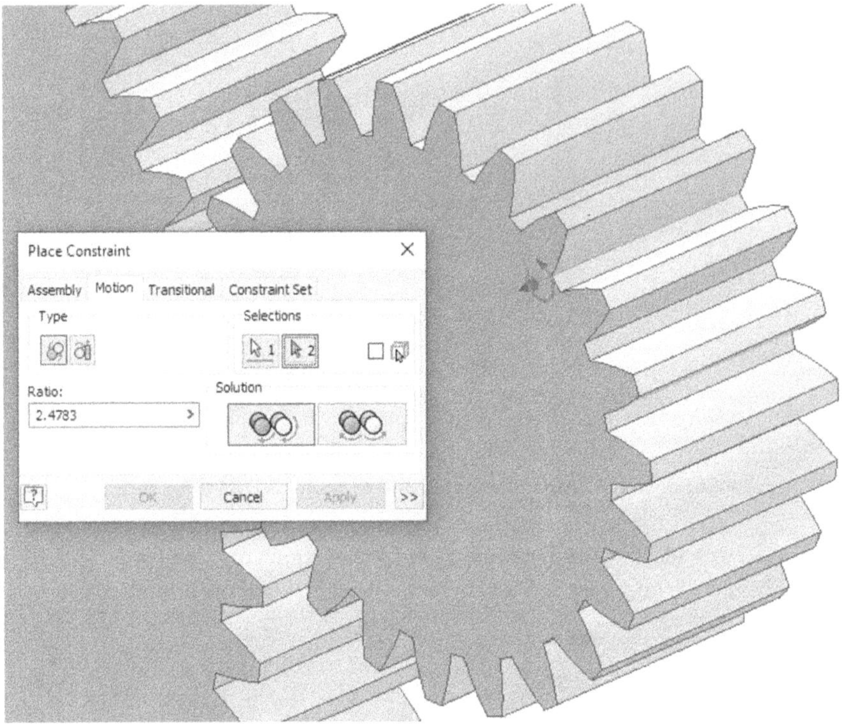

Figure 6.65: Selecting the second reference for the Motion constraint

Select **OK** to complete the **Motion** constraint.

9.   In the **Graphics** window, move your cursor over one of the gears, and then click, hold, and drag your cursor in a circular motion. Both gears turn together because of the **Motion** constraint. However, they are still not meshing correctly:

Figure 6.66: The gears rotate together but they do not mesh correctly

10. Navigate to the `Relationships` folder in the model browser, and expand + next to **Relationships**. Right-click on the **Rotation 1** Motion constraint and select **Suppress** from the menu.

11. Select **Front View** on the ViewCube.

12. Click, hold, and drag one of the gears in the **Graphics** window so that they are aligned and meshed correctly:

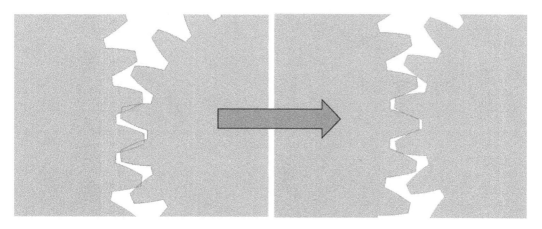

Figure 6.67: Gears meshing correctly

13. Right-click on **Rotation 1** from the `Relationships` folder in the model browser and select **Edit**:

14. In the **Edit Constraint** window, select **Reverse** as the **Solution** option.

15. Navigate to the `Relationships` folder in the model browser, and expand + next to **Relationships**. Right-click on the **Rotation 1** Motion constraint and select **Suppress** from the menu to unsuppress the Motion constraint.

16. In the **Graphics** window, move your cursor over one of the gears, and then click, hold, and drag your cursor in a circular motion. Both gears turn together because, and now the teeth mesh correctly.

17. Now, we will create an **Angle** constraint to drive the gear assembly automatically. Select **Constrain** from the **Assemble** tab.

18. Select **Angle** as the type, and for **Solution**, select **Directed Angle**.

19. For the reference, select **YZ Plane** of the first gear and **XZ Plane** of `Gear Assembly.iam`:

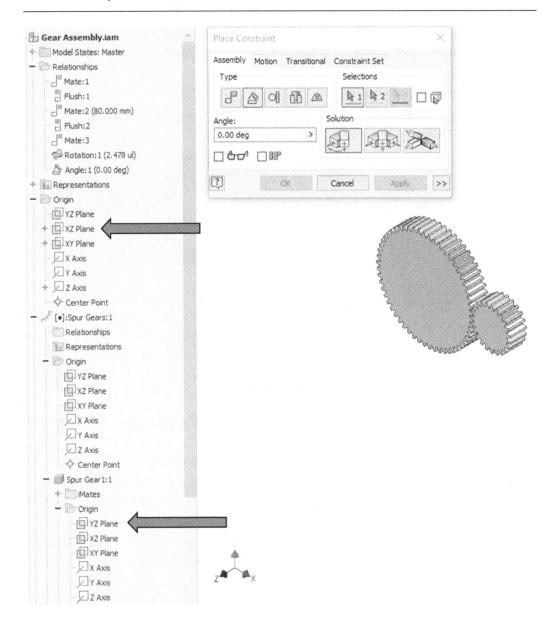

Figure 6.68: Planes to select for the Angle constraint

Select **Apply** to complete. Then, close the **Place Constraint** window.

20. In the model browser, right-click on **Angle 1 Constraint** and select **Drive**.

21. Change the **End** degree to 360 deg. Select the **Forward** button to drive the gears. Both gears will rotate and mesh correctly:

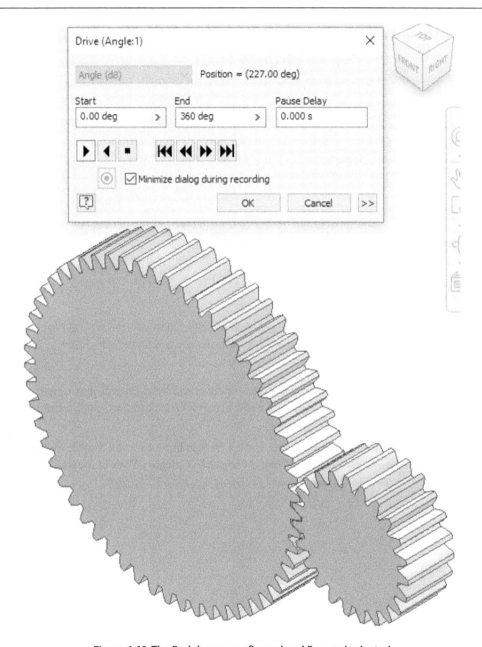

Figure 6.69: The End degree configured and Forward selected

You have applied Motion constraints to a set of gears and driven them with an Angle constraint to show the movement and correct meshing of the teeth.

Duplicating and replacing components in an assembly

Within an assembly, you can duplicate and replace components with ease. In many instances, you will need to duplicate or replicate parts or subassemblies – for example, standard components such as fasteners.

In this recipe, you will learn how to automatically place and size a fastener, and then replace and change the fastener. You will also learn to mirror and then pattern the component within an assembly. All these techniques should be combined when modeling assemblies in Inventor, as they maximize your efficiency and result in you having to recreate repeating geometry.

Getting ready

To begin this recipe, you will need to access and open `Flange Assem.iam`. This can be found in the following folder directory: `Inventor 2023 Cookbook | Chapter 6 | Flange`.

How to do it...

Ensure that you have `Flange Assem.iam` open in Inventor. In this recipe, we will begin the first steps of how to automatically place and size a fastener, replace and change the fastener, and then mirror and pattern the component within an assembly:

1. In the **Assemble** tab, under **Place**, select the dropdown and select **Place from Content Center**. This allows us to browse the **Content Center** library of standard parts within Inventor and specify the exact fastener we require.

2. Within the **Place from Content Center** window, we can browse for the content we require. In this case, we require a **Hex Head** bolt that is flanged. In **Category View** of the window, browse the following: `ISO | Fasteners | Bolts | Hex Head- Flanged`:

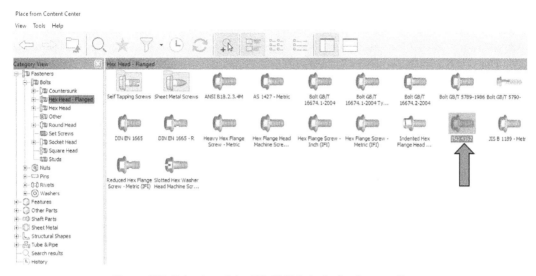

Figure 6.70: Selection of the ISO 4162 Bolt via the Content Center

Then, select the **ISO 4162 Bolt**, followed by **OK**. You can also use the filters to specify content based on a standard that you work in, such as **ISO, ANSI**, and more. A search functionality is also available, and using the *star* icon, you can mark content that you use frequently as a favorite.

3. Upon selecting **OK** in *step 2*, the **ISO 4162** bolt will appear on your cursor within Inventor. The bolt can be placed with a left click; however, if that is done, additional constraints will have to be defined to place the bolt correctly. Instead, hover the bolt over one of the tapped holes in the flange.

4. As you hover the bolt over the tapped hole, Inventor will work to read the hole information, in the model, and resize the fastener to the correct size for that hole. Once complete, the fastener will preview, and a green tick may appear, signifying that Inventor has correctly resized your chosen fastener to the hole:

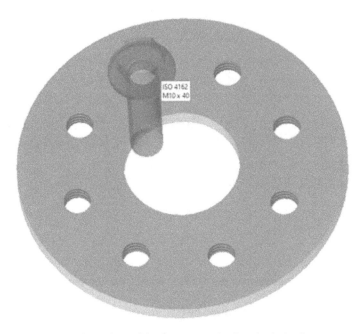

Figure 6.71: A preview of the fastener resized to the hole diameter

5.   Once the fastener has updated, as shown in *Figure 6.71*, right-click and select **Change size**:

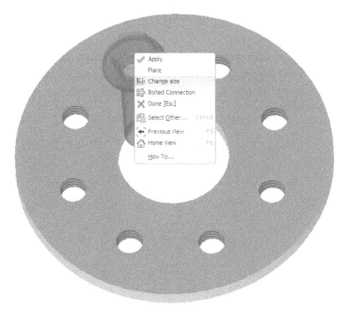

Figure 6.72: Resizing of the previewed fastener

Before placing the bolt in the flange, we need to adjust the length of the bolt.

6. In the next window, we can further define the fastener before placement. Change **Nominal Length** from 40mm to 45mm, and select **OK**. The defined **ISO 4162** bolt is placed in the flange.

7. Upon placing a component, you may need to change it at a later stage. To do this, right-click on the fastener in the **Graphics Window,** and select **Replace from Content Center**.

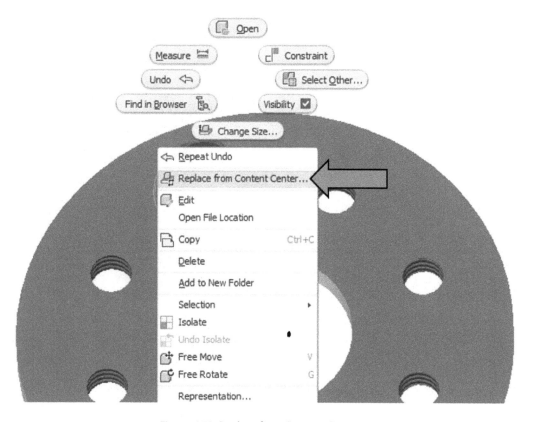

Figure 6.73: Replace from Content Center…

You can also replace components with parts you have created yourself by right-clicking on the component, selecting **Component**, and then either choosing **Replace** or **Replace All** to replace all in the series.

8. In **Place** in the **Content Center** window, select the **JIS B 1189 – Metric** bolt. Select **OK**.

9. Select **M10** as **Thread designation** and leave **Nominal Length** at *22.

10. Select **OK**, followed by **OK** again. The previously placed **ISO 4162** bolt has been replaced with the **JIS B 1189** bolt. This has been automatically resized and constrained to the assembly, so no further constraint work is required to define this.

11. There are a number of holes remaining without bolts, and there are several duplication techniques we can apply to solve this. The first is a mirror of a component within an assembly. Expand the `Origin` folder in the model browser, and highlight the **YZ**, **XZ**, and **XY** planes. Right-click and select **Visibility**:

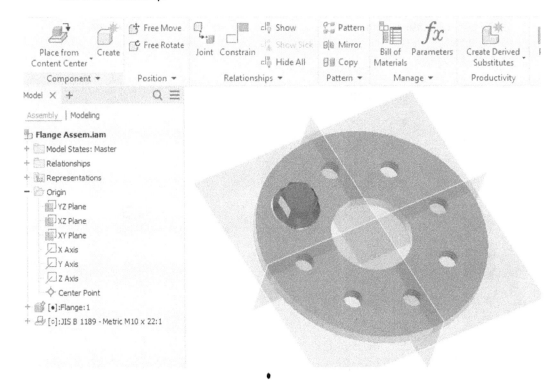

6.74: Origin planes made visible

12. In the **Assembly** tab, select **Mirror**.

13. Select **JIS B 1189** as the component to mirror, and then select **YZ Plane** as the mirror plane:

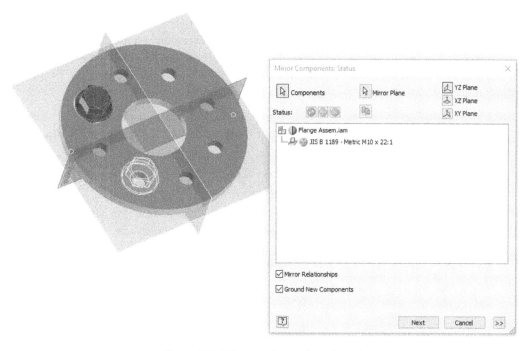

Figure 6.75: Mirror options to be selected

14. Select **Next**, followed by **OK**. The component has been mirrored into the correct position and is fully constrained:

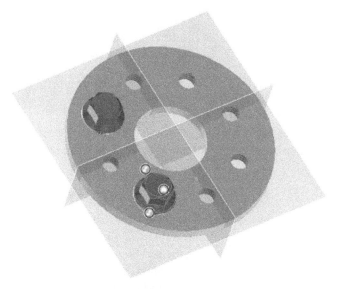

Figure 6.76: The mirrored component

Within the **Mirror** command, you can also mirror multiple components in one operation. Instead of **Mirror**, you can also use the **Copy** command, located under the **Mirror** command; this will allow you to freely place a copied part in the model, but further constraining will be required. You can also click and drag a component from the model browser into the **Graphics** window to create a copy.

15. There are still several fasteners required, and in this instance, repeated use of the **Mirror** command is not efficient. So, in the **Model Browser**, select **JIS B 1189 – Metric M10 x 22.2**, and press *Delete*.

16. We will now begin the more efficient process of using the **Pattern** commands to create and place the relevant fasteners. In the **Assemble** tab, select **Pattern**.

17. From the **Pattern** command, we can use several options to pattern and copy the selected component in the correct position. As holes inserted on the previous flange were created with a pattern, we can use this metadata in the part to assist with adding the bolt to the holes. Select **JIS B 1189 – Metric M10 x 22.2** as the component to pattern.

18. For the feature pattern, select one of the remaining holes in the flange. The **JIS B 1189** fastener will populate on these holes. Select **OK** to complete:

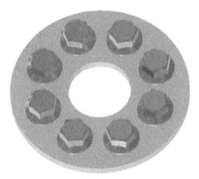

Figure 6.77: The completed assembly

You have automatically placed and sized a fastener from the Content Center, replaced and changed the fastener, mirrored it, and then used **Feature Pattern** to replicate the component within the assembly.

Creating and configuring a BOM

The BOM allows you to document and communicate the parts that make up an overall assembly for manufacture. The BOM is populated automatically, as parts are added to an assembly file, and Inventor will auto-populate this based on the metadata of the part files. In this sense, it is almost like a "live" document within the assembly file. Further customization and configuration can then be applied to the BOM to display specific information. You can also add parts to a BOM that do not have physical geometry, such as paint or grease, that may still be critical to the model, known as **virtual components**.

BOMs can also be exported as `.txt` or `.csv` files but will always remain within the assembly file itself and update as parts change in the assembly file.

In this recipe, you will learn how to do the following:

- Generate a BOM that lists components within the assembly
- Customize and edit properties in the BOM
- Create a virtual component in the BOM

Getting ready

To begin this recipe, you will need access to the `Chapter 6` folder, and open `Shaft Gear Motor.iam`. This can be found in this folder directory: `Inventor 2023 Cookbook | Chapter 6 | Shaft Gear Motor`.

How to do it...

Ensure that you have `Shaft Gear Motor.iam` open in Inventor. In this recipe, we will create and configure a BOM:

1. With `Shaft Gear Motor.iam` open in Inventor, in the **Assemble** tab, select **Bill of Materials**.

2. This generates the BOM for that assembly. If subassemblies are included in the model, they will also be included. In this case, the assembly is made up of single parts only. Within the **Bill of Materials** menu, we can make all the modifications we require to the BOM. We will start by converting the BOM to a **Parts Only** list. This is important to do if you are working with embedded subassemblies, as without this, you will not be able to see them in the BOM. Select the **Parts Only (Disabled)** tab.

3. Then right-click on **Parts Only (Disabled)** and select **Enable BOM View**. A parts-only structure is now visible:

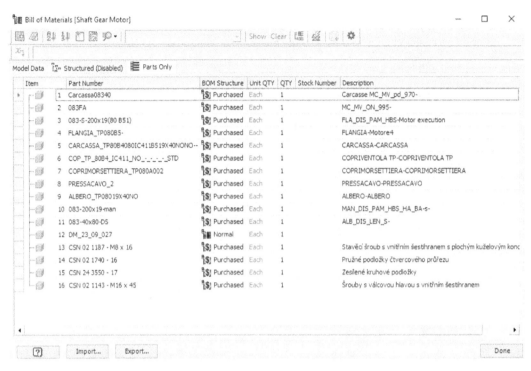

Figure 6.78: Parts Only enabled in the BOM

4. Columns and the information shown can be further customized, by dragging columns. Click and drag **QTY** and move it between **Part Number** and **BOM Structure**. The BOM will update:

	Part Number	QTY	BOM Structure	Unit QTY	Stock
1	Carcassa08340	1	$ Purchased	Each	
2	083FA	1	$ Purchased	Each	
3	083-S-200x19(80 B51)	1	$ Purchased	Each	
4	FLANGIA_TP080B5-	1	$ Purchased	Each	

Figure 6.79: Columns updated in the BOM

5. Columns can also be taken out and added. Click and drag the **Stock Number** column until a black **x** icon appears, and then release the mouse. This will remove the column.

6. We will now insert additional columns of information into the BOM. Select the **Choose Columns** button in the **Bill of Materials** window:

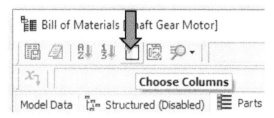

Figure 6.80: The Choose Columns button

7. In the **Customization** window that appears, double-clicking on a category will import this into the BOM, and all information will be automatically added. Browse the list and double-click **Mass** and then **Material** to add them to the **BOM** view. The columns will automatically update, based on the metadata of the parts:

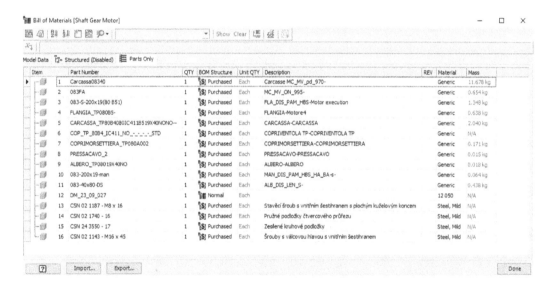

Figure 6.81: The updated BOM list with Material and Mass added

8. To show the live link between the BOM and the assembly file, we will update a material of any part and observe the **Mass** change within the BOM. Select **Done** on the BOM to close the menu.

9. In the model browser, right-click on the **PRESSACAVO_2:1** part and select **Edit**.

10. In the **Material** selection tab, browse for and select **PC/ABS Plastic**:

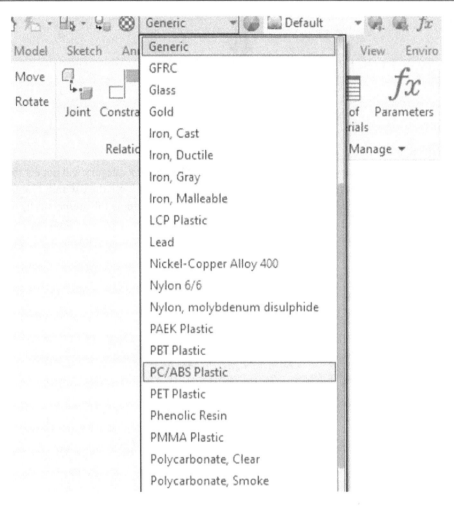

Figure 6.82: The PC/ABS Plastic material selected

11. Select **Return** in the ribbon to return to the assembly.

12. Select **Bill of Materials**, look down the list, and find the **PRESSACAVO_2:1**. You will see that the **Material** type and **Mass** have been updated to reflect the change made in the model:

| | 7 | COPRIMORSETTIERA_TP080A002 | 1 | $ Purchased | Each | COPRIMORSETTIERA-COPRIMORSETTIERA | Generic | 0.171 kg |
| | 8 | PRESSACAVO_2 | 1 | $ Purchased | Each | PRESSACAVO-PRESSACAVO | PC/ABS Plastic | 0.005 kg |

Figure 6.83: Updated Material and Mass for PRESSACAVO_2:1

13. Quantities of parts can also be changed from within the BOM. It may be that in this part, you will supply some spare versions that may not necessarily be included in the CAD model. Double-click on **QTY 1** next to **PRESSACAVO_2:1** and type 4 to change the quantity in the BOM.

14. An important aspect of the BOM we can change is the BOM structure. This is used to provide a more accurate BOM, where further categorization of components can be defined. The five options available are as follows:

   - **Normal**: Default status for all components placed in an assembly file.

   - **Phantom**: Used to simplify a design. They exist in the model but are not included in the BOM and are excluded from quantity calculations.

   - **Reference**: Used in the construction of the assembly but not used in the design. This could, for example, be a skeleton model for a frame-like structure. Reference components are excluded from all BOM calculations.

   - **Purchased**: Used to denote parts that are purchased and not manufactured.

   - **Inseparable**: These are components that must be damaged to take them apart – for example, welded parts or tight interference fitting parts.

   You are free to change parts' BOM structure in the BOM window. For the **083FA** part, double-click on the **Purchased** BOM structure in the table. Then, select **Normal** from the dropdown, which changes the BOM structure of the part:

Figure 6.84: The BOM structure changed from Purchased to Normal

15. The BOM can also be sorted and renumbered. To renumber a BOM, select the **Renumber Items** button:

Figure 6.85: The Renumber Items button in the Bill of Materials window

16. In this example, we will change the start number to 2 and the increment to 2. Select **OK** to complete:

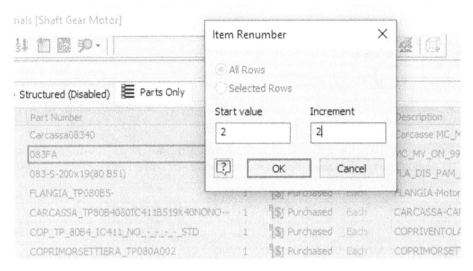

Figure 6.86: Item renumbering in the Bill of Materials window

The BOM will renumber as specified.

17. The BOM can also be sorted and reorganized with the **Sort Items** command. Select the **Sort Items** command:

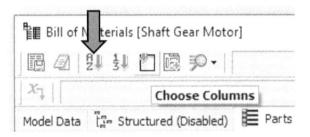

Figure 6.87: The Choose Columns command

18. Change the sorting fields, as shown in *Figure 6.88*, using the drop-down menu, and select **OK**. This will sort the BOM as per the sorting specifications:

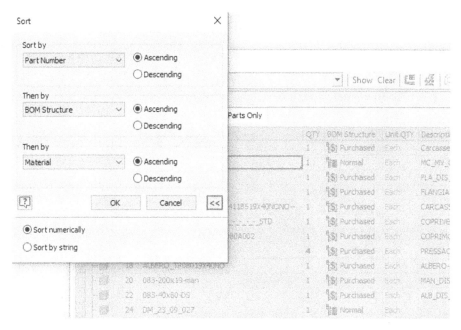

Figure 6.88: Sorting columns in the BOM

19. Close the **Bill of Materials** window, as we will now create a virtual component called **Grease**. This is a component that's important to the design, which we need in the CAD but is impossible to model.

20. In the **Assembly** tab, within the component panel, select **Create**.

21. Create a new component called `Grease`, select **Virtual Component**, and select **Purchased** under **Default BOM Structure**. Select **OK** to complete:

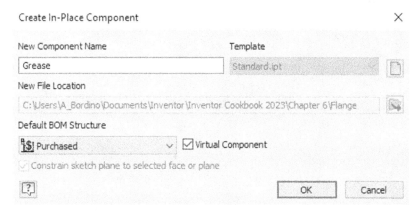

Figure 6.89: Defining the name and default BOM structure of the virtual component

22. We now need to create a parameter to define the base quantity of **Grease**, measured in millimeters. Right-click on the **Grease** component in the model browser. Select **Component Settings**.

23. Select the **Edit Parameters** button:

Figure 6.90: The Edit Parameters button in Grease Component Settings

24. Select **Add Numeric** in the **Parameters** dialog box.

25. Create a new parameter called `Base_Quantity`, set the units to `ml`, and the equation to `1 ml`. Select **Done** to complete:

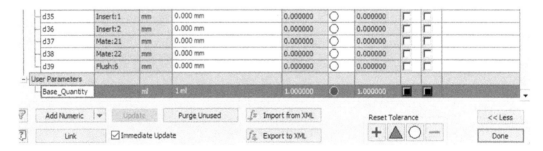

Figure 6.91: A new parameter created for the virtual Grease component

26. We will now change the base quantity to the correct unit. Change **Base Quantity** to `Base_Quantity (1ml)`, and select **OK**:

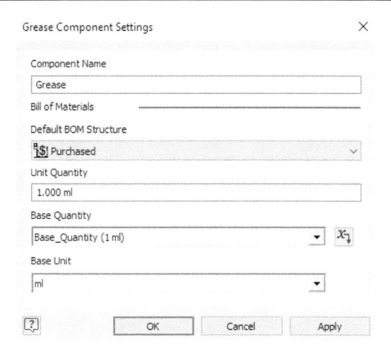

Figure 6.92: Adding Base_Quantity (1 ml) to the virtual component

27. We will now change **Grease** so it has the correct quantity. So, select **Bill of Materials**, then in the **Grease** item, change the quantity from **1ml** to 100.000 ml:

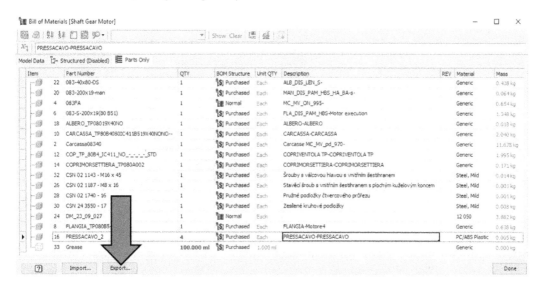

Figure 6.93: The quantity of the virtual grease component changed from 1 ml to 100.000 ml

28. A BOM can also be exported as a `.csv` or text file. To do this, select **Export...**.

29. In the Windows menu, browse and save your BOM as an `.xml` file, named `TEST BOM`.

You have generated and customized a BOM for components within an assembly, edited properties, added a virtual component, and exported the BOM in an `.xml` file.

Detailing and customizing a BOM in a drawing file

In this recipe, you will create a **General Arrangement** (**GA**), a drawing of an existing assembly, add and customize a BOM to the drawing sheet, and auto-balloon all components within the drawing.

Getting ready

To begin this recipe, you will need access to the `Chapter 6` folder and `Shaft Gear Motor.iam`. This can be found in this folder directory: `Inventor 2023 Cookbook|Chapter 6|Shaft Gear Motor`.

How to do it...

Ensure that you have `Shaft Gear Motor.iam` open in Inventor. To create a GA drawing, and add and customize a BOM to the drawing sheet, we will follow these steps:

1. Select **File**, **New**, **Metric**, `ISO.idw`, and then select **Create**.

2. In the new `ISO.idw` file, select **Base View**, and browse `Shaft Gear Motor.iam`. Use the ViewCube to rotate the base view so that the ViewCube is displaying the top view, as shown in *Figure 6.94*. Change **Scale** to `1:4`:

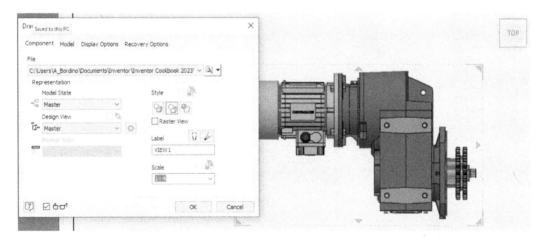

Figure 6.94: The top view of the shaft gear motor

3. Move the cursor toward the top right of the drawing screen and left-click to place a projected ISO view of the assembly. Right-click and select **OK** to complete:

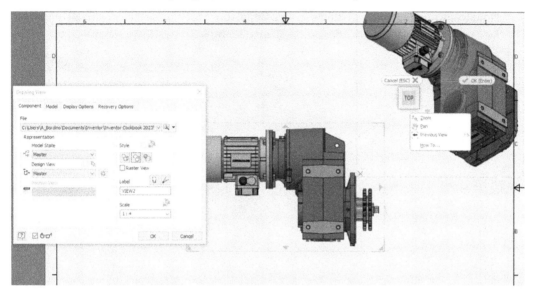

Figure 6.95: The ISO view added

4. Click and drag the two placed views so they resemble *Figure 6.96*:

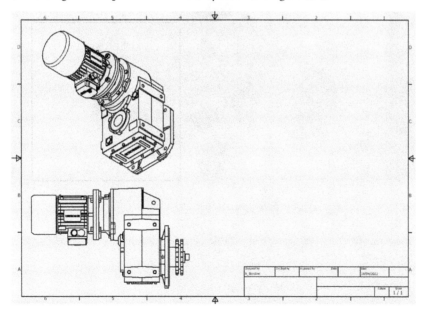

Figure 6.96: Drag the placed views in this position on the drawing sheet

5. Select the base view and hit *Delete*. Select **OK** to complete. Drag the ISO view to the bottom left of the drawing sheet.

6. Double-click the ISO view. Using the ViewCube, move the view to the position shown in *Figure 6.97* to re-orientate it. Then, select **Shaded** for **Style**. Select **OK** to complete:

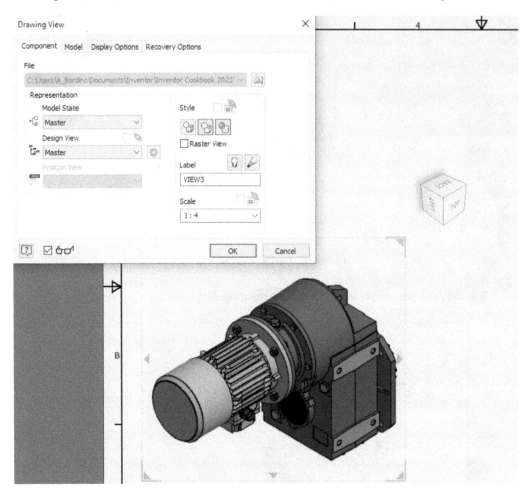

Figure 6.97: The completed ISO view

7. Select the **Annotate** tab. Navigate to the **Table** commands and select **Parts List**.

8. Select **ISO View** on the drawing sheet as the referenced view.

9. Select **Parts Only** from the **BOM View** options:

Parts List

Source

🔲 Select View

C:\Users\A_Bordino\Documents\Inventor\Inventor C ∨   🔍

Master   ∨

BOM Settings and Properties

BOM View	Numbering	Min.Digits
Parts Only ∨	Numeric ∨	1 ∨

Table Wrapping

Direction to Wrap Table:

◉ Left   ○ Right

☐ Enable Automatic Wrap

   ○ Maximum Rows   10

   ◉ Number of Sections   1

[?]   OK   Cancel

Figure 6.98: The Parts List options to select

10. Select **OK**. A preview BOM will appear on the cursor. Left-click at the top right of the drawing sheet to place this. The BOM will be placed here.

11. Double-click on the newly placed BOM in the drawing.

12. As in the *Creating and configuring a BOM* recipe, you can customize a BOM in the drawing environment the same way you did in the assembly file. Select **1** in the **QTY** column for the **083FA** part and change it to 5:

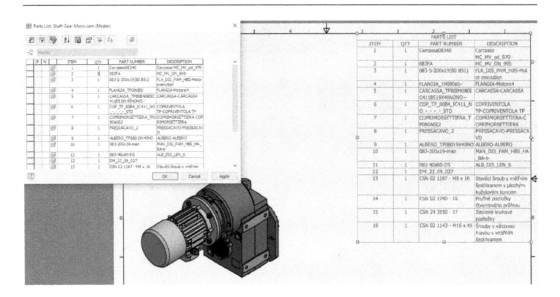

Figure 6.99: Customizing the BOM in the drawing environment

13. Select the **Column Chooser** command at the top left of the **Parts List** dialog box.

14. Browse for **Mass** and **Material**, and select them. Then, select **Add** to move them into the BOM. Select **OK** to complete, followed by **OK** again. The parts list or BOM is updated on the drawing sheet:

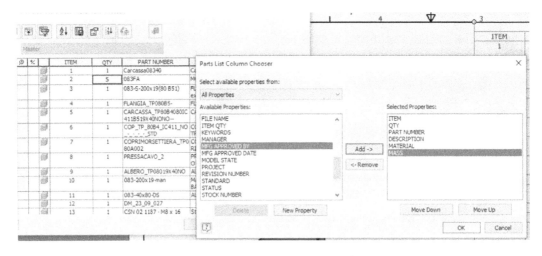

Figure 6.100: Choosing additional columns for the BOM

15. To annotate and auto-balloon, in the **Annotate** tab, select **Auto Balloon**.

16. Select the ISO view on the drawing sheet.

17. You can now select the specific items to balloon from the assembly. In this case, draw a box around the whole view to select all parts. Inventor will ignore duplicate parts in the auto balloon by default:

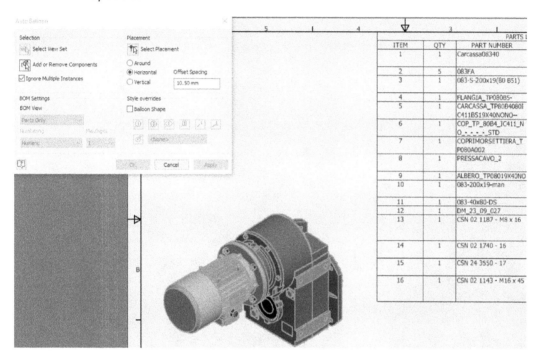

Figure 6.101: All parts selected from the assembly for the auto balloon

18. Select the **Select Placement** button.

19. Select **Around** from the list. Move your cursor toward the drawing view on the drawing sheet. You will see that previews of the balloons populate. When ready, left-click and select **OK** to place:

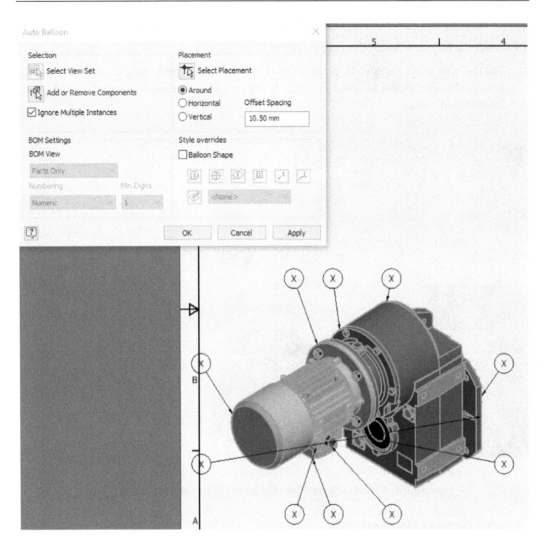

Figure 6.102: A preview placement of the auto balloon

If balloons are placed in an undesirable location, they can be moved freely after placement with a click and drag.

20. The auto balloon has now correctly numbered the parts of the assembly on the drawing. If changes are made to the file, the BOM and the balloons will update if required:

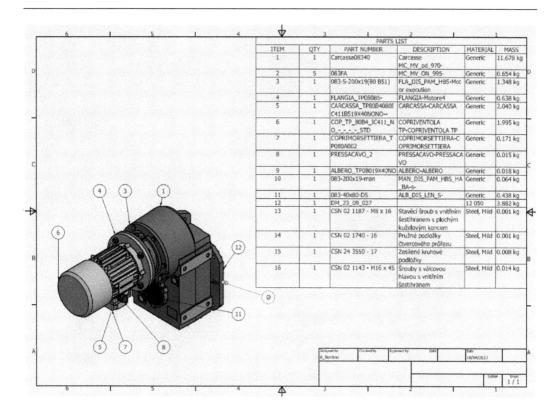

PARTS LIST					
ITEM	QTY	PART NUMBER	DESCRIPTION	MATERIAL	MASS
1	1	Carcassa08340	Carcasse MC_MV_pd_970-	Generic	11.678 kg
2	5	083FA	MC_MV_ON_995-	Generic	0.654 kg
3	1	083-5-200x19(80 B51)	FLA_DIS_PAM_HBS-Mot or execution	Generic	1.348 kg
4	1	FLANGIA_TP080B5-	FLANGIA-Motore4	Generic	0.638 kg
5	1	CARCASSA_TP80B4080I C411BS19X40NONO--	CARCASSA-CARCASSA	Generic	2.040 kg
6	1	COP_TP_80B4_IC411_N O_-_-_-_STD	COPRIVENTOLA TP-COPRIVENTOLA TP	Generic	1.995 kg
7	1	COPRIMORSETTIERA_T P080A002	COPRIMORSETTIERA-C OPRIMORSETTIERA	Generic	0.171 kg
8	1	PRESSACAVO_2	PRESSACAVO-PRESSACA VO	Generic	0.015 kg
9	1	ALBERO_TP08019X40NO	ALBERO-ALBERO	Generic	0.018 kg
10	1	083-200x19-man	MAN_DIS_PAM_HBS_HA_BA-s-	Generic	0.064 kg
11	1	083-40x80-DS	ALB_DIS_LEN_S-	Generic	0.438 kg
12	1	DM_23_09_027		12 050	3.882 kg
13	1	CSN 02 1187 - M8 x 16	Stavěcí šroub s vnitřním šestihranem s plochým kuželovým koncem	Steel, Mild	0.001 kg
14	1	CSN 02 1740 - 16	Pružné podložky čtvercového průřezu	Steel, Mild	0.001 kg
15	1	CSN 24 3550 - 17	Zesílené kruhové podložky	Steel, Mild	0.008 kg
16	1	CSN 02 1143 - M16 x 45	Šrouby s válcovou hlavou s vnitřním šestihranem	Steel, Mild	0.014 kg

Figure 6.103: The completed Auto Balloon operation

You have created a new drawing of an existing assembly, placed an isometric view, added a BOM and customized it, and then used **Auto Balloon** to detail the components.

Model credits

The model credits for this chapter are as follows:

- **Brake** (by Gabriel Medeiros): `https://grabcad.com/library/brake-51/details?folder_id=10984917`

- **Vice** (by Ahmed Habashy): `https://grabcad.com/library/vice-86/details?folder_id=4922098`

- **SHAFT GEAR MOTOR WITH ACCESSORIES** (by Dařena Miroslav) `https://grabcad.com/library/shaft-gear-motor-with-accessories-1`

7

Model and Assembly Simplification with Simplify, Derive, and Model States

When working with assemblies and individual parts, creating simplified models with a reduced Level of Detail is often required as a part of the manufacturing workflow. Within Inventor, there are several workflows and tools that can be employed to achieve this. The **Simplify** tool (**Shrinkwrap** in versions 2021 and earlier) allows you to take your existing parts and assemblies and simplify them into a state that not only is easier to use but also allows you to protect your intellectual property. The **Derive** tool enables you to reference existing parts and assemblies as base components for modifications. **Model States** (**Level of Detail** in versions 2021 and earlier) allow the configuration and simplification of parts and assemblies and are an alternative to iParts and iAssemblies (see *Chapter 3, Driving Automation and Parametric Modeling in Inventor*, for more details).

In this chapter, you will learn about the following:

- Simplifying assemblies
- Deriving components
- Creating model states in parts
- Creating model states in assemblies
- Using model states in drawings

This chapter also aims to give context to each of the methodologies and how and why they should be used in your workflows.

Technical requirements

To complete this chapter, you will need access to the practice files in the Chapter 7 folder within the Inventor Cookbook 2023 folder.

Simplifying assemblies

In this recipe, you will use the **Simplify** tool to create a simplified model of a complex assembly, reduce processing time, and protect intellectual property.

The advantages of using the **Simplify** tool in your workflows are as follows:

- Reduces the file size and complexity of large assemblies
- Reduces the opening times of complex assemblies
- Filters relevant design data
- Protects intellectual property by removing internal components
- Improves performance

The **Simplify** command enables you to create a part from an assembly by specifying the components and features you want to remove individually or by using one of the built-in presets within the **Simplify** tool.

Getting ready

To begin this recipe, you will need to open the Shaft Gear Motor2.iam file located in the Chapter 7 folder within the Inventor Cookbook 2023 folder.

How to do it...

To begin the following steps, ensure that you have the Shaft Gear Motor.iam assembly file open in Inventor:

1. Navigate to the **Assemble** tab and select the **Simplify** tool as shown in *Figure 7.1*:

Figure 7.1: Simplify tool

2. The **Simplify** tool now opens and the following dialog appears. *Figure 7.2* details this with reference to the many options available and what they are capable of:

Figure 7.2: Simplify tool dialog box

A to **I** show the various areas of the tool that we can use to create simplified parts from complex assemblies:

- **A** – This is the presets area. From here, you can define whether you want to remove small parts, remove medium-sized parts, or envelope all the parts in the assembly as separate bounding boxes or envelopes. Presets can also be saved for future use.

- **B** – Allows you to load in any linked **Model State**, **View**, or **Position** representations for an assembly and **Simplify** from these.

- **C** – Replace each top-level component or all parts with envelopes.

- **D** – Exclude certain parts by size and define them with the **Max. Diagonal** field. There is also a measuring tool located here so that you can perform checks on the model if required.

- **E** – Enables you to manually select parts to exclude from the model. You can also select **All Occurrences**, which is especially useful for excluding content such as nuts, bolts, or other repeating parts.

- **F** – Define the type of file to create and the template to use for the new simplified part. You can also set the name of the newly created part and define the location and **BOM Structure**.

- **G** – This section allows you to define what style you would like the new simplified component to be, from a **Single Solid Body** to a **Single Composite Surface** feature.

- **H** – **Advanced Properties** allow you to rename the part and break the link between the parent assembly. With this unchecked, the changes to the parent assembly will be reflected in the linked simplified part.

- **I** – These checkboxes allow you to fill in all internal voids to protect intellectual property, remove internal parts, and color override from the source components.

From the presets dropdown (**A**) in *Figure 7.2*, select ***Remove the least detail (small parts and features)**. This will update the model.

3. In this instance, a bounding box or envelope is not required. Instead, we will exclude some smaller parts by size. Select the **Exclude parts by size** checkbox labeled **D** in *Figure 7.2*.

4. For **Max. Diagonal**, type 230mm. The assembly updates to reflect this by excluding all parts within the 230 mm max diagonal threshold. **9/16 Parts** should now be excluded from the simplified model.

5. Select the **9/16 Parts** dialog. You can now select parts in the graphics window to exclude them manually. In this case, select the components shown in *Figure 7.3*. When selected, they will disappear:

Figure 7.3: Selecting components to remove from the simplified model

6. Rename the new component from Shaft Gear Motor2_Simplify_1 to SGM Simplify.
    Leave **Location** and **BOM Structure** as their defaults.

7. Under **Style**, select the option highlighted in the figure, which creates a **Single Solid Body** part
    with seams between planar faces.

8. Under **Advanced Properties**, ensure that the following is selected as shown in *Figure 7.4*:

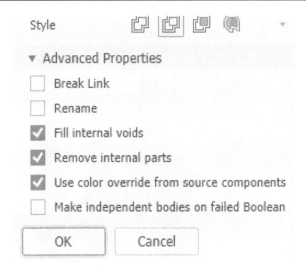

Figure 7.4: Advanced properties to select

9. Select **OK** to complete. The simplified part will be generated in a new Inventor tab. *Figure 7.5* shows a comparison of the two models:

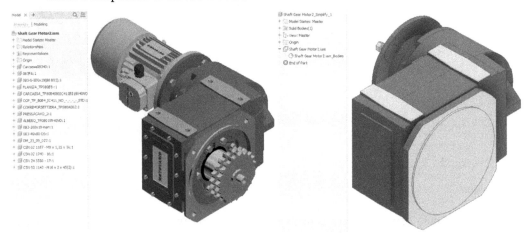

Figure 7.5: The original assembly compared to the simplified model

10. Close all files and do not save.

You have successfully created a simplified part file from a complex assembly, excluding and hiding certain parts, removing unnecessary features, and protecting IP by removing internal geometry. You have also exported this as a **Single Solid Body** feature.

There are many more ways that this assembly could have been simplified. This will depend on the scenario, project constraints, the design itself, and the intention and end use of the simplified assembly part. This example has used a small assembly to maximize the processing time, but a much larger assembly of many thousands of parts and sub-assemblies could also be simplified.

> **Express Mode**
>
> When working with extremely large and demanding assemblies consisting of thousands or tens of thousands of parts, running the Inventor assembly in Express Mode will reduce processing power. Your computer's RAM, technical specifications, and the geometry type and complexity of the file will also be limiting factors.

Deriving components

In this recipe, you will use the **Derive** tool to create a new derived part from an existing assembly, with existing information such as sketches, work features, and parameters.

The original part from the **Derive** operation is known as the base component. Derived parts can be derived from **base components** of parts, assemblies, sheet metal parts, and weldments. When deriving a component, components of an assembly can be excluded or suppressed and even a mirror of the base component can be created.

Derived parts and assemblies are useful for controlling changes to models and creating a simplified version for parts or assemblies, much like the **Simplify** functionality (see the *Simplifying assemblies* recipe). Modifications to the original model from which the base component derives sync automatically across both files. A derived component usually has less detail than the base component and therefore is another way to reduce memory and loading times for certain workflows.

One of the common uses of a derived part is to explore design alternatives and manufacturing processes. Here are examples of other uses:

- Creating a casting blank
- Detailing how a weldment can be configured in several ways, including post-assembly operations
- Subtracting or joining parts to create a single part
- Deriving parts from multi-body parts
- Deriving parameters from an assembly or part
- Creating a simplified version of a complex part or assembly
- Reducing the load times of a complex model

Getting ready

To begin this recipe, you will need to open the `Shock_Absorber_Assy2.iam` file located in the `Chapter 7` folder within the `Inventor Cookbook 2023` folder.

How to do it...

Ensure that you have `Shock_Absorber_Assy2.iam` open. You will not be actively using this file and the assembly that will act as a base component does not need to be open to use the **Derive** command in a new document. However, opening this file provides context as to what assembly we will be deriving.

The first step will be to begin the creation of a new Inventor part file to execute the **Derive** operation in:

1. Create a new Inventor `Standard (mm) .ipt` file.

2. With the new file open, we will now create a derived and simplified version of `Shock_Absorber_Assy2.iam`. Navigate in the ribbon to the **Manage** tab in the new part file and select **Derive**:

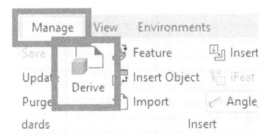

Figure 7.6: The Derive command in the ribbon

3. Inventor now prompts you to identify the part or assembly on which you wish to use the **Derive** tool. Browse the following: `Inventor | Inventor Cookbook 2023 | Chapter 7 | Shock Absorber 2 | Shock_Absorber_Assy2.iam`. Then, click **Open**:

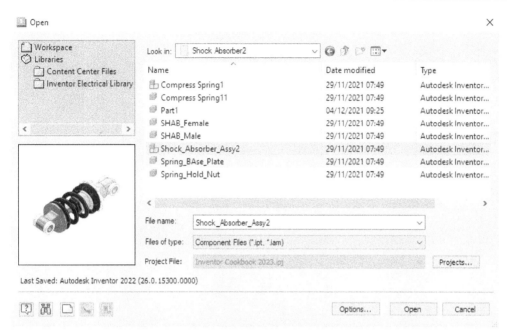

Figure 7.7: Selecting the assembly to derive from

4. A preview of the assembly now opens in the part file and the **Derived Assembly** window opens to allow you to configure and specify how the derived component should be. On the **Derive style** options, select the option highlighted in the figure, which creates a derived **Single Solid Body** part with no seams between planar faces:

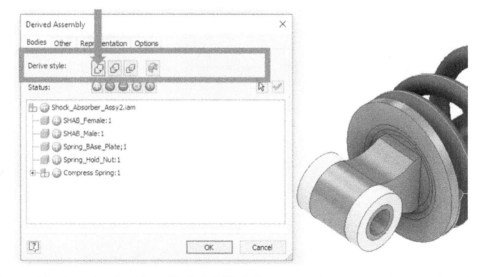

Figure 7.8: Derive style selected as a Single Solid Body feature merging out seams between planar faces

This will allow us to create a **Single Solid Body** feature from the assembly. There are other options to maintain separate solid bodies or create composite surfaces, but in this case, we only require a **Single Solid Body** feature as our final output.

5. Click on the green + symbol next to the SHAB_Female: 1 part twice until the icon changes to a green circle with a white square. The preview should display as *Figure 7.9*:

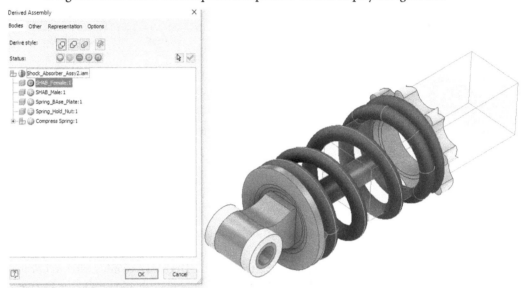

Figure 7.9: Bounding box created of one of the components

This will replace the component in the assembly with a bounding box instead.

6. We will now exclude some components entirely from the model. Click on the green + symbol next to Spring_BAse_Plate_1 once until the green + symbol changes to a red stop sign. In the overview, the component will turn red, meaning it will be excluded from the derived version, as shown here:

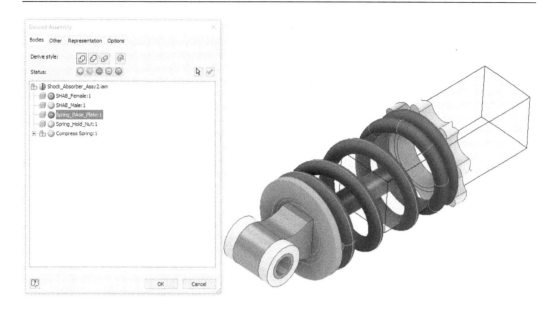

Figure 7.10: Spring_BAse_Plate_1 removed from the derived part

7. Navigate and select the **Other** tab. From here, you can also choose to bring across certain parameters, work planes, and more into the derived component. In our example, this is not required. Select the **Representation** tab. Here, you can start to define a derived component from an existing **Position View**, **Detail View**, or **Model State** representation. This assembly has no **Position View** or **Model State** representations except for the default, **'Primary'**.

8. Select the **Options** tab in the **Derived Assembly** window.

9. Here, we can further simplify the derived component by eliminating parts based on size or visibility. As this is a relatively simple assembly, there is no real need to use this functionality.

    Next, for **Hole patching**, select **All**. This will patch any holes in the assembly.

10. In the **Scale factor** area, you will see that this is set to the default, `1.000ul`, meaning the derived component will be 1:1 of the original. Change this to `.500ul`:

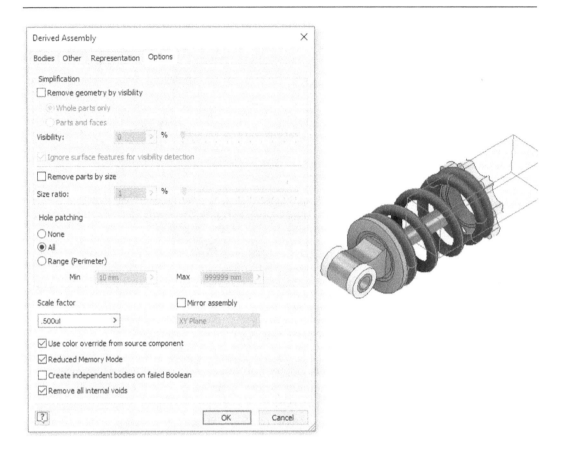

Figure 7.11: Hole patching selected and the scale factor changed to .500 ul

This is also an effective way of securing the IP of designs further when collaborating with external stakeholders on a project in some scenarios.

11. Select **OK** to complete the **Derive** operation. The operation is completed and the derived part is shown in *Figure 7.12*:

Figure 7.12: Derived component from the assembly completed

The derived component has successfully been created. You have simplified and derived a single part that is representative of an assembly using **Derive**. Although this example used a small assembly, the operation can also be done on much larger complex assemblies and single parts as well.

Creating model states in parts

Model states are incredibly versatile and effective ways of simplifying parts and assemblies and creating configurations of parts or assemblies. In this respect, they are an alternative to iParts and iAssemblies.

Model states allow for the creation of multiple representations of a part or assembly within a single file. This workflow makes it easier and more efficient to create and manage configurations within a design, and as all variations exist within one file, there is no risk of accidentally breaking the association between files. Model states are also an effective way of not only creating configurations of a product but also detailing assembly stages or manufacturing stages within a model.

Model states in parts can be used to create or represent the following:

- Manufacturing or production stages
- Manufacturing preparation stages
- Ranges or product families
- Simplification levels
- The range of movement of parts

All changes to model states are conducted within the design file and within **Model Browser**, so there is no lengthy process of configuration with tables as there is with iParts or iAssemblies. Model states can also be edited once created.

When creating model states, you can modify the following configurable attributes directly in the part or assembly file:

- Suppress or unsuppress a part of an assembly's features
- Suppress or unsuppress components (only available in assemblies)
- Assign materials (only in parts)
- Parameters
- iProperties
- Constraints (only in assemblies)
- Rename model states

> **Model states with Autodesk Vault**
>
> It is important to note that model states are stored in a single file instead of multiple files as with iParts or iAssemblies. If using Autodesk Vault Professional, additional functionality is available for model states if you are using an item-centric workflow. If you use Vault Basic or Vault Workgroup, then files that contain model states will display master model properties. If your workflow requires separate files to undergo a release process, then iParts or iAssemblies would be more suitable.

Getting ready

To begin this recipe, you will need to open the `Handle.ipt` file located in the `Screwdriver Model States` folder within the `Chapter 7` folder of the `Inventor Cookbook 2023` folder.

How to do it...

Before beginning the following steps, ensure that you have the `Handle.ipt` file open. In this recipe, we will create various model states of configurations for a handle for a screwdriver:

1. With `Handle.ipt` open, navigate to **Model Browser**. Expand the **Model States** folder. As shown in *Figure 7.13*, the only model state available is the default – **[Primary]**. It is here that we can configure the variations:

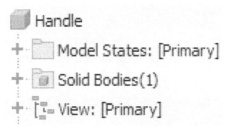

Figure 7.13: Model States by default in the file

2. Right-click on the **Model States: [Primary]** folder and select **New**. This creates a new model state. Repeat this until there are three model states in total, as shown in *Figure 7.14*:

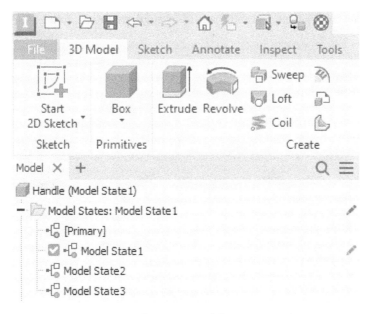

Figure 7.14: Three new model states created

3. The three model states have been created but the configurations have not yet been applied. **Model State1** has the gray and white check mark next to it, indicating that it is the active model state. A blue pencil icon is also present next to the active model state. Any changes made to the active model state will be reflected in that model state. To change the model state, simply double-click the model state's name. Double-click and activate **Model State1**.

4. It is important to name the model states correctly so that users can ascertain what they're for. This also makes linking model states in assemblies much easier. Select **Model State1** twice so that you can rename it. Rename all model states as per the names given in *Figure 7.15*:

Figure 7.15: Renaming the Handle model states

5. The model will not update in length yet, as only the name has been defined; we will now work to change length parameters to differentiate between the various model states. Ensure that Handle_Long is the active model state.

6. Select the **Parameters** command and change the **Length** parameter from 150 mm to 250 mm:

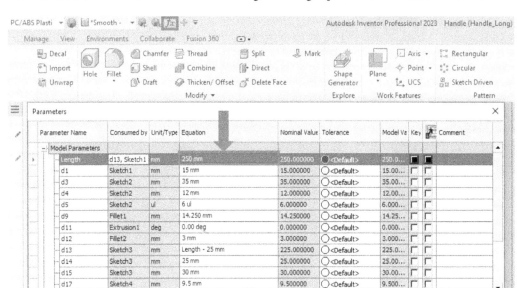

Figure 7.16: Length parameter changed for the Handle_Long model state

7. Select **Done** to complete. The model updates to reflect the change in the model state.

8. Select the **Appearance** dropdown and select **Clear Override**.

9. Select the **Material** dropdown and select **Thermoplastic Resin** as the option.

10. Activate the Handle_Medium model state. In this case, the **Length** parameter is at the desired size, but we need to change **Material** from **Thermoplastic Resin** to **PC/ABS Plastic**.

11. Select Handle_Medium in the top level of **Model Browser** so that the whole model is highlighted.

12. Select the **Material** option, browse for **PC/ABS Plastic**, and select it:

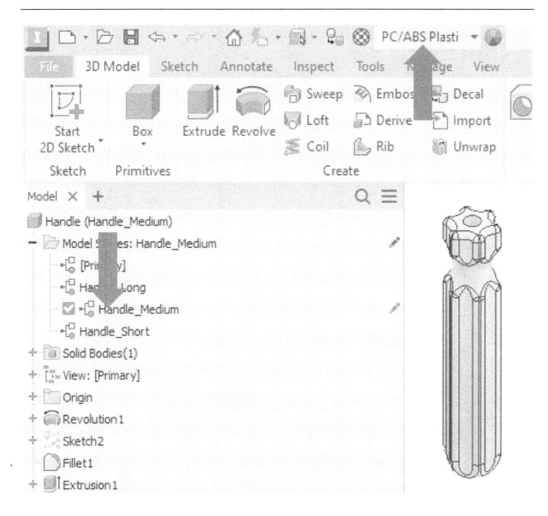

Figure 7.17: Handle_Medium material changed to PC/ABS Plastic

13. Activate the `Handle_Short` model state. Select **Parameters** and then change the **Length** parameter to `120` mm.

14. Change **Material** from **PC/ABS Plastic** to **Polypropylene**.

15. Right-click on **Fillet1** and select **Suppress**:

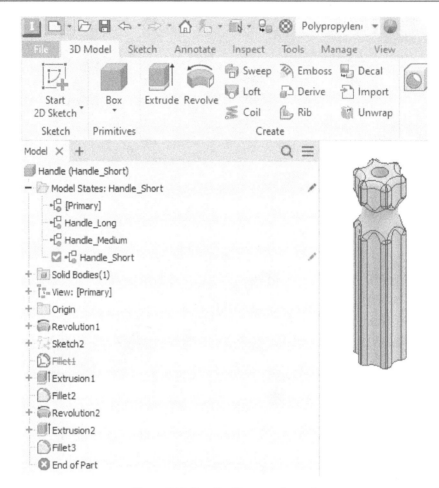

Figure 7.18: Handle_Short configured

16. By double-clicking on the model states, you will see that `Handle.ipt` updates to reflect the suppressed features and changes to geometry.

17. Select the **Save** icon to save the file.

Within the `Handle.ipt` file, you have created several model states that change the model and configure the length, as well as **Material** properties, and suppressed a feature.

Creating model states in assemblies

In this recipe, you will use a different version of the `HandleMS.ipt` screwdriver handle and combine this with other parts featuring model states to create model states and a configuration of the product in an assembly file.

Getting ready

To begin this recipe, open `ScrewDriverModelStateAssem.iam`, located at `Inventor Cookbook 2023 | Chapter 7 | Screwdriver Model States`.

How to do it...

Before starting these steps, ensure that you have the `ScrewDriverModelStateAssem.iam` file open. We will link the model states for the `HandleMS.ipt` and `Shaft.ipt` files so that three total variations of the screwdriver are created, a *long*, *medium*, and *short* variant:

1. With `ScrewDriverModelStateAssem.iam` open, navigate to the top level of **Model Browser**. We need to create three corresponding model states at the top level of the assembly to drive the configurations. Right-click on **Model States** and select **New**. Repeat this twice to create three model states in total:

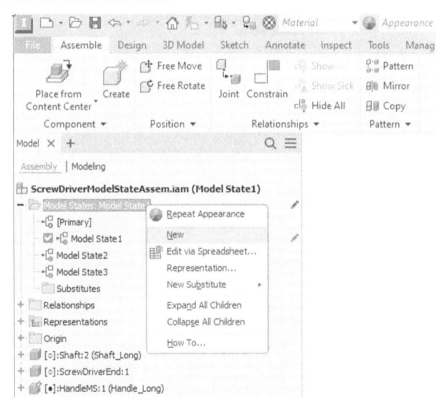

Figure 7.19: Model states created at the top level of the assembly

2. By clicking on the name of the model states, type and change the names so that they reflect the names as shown in *Figure 7.20*:

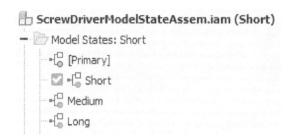

Figure 7.20: Model states with new names

3. These model states will drive the assembly, so it is important that they are named appropriately. Double-click on the **Long** model state to activate it.

4. Navigate **Model Browser**. Expand the files for both the HandleMS and Shaft components:

Figure 7.21: Expanded model states of the components in the assembly

5. Double-click on HandleMS and Shaft so that both of the **Long** options are active as per *Figure 7.21*. With the **Long** top-level model state active, and the other **Long** model states active within it, Inventor will now select the long shaft and handle for the long configuration.

6. Repeat *step 5* for the **Medium** and **Short** model states. Ensure that Handle_Short and Shaft_Short are selected under the **Short** top-level model state, and Handle_Medium and Shaft_Medium are selected under the **Medium** top-level model state.

7. At the top level of the assembly in the top level of the model states, cycle between **Long**, **Medium**, and **Short**. The screwdriver should update as per *Figure 7.22*:

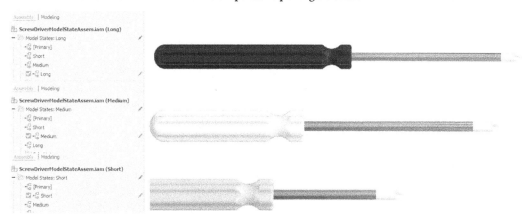

Figure 7.22: Model states controlling the configurations of the assembly

Using model states in two separate part files, you have combined these with model states in a top-level assembly to drive the product configuration of a screwdriver. The size of the components changes as the material and feature suppression change.

Substitute model states

With model states, you can also replace components in an assembly with a simpler one that represents it, known as a substitute. The substitute contains all original BOM information as well as the physical information and iProperties. With this operation, you can also simplify the geometry.

Using model states in drawings

In this recipe, you will learn how model states can be used in drawings. You will deploy several model states into a drawing with different size variants and then produce a table detailing the configurations.

Getting ready

To begin this recipe, you will need to create a New ISO Metric (mm) drawing file and have this open.

How to do it...

Before starting, ensure that you have a new ISO (mm) drawing file open. We will begin by placing a view of HandleMS and selecting a model state to display in the drawing view:

1. Select **Base View** from the top left of the ribbon.

2. Browse for HandleMS.ipt and select this.

3. Under **Model States**, ensure that Handle_Long is active and leave all other settings as default as per *Figure 7.23*. Rotate the view using **View Cube** so that the handle is placed horizontally on the sheet. Select the shaded **Style** option. Select **OK** to place the **Drawing View** feature:

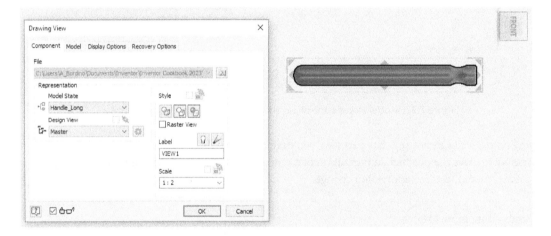

Figure 7.23: The Handle_Long model state view placed

4. Repeat *steps 1* to *3* for the **Medium** and **Long** model states of HandleMS. The drawing sheet should resemble *Figure 7.24*:

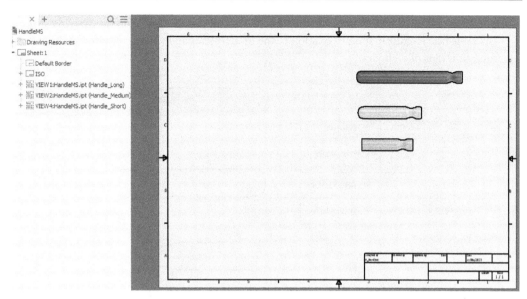

Figure 7.24: All model states for HandleMS shown on the drawing sheet

5. Select the **Annotate** tab. Select **General Table**.

6. Select **VIEW 1: HandleMS.ipt (Handle_Long)**.

7. Select **Column Chooser**, followed by **Add ->** three times, to move all attributes across to the general table. These are attributes that define the configuration of the models:

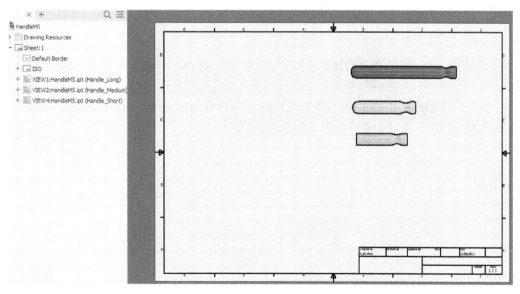

Figure 7.25: All attributes of the model states selected for the table

8. Select **OK**, followed by **OK** to create the table. Click and place the table on the drawing sheet. The table lists all information in relation to the model states and what is required to drive the configurations of each view:

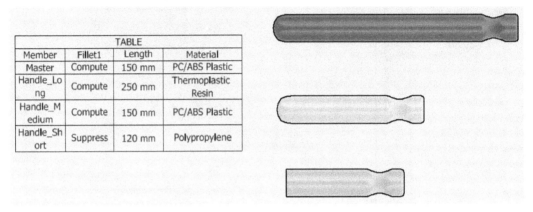

TABLE			
Member	Fillet1	Length	Material
Master	Compute	150 mm	PC/ABS Plastic
Handle_Long	Compute	250 mm	Thermoplastic Resin
Handle_Medium	Compute	150 mm	PC/ABS Plastic
Handle_Short	Suppress	120 mm	Polypropylene

Figure 7.26: Attributes for the HandleMS configuration displayed on the drawing with the corresponding model states

You have placed various drawing views of the same part with different model states active and produced a table that details the differences between each configuration. Although completed with a part file, the same could be done with an assembly file.

Model credits

The model credits for this chapter are as follows:

- **SHAFT GEAR MOTOR WITH ACCESSORIES** (by Dařena Miroslav): `https://grabcad.com/library/shaft-gear-motor-with-accessories-1`

- **Shock Absorber** (by Suriya SB): `https://grabcad.com/library/shock-absorber-370/details?folder_id=10235721`

8

Design Accelerators – Specialized Inventor Tool Sets for Frames, Shafts, and Bolted Connections

Inventor has a wide range of tools to accelerate design engineering tasks and generate components automatically with minimum required input from the user. This allows users to configure examples of structural frames, shafts, gears, and bolted connections much more quickly and efficiently. If creating any of these features, the relevant design accelerator or environment should be used, as creating these parts individually using standard modeling practices would be extremely time intensive. By using the **Design Accelerator** or **specialized toolset**, future design changes and edits can be made more efficiently, and the parts can be made more quickly.

In this chapter, you will learn about the following:

- Creating frames using the **Frame Generator** environment
- Detailing the BOM for frames in the drawing environment and producing cut lists
- **Structural Shape Author** – creating your own custom frame members and publishing them to the Content Center
- Creating gears in Inventor with the **Spur Gear Design Accelerator**
- Creating shafts in Inventor with the **Shaft Design Accelerator**
- Applying automatic **bolted connections**

This chapter also aims to give context to each of the methodologies and how and why they should be used in your workflows. There are many more design accelerators contained within Inventor, but to cover them all would take far more than one chapter. This book aims to cover the most widely used accelerators and environments.

By the end of this chapter, you will learn how to use and apply the design accelerators including Frame Generator, shafts, and bolted connections.

Technical requirements

To complete this chapter, you will need to access the practice files in the Chapter 8 folder within the Inventor Cookbook 2023 folder.

Creating frames using the Frame Generator environment

Frame Generator allows for the creation of internal frame and external frame assemblies for machines from the Assembly and Weldment environments. There are multiple ways in which a frame can be created from Frame Generator; in this example, we will define a structural skeletal model as the basis of the frame and then proceed to add members. Additional 2D and 3D sketches can be used to create additional members, as required.

Frame Generator uses existing frame members from the Content Center. Once selected, you can then define the size, material, and appearance parameters. Frame members can then be generated onto existing sketch lines or edges.

Once placed, there are various end treatment functions such as miters, cuts, and joints that can be applied.

This recipe focuses on creating frames from scratch using the **Frame Generator** environment. Once created, the frame will be edited and modified with various end treatment options, such as a miter.

Getting ready

To begin this recipe, you will need to create a New Standard (mm) Assembly .iam file and have it open.

How to do it...

To begin, ensure that you have a New Standard (mm) Assembly .iam file open in Inventor. The first step for the frame creation is to make a skeletal model to act as a reference. This model will drive our frame in terms of overall size and area:

1.   Within the **Assemble** tab, select **Create**.

2. In the **Create In-Place Component** window, change the name of the part to S Frame1.ipt. Define the template as Metric\Standard(mm).ipt and ensure that the location of the file is set to the Chapter 8 folder within Inventor Cookbook 2023:

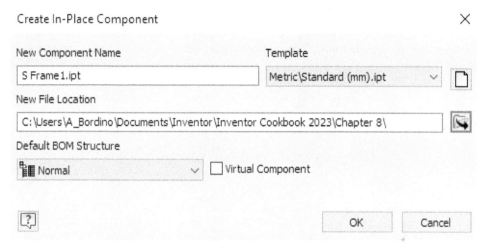

Figure 8.1: The Create In-Place Component Window, showing the values to change in step 2

3. Select **OK** to complete.

4. Expand the **Origin** folder in the **Model** browser and select the XY plane to create the initial sketch.

5. Select **Start 2D Sketch** from the top left of the ribbon and then select the XY plane either from the graphics window or the **Model** browser.

6. Select **Rectangle** and then **Two Point Center Rectangle**. Create the sketch shown in *Figure 8.2* from the origin. The rectangle needs to be 1000 mm x 500 mm. This will form the basis of our structural frame skid:

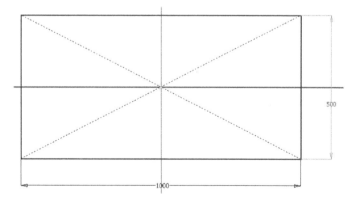

Figure 8.2: Two-point center rectangle to create on the XY plane

7. Select **Finish Sketch**.

8. Select **Extrude** and extrude the rectangle created in *steps 6* and *7* by 1500 mm in one direction. Select **OK** to complete.

9. Select **New Sketch** and create a sketch on one of the large 1,000 x 500 sides of the extruded rectangle.

10. Create a line sketch on the face as shown in *Figure 8.3*. Select **Finish Sketch** and then repeat this operation on the other side. These additional lines will form the structure of cross members in the frame:

Figure 8.3: Sketch line to create on the face of the recently extruded solid

11. Select the **Material Selection** dropdown. Then, select **Autodesk Material Library** and choose **Acrylic Clear** as the **Material** option. This is because this part will act as a skeletal reference model for the placement of members. Applying a clear material allows the user to see all the visible edges when selecting and applying frame members. If the wrong member is selected, hold *Shift* and select the incorrect member to remove it.

12. Select **Return** to return to the top level of the assembly. Change **Visual Style** to **Shaded with Edges**. This will also make the placement of the members much easier:

Figure 8.4 Skeletal reference model of the frame

13. Select the **Design** tab and then select **Insert Frame**:

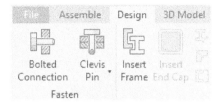

Figure 8.5: The Insert Frame command

14. Inventor will now prompt you to save the assembly before continuing. Select **Yes**, name the file FrameSubAssem1.iam, and save it in the Chapter 8 folder. Select **Save** and **OK**.

15. Frame Generator now reveals the **Insert Frame** panel. Here, we can now define the types of frames we wish to place. Select the edges of the reference model as shown in *Figure 8.6*. As you select the sketch lines of your model, they will populate and preview with default frame members, as shown in *Figure 8.6*:

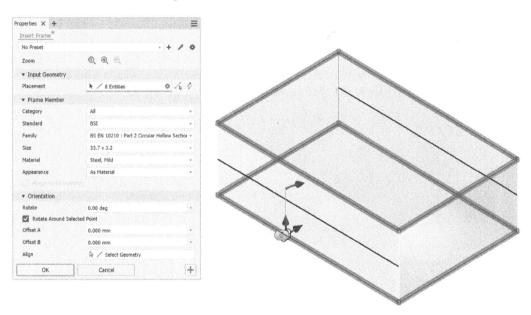

Figure 8.6: Placement of frame members

16. Zoom into one of the members and you will notice that the graphical reference shown in *Figure 8.7* is present. This allows for the placement and definition of the frame to be edited based on the reference sketch line or edge. Frames can also be rotated in position. In this instance, leave all the settings as their defaults:

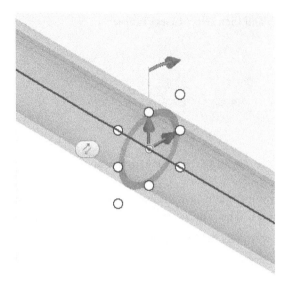

Figure 8.7: Graphical interface that can be used to alter frame member placement

17. The next section of the command allows the **Category**, **Size**, **Family**, and **Material** options to be defined for these members. Select **All** from the **Category** drop-down list and then select **Square/Rectangular Tubes**. We will now define the frame members as square sections of steel as opposed to round bars, which they are by default:

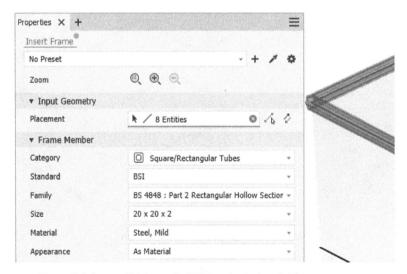

Figure 8.8: Square/Rectangular Tubes selected as the frame category

18. Select **Standard** and change it from the default value to **ISO**. The families that are available then change to ISO frame members as a result.

19. Change the **Family** option for the members to **ISO 4019 Square…** and the **Size** option to **50 x 50 x 2**. This is shown in *Figure 8.9*. The frame members in the preview update in real time within the graphics window. You will notice in the preview that the members are currently blending into each other and are not mitered or cut correctly, this will be changed and corrected after placement:

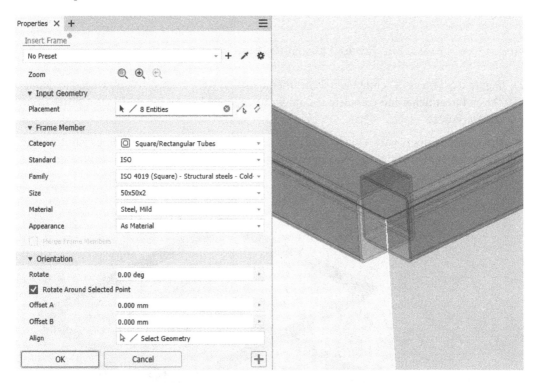

Figure 8.9: Family and size of members changed

20. Further **Orientation** of the members can be defined, but in this instance, the members are defined correctly for our design, so select **OK** to complete the generation of the members.

    Inventor now prompts you to name the new members – leave the fields as default and select **OK**, followed by **OK** again. If required, individual names can be applied to individual frame members:

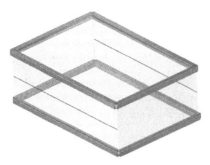

Figure 8.10: Completed frame members

21. *Figure 8.10* shows the completed frame members that were created. We now need to define the end treatments and complete the other missing members of the skid. Select **Miter** from the **Design** tab.

22. Select all members with a click, hold, and drag of the mouse in the graphics window. All selected members should go blue upon selection, indicating you have included the member as part of your selection:

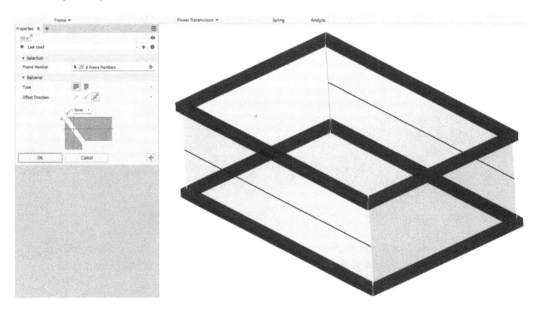

Figure 8.11: Members selected by the Miter command

23. Change **Miter Offset** to 5mm for the placement of a weld. Select **OK** to complete.

24. All joining corners of the frame are now automatically mitered, as previously defined in *steps 22* and *23*:

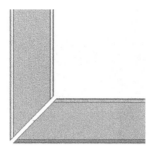

Figure 8.12: Miter defined and applied to all joining frame members

25. Select **Insert Frame**. Select the four members as shown in *Figure 8.13*:

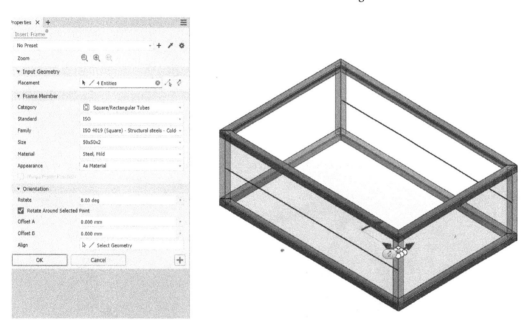

Figure 8.13: Four additional frame members selected

26. Select **OK** followed by **OK** again to complete the step.

27. Select **Trim/Extend** to cut the vertical frame members so that they interface with the top miter frame correctly.

28. Select the bottom face shown in *Figure 8.14* as the first reference. This is the face we want the frame member to be cut to:

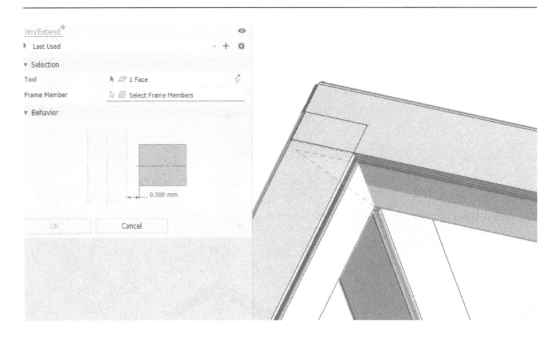

Figure 8.14: The Trim/Extend command in progress

29. For the second frame members, select all vertical frame members, as shown in *Figure 8.15*:

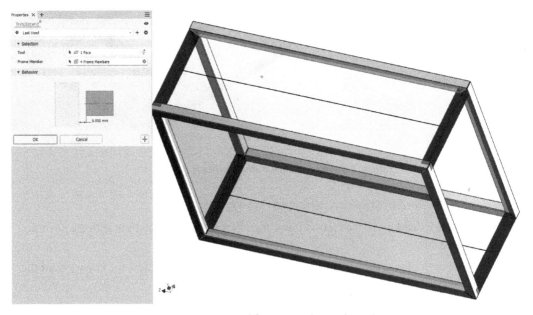

Figure 8.15: Vertical frame members selected

30. Select **OK** to complete.

31. One side of all the vertical frame members has been trimmed to the bottom face of the miter sections, but the other face is still interfacing in the wrong way (see *Figure 8.16*). The **Trim/ Extend** function from *steps 26* to *28* must be repeated on the other side:

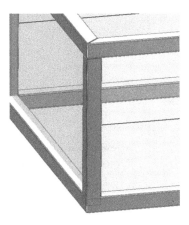

Figure 8.16: Correct interface of members at the top of the frame
and an incorrect one at the bottom of the frame

32. Repeat *steps 26* to *30* on the other side of the vertical members. The result should be that of *Figure 8.17*:

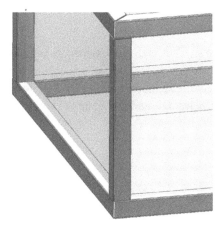

Figure 8.17: Trim/Extend repeated for the other side of the vertical members

33. A design change needs to be actioned on the four vertical frame members. Select all members shown in *Figure 8.18* in the **Model** browser and select **Edit with Frame Generator**:

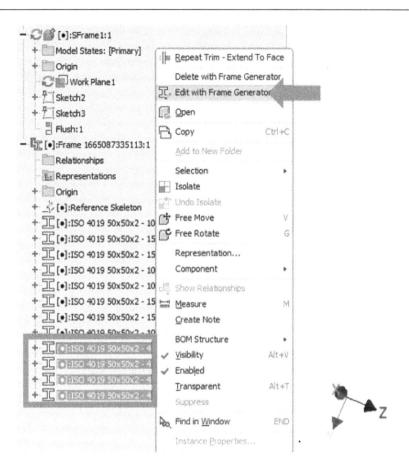

Figure 8.18: Edit with Frame Generator selected

34. Change the **Family** option of the members to **ISO 12633-2 (Rectangular)**. The model updates and the changes are made. Select **OK** and **OK** again to complete. **All End Treatments** should be preserved:

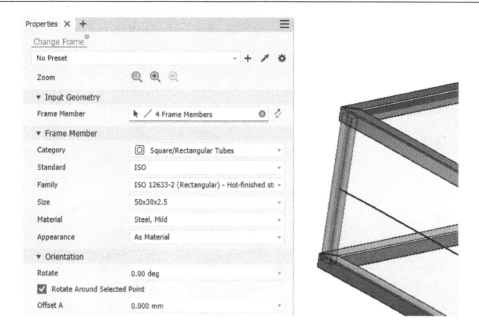

Figure 8.19: Design change actioned, ISO 12633-2 (Rectangular) selected

35. Select **Insert Frame Generator**. Select the two remaining horizontal sketch lines in the model to create the members. Change the **Family** option to **ISO 12633-2 (Square)** and change the **Size** option to **40x40x3**. This is shown in *Figure 8.20*:

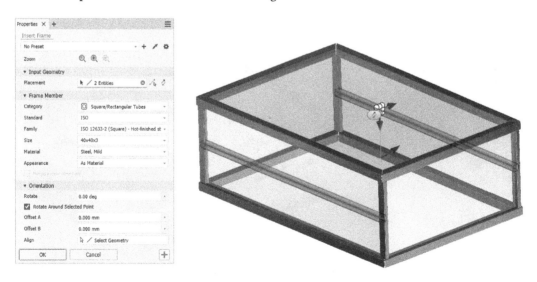

Figure 8.20: Members selected and defined

36. Select **OK** followed by **OK** again to complete. We will now notch these new members into the existing framework.

37. Select **Notch**. Then select the two members from the frame shown in *Figure 8.21*.

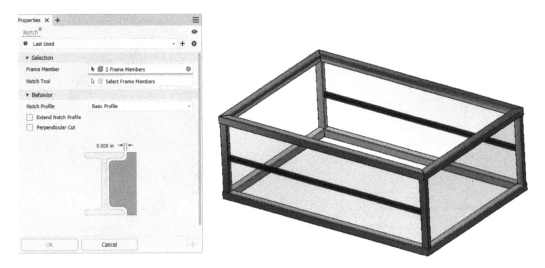

Figure 8.21: Members selected for the Notch command

38. Select the frame members from the command and then select the four vertical members for the notch to interface with. *Figure 8.22* represents this:

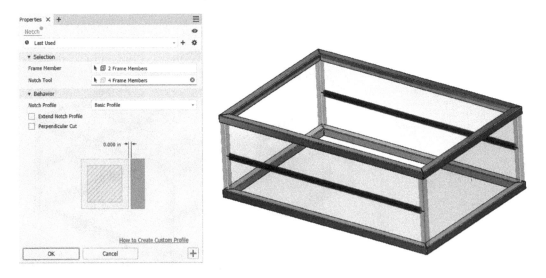

Figure 8.22: Four vertical members selected for the Notch command

39. Select **OK** followed by **OK** to complete.

40. We will now extend the frame to create legs for the skid and use Frame Generator to generate end caps for the profiles.

41. Right-click **SFrame1.ipt** in the browser and select **Edit**.

42. Select the **Manage** tab and then select **Parameters**.

43. Select the **d0** parameter and change the value from 1000mm to 1500mm. Select **OK** and then select **Return** to return to the top-level assembly. The frame updates and changes back to its original state. All of the existing frame definitions and end treatments are maintained:

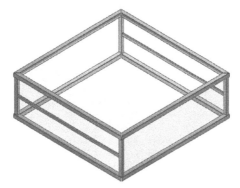

Figure 8.23: Updated frame model

Using Frame Generator, you have created a frame skid structure with various frame members and end treatments. You have also updated the frame based on the parametric skeletal reference model that drives the assembly.

Detailing the BOM for frames in the drawing environment and producing cut lists

Detailing a BOM for components created with Frame Generator is useful to communicate cut lengths and other manufacturing information. In this recipe, you will learn how to adapt and edit a BOM for frame-generated parts.

Getting ready

To begin this recipe, open FrameBOMGA.idw from the Chapter 8 folder in Inventor Cookbook 2023.

How to do it...

With `FrameBOMGA.idw` open in Inventor, you will notice several views of a framework have already been placed. We will now generate the default BOM:

1. Select the **Annotate** tab and then select **Parts List**. For the BOM view, change the option from **Structured** to **Parts Only**. This provides a true list of all parts in all assemblies and sub-assemblies within the overall model.

2. Select one of the views and place this on the drawing as per *Figure 8.24*:

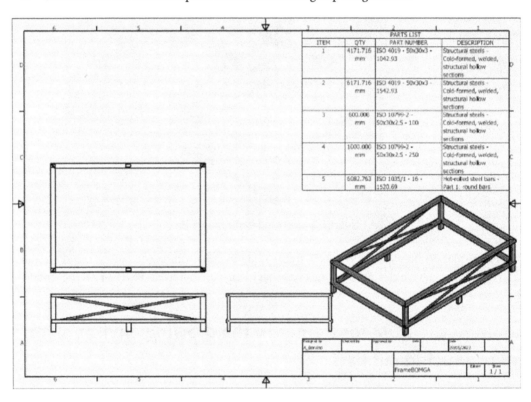

Figure 8.24: Default BOM placed for the frame product

3. The issue with this BOM structure is that it has produced an itemized list of every single framework and this is not often the best way of displaying this for manufacturing purposes. We will now adapt the BOM so that it is more appropriate for cutting lists and grouped in length for purchasing.

    Select the **Manage** tab and select **Styles Editor**:

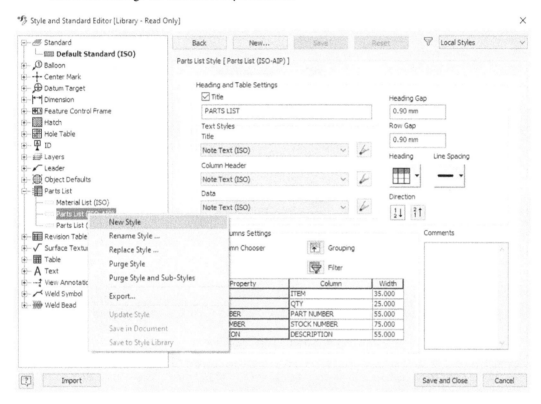

Figure 8.25: The expanded parts list and creating a new style

4. Expand **Parts List**, right-click on **Part List (ISO)**, and select **New Style**, as shown in *Figure 8.25*.

5. Rename your new local style Frame BOM. You can also select whether you want to add this to your template at this point. Select **OK**.

6. Select **Column Chooser**:

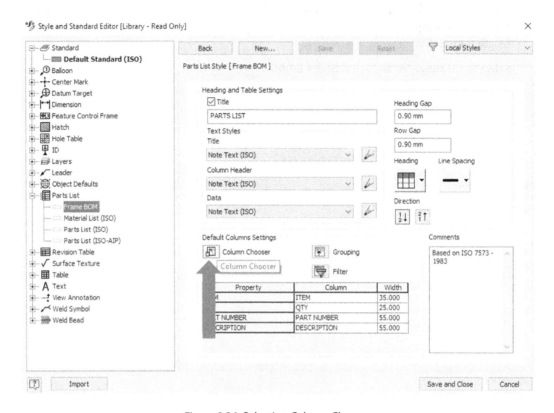

Figure 8.26: Selecting Column Chooser

7. On the right-hand panel, scroll down, select **UNIT QTY**, and then choose **Add**. Select **STOCK NUMBER** from the right panel and then **Add**. Select **UNIT QTY** from the right-hand panel and choose **Remove**. Your menu should have the selected properties on the right as shown in *Figure 8.27*. Finally, select **OK**:

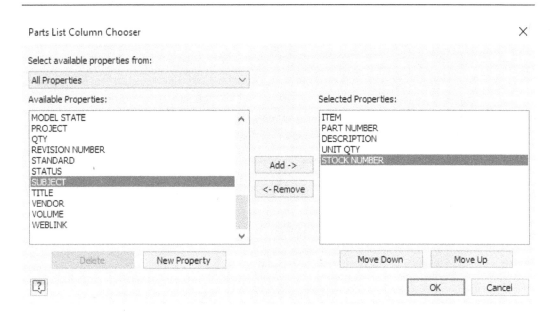

Figure 8.27: Selected Properties required

8. Select **UNIT QTY** as shown in *Figure 8.28*:

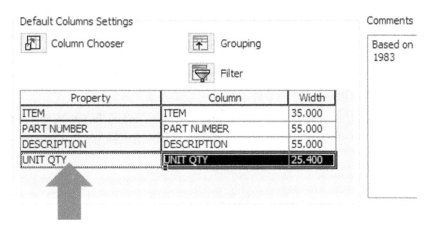

Figure 8.28: UNIT QTY to select

9. Select the **Substitution** tab and then select **Enable Value Substitution**. Under **When exists, use value of**, select the drop-down menu, and select **Browse Properties**. Select **UNIT QTY** and select **OK**.

10. Under **When rows are merged, value used is**, select the dropdown and select **Sum of Values**:

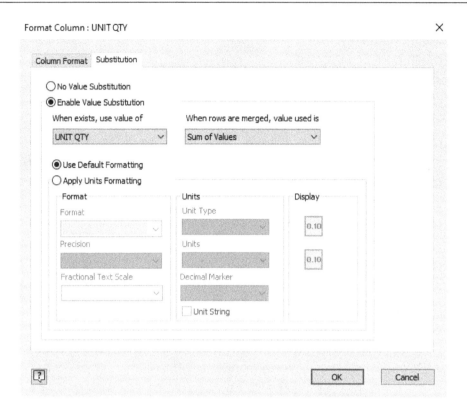

Figure 8.29: UNIT QTY and Sum of Values selected

This will now mean the BOM will group all **UNIT QTY** properties of specific cut lengths and multiply them to a total in the BOM.

11. To further improve the display, select **Apply Units Formatting**, and on the screen, apply a **Precision** setting of 1.1, set **Units** to m, and change **Decimal Marker** to .Period. Then, select **OK**:

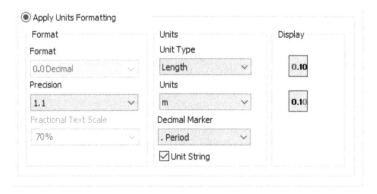

Figure 8.30: Units Formatting to apply

12. Select **Grouping**:

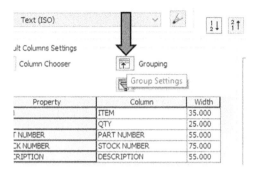

Figure 8.31: Grouping

13. Select **Group** and then select **First Key** as DESCRIPTION. Select **OK**.

14. Select **Save** followed by **Save & Close.**

15. The BOM previously placed has not been updated based on the stages – we will need to reapply a new BOM to reflect the changes made. Select the **Annotate** tab.

16. Select **Parts List**. Then, on the right of the ribbon, select **Frame BOM** as the parts list standard, as per *Figure 8.32*:

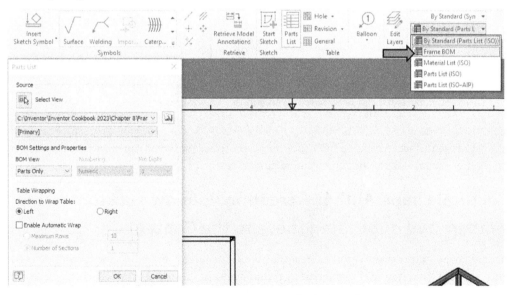

Figure 8.32: Frame BOM standard selected

17. Go to **Select View**, then select one of the drawing views, and select **OK**. Click anywhere on the sheet to place the new BOM:

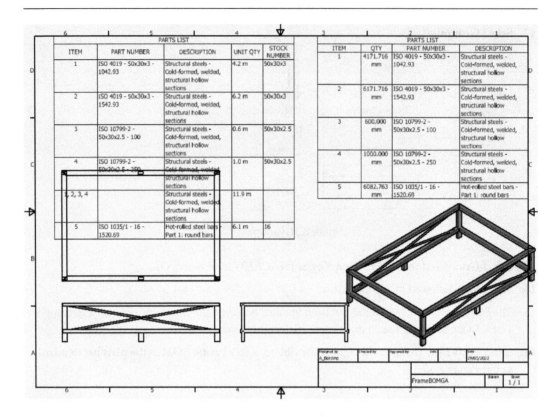

Figure 8.33: New frame BOM placed

18. The new BOM lists the unit quantity of the specific lengths and types of steel for purchasing. The original parts list can now be removed and the drawing rearranged.

By editing the styles and standards, we have created a new BOM configuration that can be saved to your template so that when creating BOMs for frame-generated parts, you can create an accurate cut list and a more suitable BOM for purchasing requirements.

Structural Shape Author – creating your own custom frame members and publishing these to the Content Center

Structural Shape Author enables you to create custom content for components for use within **Tube & Pipe**, **Frame Generator**, and as standalone components. In this recipe, you will create a custom aluminum extrusion, define its parameters, and then publish it to your Content Center for use with the frame generator.

Getting ready

To begin this recipe, you will need to **Run Inventor as Administrator** from Windows. Open `AluEx1.ipt` from the `Chapter 8` folder in `Inventor Cookbook 2023`.

How to do it...

With `AluEx1.ipt` open in Inventor, you will see that `AluEX1.ipt` is a simple U-channel aluminum extruded profile. We will perform a series of operations to define and then author this component so that it can be used within the Frame Generator environment for future use:

1. Expand **1** in the **Model** browser, right-click on **Sketch1**, and select **Edit Sketch** (the sketch is seen in *Figure 8.34*). You will notice that the profile is already defined and key parameters have already been applied to control the **Length**, **Height**, **Width**, and **Thickness** settings for the extrusion. It is necessary to have these parameters to create custom content:

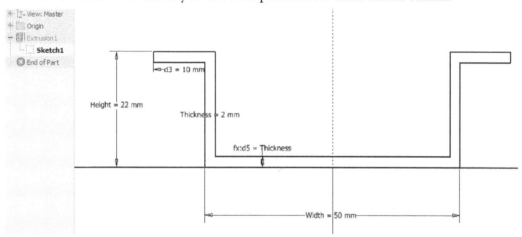

Figure 8.34: Sketch1 of the U-channel extrusion

Select **Finish Sketch** to return to the part.

2. Select the **Manage** tab and then select **Create iPart.**

3. Right-click on the **Length** cell and select **Custom Parameter Column**. The length will be variable and when we publish to the Content Center later, this will make the custom extrusion easier to use when not using Frame Generator:

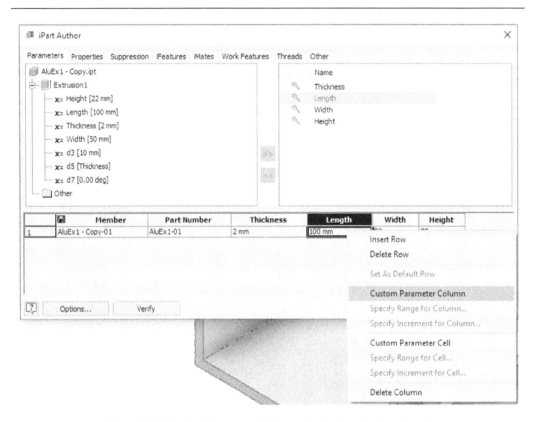

Figure 8.35: Custom Parameter Column selected on the Length cell

4.   Select the key next to **Width** and **Height** so that they are classified as key parameters. This will make the process of selecting within Frame Generator easier once published:

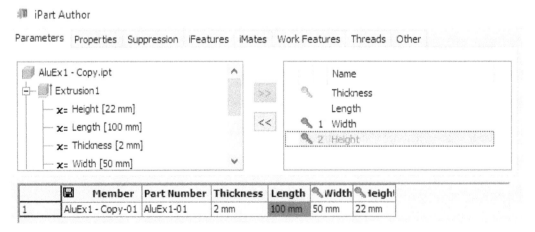

Figure 8.36: Width and Height selected as key parameters

5. Select the **Properties** tab in **iPart Author**, expand **Project**, and select **Stock Number**. Select >> to move **Stock Number** to the iPart.

6. Select **Description** and then select >> to move the **Description** property to the iPart.

7. Rename the values of **Description** and **Stock Number** as detailed in *Figure 8.37*. Select **OK** to complete:

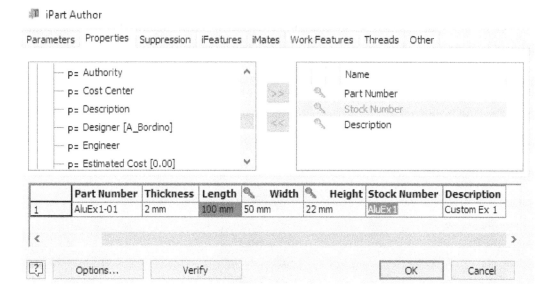

Figure 8.37: Description and Stock Number properties added and defined

8. Select **Structural Shape** author from the **Author** panel in the **Manage** tab:

Figure 8.38: The Structural Shape author command

9. We now need to select the **Category** setting for the Content Center. Select the dropdown that currently displays `*required`. Select **Other**. The **Base Extrusion** and **Default Base Point** settings are correct and can be left as their defaults.

> **Notch profiles**
>
> You can also define a Notch profile; this is normally required for more complicated profiles with multiple internal voids. To create this, edit the original sketch for the profile, and create a bounding sketch box around the profile. Within the optional selection, you can then click and define the areas that would not be notched, as shown in *Figure 8.39*.

10. The parameters must now be mapped. Select **Parameter Mapping** from the **Structural Shape Authoring** menu.

11. Within **Table Columns**, we can select the relevant iPart to map parameters. For **Base Length**, select **Length** from the dropdown.

12. For **Shape Height**, select **Height** from the dropdown.

13. For **Shape Width**, select **Width** from the dropdown.

14. For **Shape Thickness**, select **Thickness** from the dropdown:

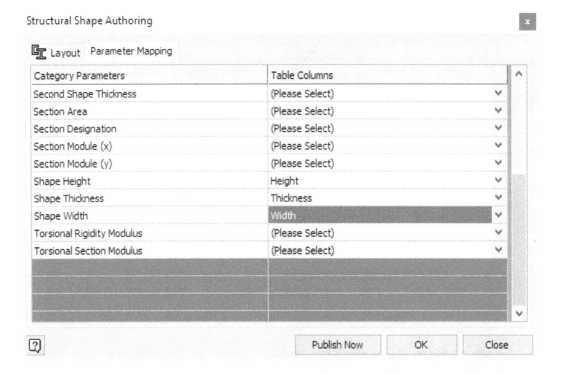

Figure 8.39: Category Parameters mapped

15. Select **OK** to complete, followed by another **OK**. The part has now been authorized and needs to be placed in the Content Center.

16. The next step is to publish this authored part to the Content Center. This can be done either with **Desktop Content** or **Autodesk Vault Server** content. In this example, we will use **Desktop Content**. Navigate to the Inventor home screen.

17. Ensure that `Inventor 2023 Cookbook` is selected as the active project and select **Settings**.

18. Select **Configure Content Center Libraries**:

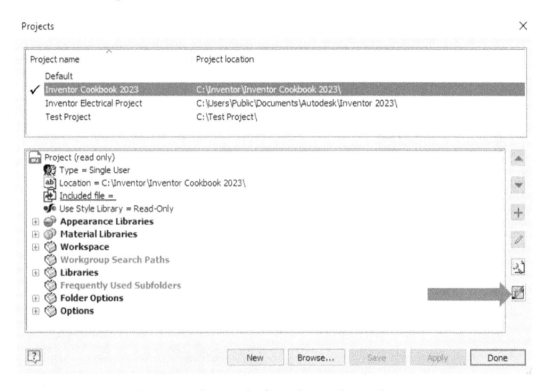

Figure 8.40: Choosing Configure Content Center Libraries

19. Select **Create Library**:

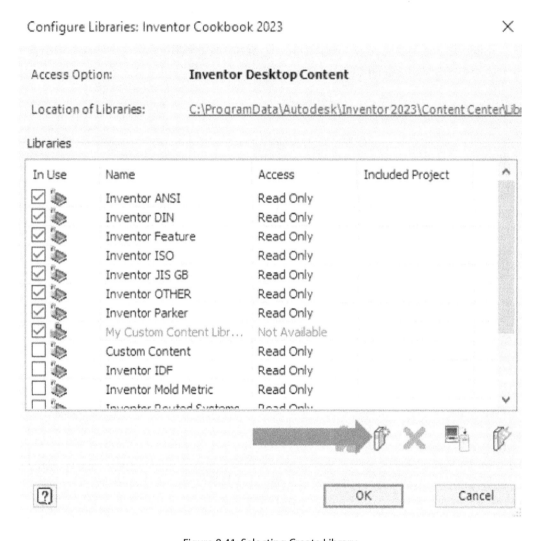

Figure 8.41: Selecting Create Library

20. Name your library My Custom Content Library. Select **OK**. The new library created will be set to **Read/Write**, which is essential to proceed. Select **Done** and **Save**.

21. We can now proceed to publish the part. Within the AluEx1.ipt file, navigate to the **Manage** tab, then to the **Content Center** panel, and select **Publish Part**:

Figure 8.42: Publish Part

22. Select **Next**. Then select **Next** again for the **Category** option. You have already mapped the parameters previously, so once more, select **Next**.

23. Select **Length** and then select → to bring this to the key columns. Then, select **Next**:

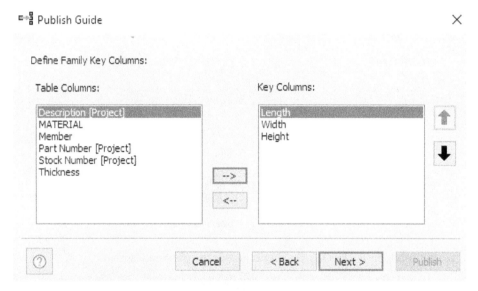

Figure 8.43: Publish Guide with Length selected as a key parameter

24. Select **OK** on the warning that displays. This is a warning to you that when placing multiple instances of this part, external to Frame Generator, the names of these parts will not be unique. This is okay in this instance and for use within Frame Generator. If required, you can go back and modify this in the iPart at a later stage.

25. Define **Family Properties** as per *Figure 8.44* by selecting **Standard Organization** as **CUSTOM**. This will make finding custom content much easier within our library. Then, select **Next**:

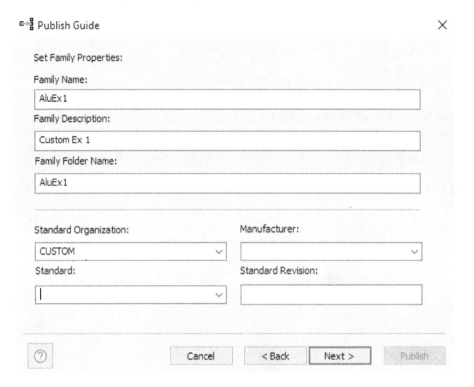

Figure 8.44: Family Properties to set

26. Select **Publish** followed by **OK**. The part is now in the Content Center.

27. Create a brand new .iam file. Select **Place from Content Center**, browse **Structural Shapes**, and then find **Other**. The published part is present and available for use:

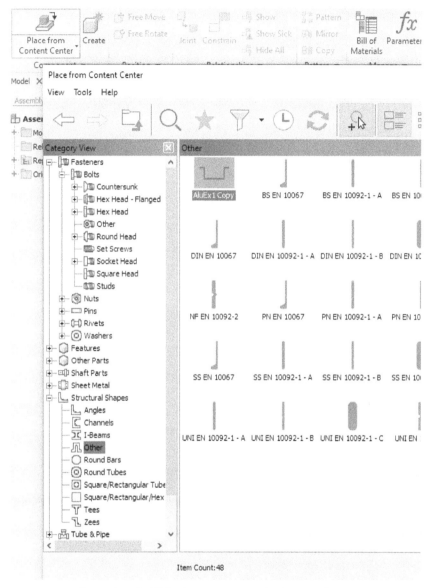

Figure 8.45: Published custom part in the Content Center

28. Select **Cancel** to close the **Place from Content Center** window.

29. Select the **Design** tab and then select **Insert Frame**. Save the assembly if prompted as Assembly1 in the Chapter 8 folder. Set **Category** to **Other** and set **Standard** to **CUSTOM**. The custom profile is available to select and use within Frame Generator:

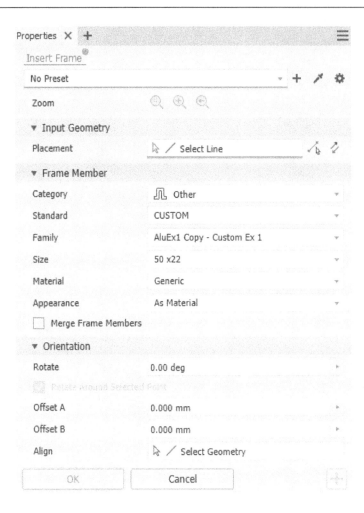

Figure 8.46: Custom U-channel aluminum extrusion available for use in Frame Generator

Using **Structural Shape Author**, you have created a custom aluminum profile, defined the parameters, and published this to the Content Center for use in Frame Generator.

Creating gears in Inventor with the Spur Gear Design Accelerator

Inventor's Design Accelerators allow you to specify, calculate and generate gearing for a variety of different solutions, such as spur, worm, and bevel. In this recipe, we will calculate and generate two meshing spur gears on two shafts that have already been defined in an assembly.

Getting ready

To begin this recipe, open `SpurGear.iam` from the `Chapter 8` folder in `Inventor Cookbook 2023`.

How to do it...

Ensure that `SpurGear.iam` is open in Inventor. Within this assembly file, two shafts have been created and are visible in the assembly. A plate that the shafts interface with is also present. The wider context of the design is not contained within this assembly.

We will start by defining the location of the two meshing spur gears on the existing shafts:

1. Select the **Design** tab. Navigate to the **Power Transmission** panel and select **Spur Gear**:

Figure 8.47: Spur Gear selected

2. The **Spur Gears Component Generator** tool now opens. From here, we can define and calculate all aspects of the gears that we desire for this design. In the **Design Guide** section in the top left of the menu, select **Module and Number of Teeth** as the option, as the shafts' centers are already pre-defined.

3. Set **Desired Gear Ratio** to `1.8000 ul`. Then, select **Calculate**. The **Module** value will change to `3.000 mm`:

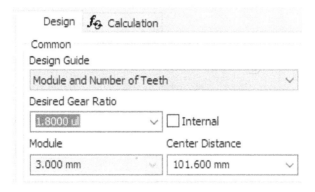

Figure 8.48: Design Guide and Desired Gear Ratio selected

4. For the **Center Distance** value, ensure that this is set to 101.60 mm. Inventor should pick this up automatically, but if not, you can manually override. This is the center distance between the two shafts.

5. The gears can now be defined in terms of geometry. Under **Gear 1**, select the dropdown for **Component**. As we have existing shafts, we can use these as a reference.

6. Select the red arrow next to **Cylindrical Face** under **Gear 1**:

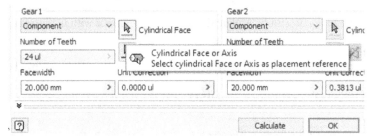

Figure 8.49: Cylindrical Face arrow selected

7. Then, select the geometry of the shaft shown in *Figure 8.50*:

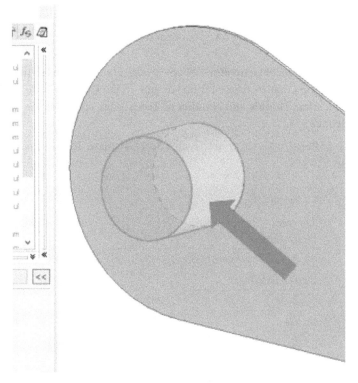

Figure 8.50: Geometry to select

8. Select the red arrow next to **Start plane** under **Gear 1**:

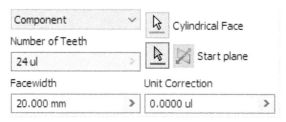

Figure 8.51: The Start plane button

9. Now, we can define from the shaft geometry where the gear should start. Select the geometry shown in *Figure 8.52*:

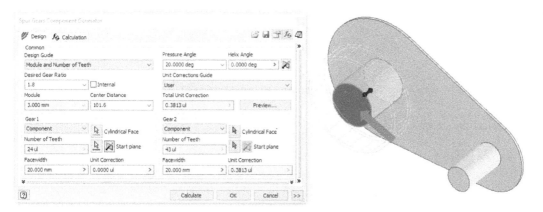

Figure 8.52: Geometry selected and the preview of gear displayed

10. Select **Flip Start Plane** if the gear preview is displayed on the other side of the shaft. The end preview should resemble that of *Figure 8.52*:

Figure 8.53: Flip Start plane

11. Under **Gear 2**, select **Component** and then select **Cylindrical Face,** and select the geometry shown in *Figure 8.54* as a reference:

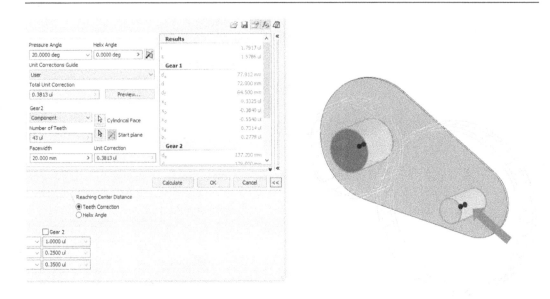

Figure 8.54: Geometry to select for Gear 2

12. Select the start plane as per *Figure 8.55* for **Gear 2**:

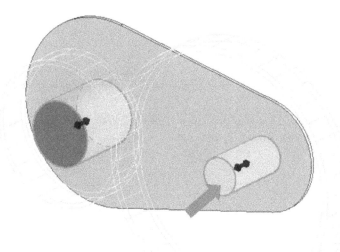

Figure 8.55: Start plane for Gear 2 selected

13. Select **Flip Start Plane** if required. The gears must be aligned as shown in *Figure 8.56*:

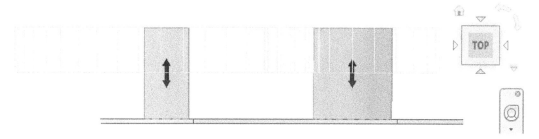

Figure 8.56: Gears preview shows that the gears are aligned

14. Select the **Calculation** tab from within the **Spur Gears Component Generator** window.

15. From here, you can specify the required power, lubrication, and estimated life cycle of the gears. In this instance, leave all as their defaults and select **OK** followed by **OK** again:

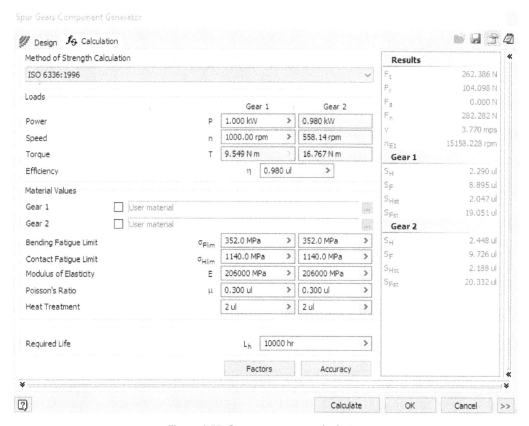

Figure 8.57: Gear component calculation

The gears are generated within the assembly and mesh correctly, as per our previous calculation and inputs:

Figure 8.58: Completed and calculated meshing spur gear set

16. Modifications can be made at this point. In the graphics window, right-click on the larger of the two gears and select **Edit using Design Accelerator**:

Figure 8.59: Edit using Design Accelerator

17. Under **Helix Angle**, type 15 degrees for the **Angle** value. This will transform the spur gears into helical spur gears. Select **OK**.

18. The assembly updates and the gears are updated to helical gears as per the previous **Helix Angle** input.

Figure 8.60: Assembly updated to helical gears

Using the **Spur Gear Design Accelerator** tool, you have successfully defined, calculated, and generated two intermeshing helical spur gears, and placed them on an existing shaft, you have then modified the gears with a helix angle.

Creating shafts in Inventor with the Shaft Component Generator tool

To create shafts in Inventor, the most efficient way is to use the **Shaft Component Accelerator** tool. This allows users to specify all aspects of the shaft, generate the geometry, and apply any necessary calculations. Upon specification, the shaft is assembled from single sections (cylinder, cone, and polygon). Additional end features such as chamfers can also be applied to the generator. By defining a shaft using the **Shaft Component Generator** tool, future design changes and edits are much easier to facilitate, as the generator can be used to redefine the placement and geometry of the shaft.

Once a shaft is defined and placed, it can also be saved into your templates for future use.

In this recipe, you will define a shaft and place a ball bearing onto it, using both the **Shaft Component Generator** tool and the **Bearing Generator** tool.

Getting ready

To begin this recipe, open Shaft Assem.iam from the Chapter 8 folder in Inventor Cookbook 2023.

How to do it...

With `Shaft Assem.iam` open in Inventor, a simple housing box is shown, which we will generate a shaft for. The shaft will be placed central to the circular cut-out on the model. Once a shaft has been generated, we will then specify a ball bearing to be placed on the model:

1. Select the **Design** tab within the assembly file.

2. Select the **Shaft** command from the **Power Transmission** panel in the ribbon:

Figure 8.61: The Shaft Component Generator command

3. The **Shaft Component Generator** tool now opens. Automatically, a default shaft will appear on the cursor – left-click to place this preview anywhere. As we change values in the generator, the preview will update live as a representation of what the shaft will look like. We will place the shaft in its correct location once we have defined the cylinders:

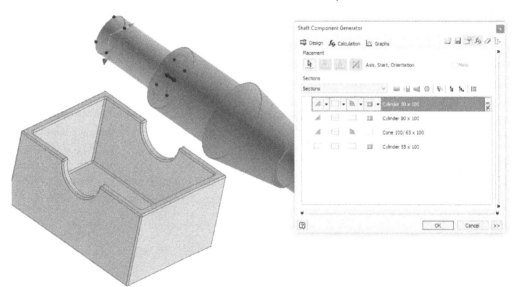

Figure 8.62: Default preview shaft placed in the assembly

4. The commands highlighted in *Figure 8.63* allow you to define additional cylinders, polygons, and cones, or to select geometry in the assembly to reference for the shaft generation – this would be an alternative way of placing the shaft compared to the method we will use, which is to define it first and then use joints in the assembly you locate:

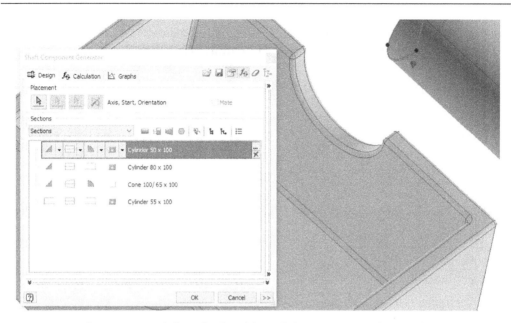

Figure 8.63: The Sections tool allows for the additional placement and definition of shaft sections

5. It is best to work from one side of the shaft to the other and define each key segment. Select the **End Section** dropdown next for the first shaft. Select **Thread**:

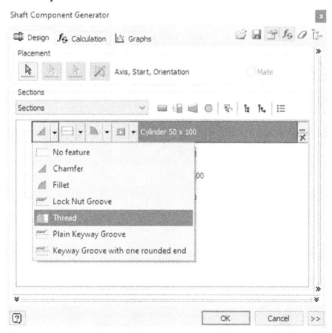

Figure 8.64: Thread selection for the end of the shaft

6. We can now define how we want the end thread for this section of the shaft to be defined. Set the **Thread** and **Shaft** dimensions to the values shown in *Figure 8.65*:

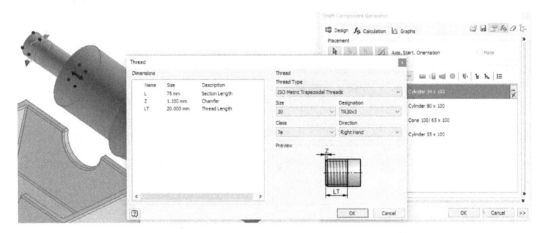

Figure 8.65: Thread and Shaft section values to define

Select **OK** to complete. This section of the shaft updates in the preview.

7. Select the dropdown for the next cylinder and select **Chamfer**:

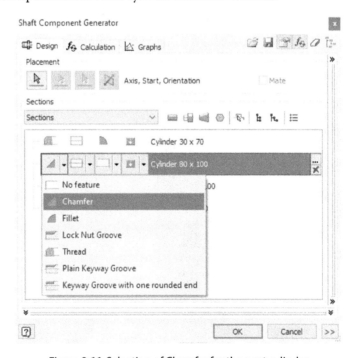

Figure 8.66: Selection of Chamfer for the next cylinder

8. Define the **Chamfer** value as 3   mm and select the green tick to apply:

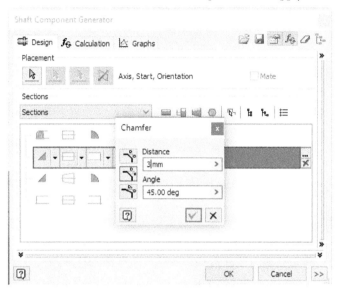

Figure 8.67: Shaft Chamfer distance changed to 3 mm

9. Double-click on **Cylinder 80 x 100**.

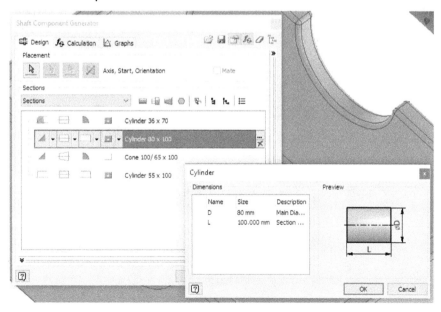

Figure 8.68: Cylinder size selection

We will now change the size of this section.

10. In the next window, change the **Length** and **Diameter** options to the values shown in *Figure 8.69*. Select **OK** to complete. The preview of the shaft changes to reflect this:

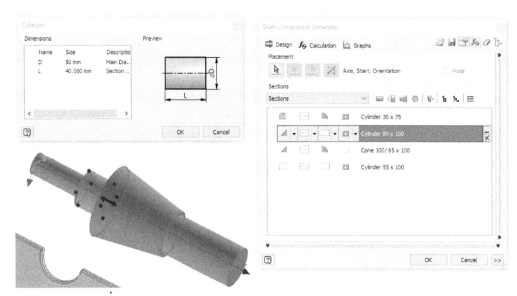

Figure 8.69: Values to change for the second cylinder of the shaft

11. For the third cylinder, select the dropdown for **Cone**, change it, and set this to **Polygon**. Select **OK** on any warning that appears. The cone is changed to a polygon that we can define:

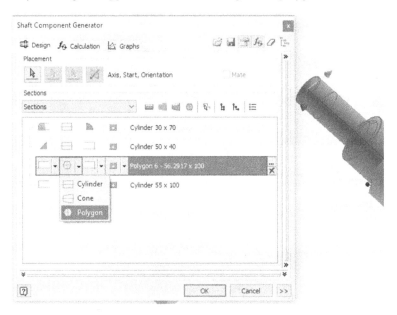

Figure 8.70: Polygon selected

12. Double-click on **Polygon 6-56.2917 x 100**. In the next window, define D_{in} as 65 mm. Select **OK** to complete:

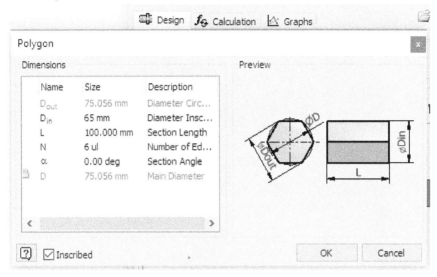

Figure 8.71: Polygon D_{in} defined as 65 mm

13. Select **OK** to close the window and complete the generation of the shaft. Select **OK** again when prompted to name the file.

14. Left-click to place the shaft anywhere within the assembly. Now that the shaft has been defined, at any point, by either right-clicking on the shaft or selecting **Edit using Design Accelerator**, or by expanding the shaft in the **Model** browser, you can make design changes to the shaft:

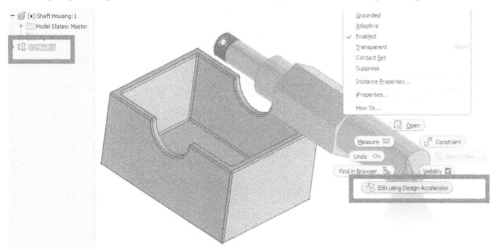

Figure 8.72: How to edit the defined shaft

15. Select the **Joint** command.

16. Set **Type** to **Rotational** and select the geometry shown in *Figure 8.73* as the first reference:

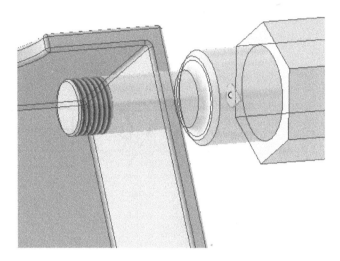

Figure 8.73: The first reference to select for the rotational joint

17. Select the edge shown in *Figure 8.74* as the next reference:

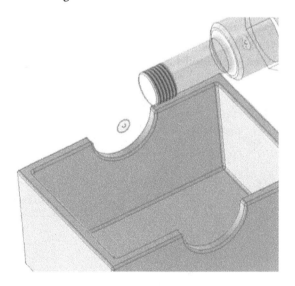

Figure 8.74: Next edge to select

18. Select **Flip Component** to ensure that the shaft aligns correctly as shown in *Figure 8.75*:

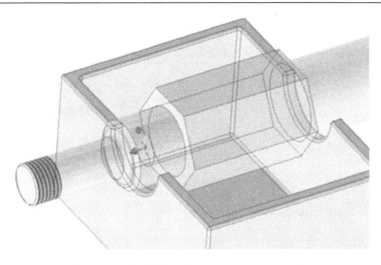

Figure 8.75: Shaft flipped to the correct orientation

19. Before we place the bearings, the clearance between the shaft and housing is too tight on either side, so we will now edit the shaft to accommodate a greater range of bearings. Right-click on the shaft and select **Edit using Design Accelerator**.

20. For **Cylinder 2**, set **Diameter** to 3 5 mm:

Figure 8.76: Change to the diameter of Cylinder 2

21. For **Cylinder 3** also set **Diameter** to 3 5 mm:

Figure 8.77: The diameter of Cylinder 3 changed to 35 mm

22. Select **OK**. There is now ample clearance to place a bearing on either side.

23. Select the **Design** tab from the ribbon. Then, select **Bearing** from the **Power Transmission** area of the ribbon.

24. Select the drop-down menu next to **Angular Contact Bearings** and then select **All Ball Bearing Categories**:

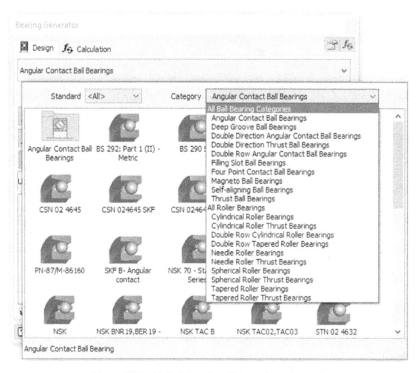

Figure 8.78: All Ball Bearing Categories selected

25. Select the red arrow next to **Cylindrical Face** and select the geometry of the bearing shown in *Figure 8.79*:

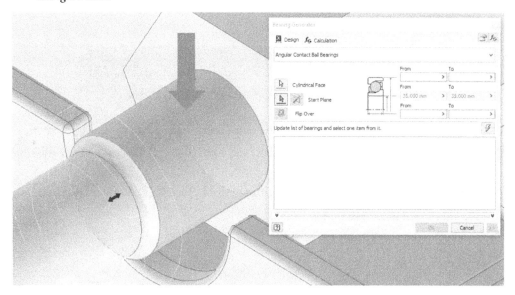

Figure 8.79: The Cylindrical Face reference for the bearing

26. Select the red arrow next to **Start Plane** and then choose the geometry shown in *Figure 8.79*. This is the face the bearing will be flush to:

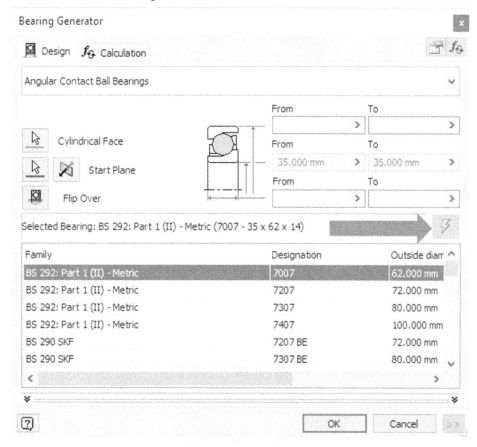

Figure 8.80: Reference Start Plane selected

27. Now, we can filter the Content Center to show the applicable bearings that fit these design requirements. Select the **Update** button shown and highlighted by the arrow in *Figure 8.80*.

28. Select **BS 292:Part 1 (II) Metric 7007 62mm Bearing**. Click on the **Calculation** tab and, from here, you can specify a bearing further in terms of the loads, speeds, lubrication, and desired life expectancy. In this case, select **OK**, followed by **OK** again.

The bearing is placed and constrained onto the shaft in the correct position:

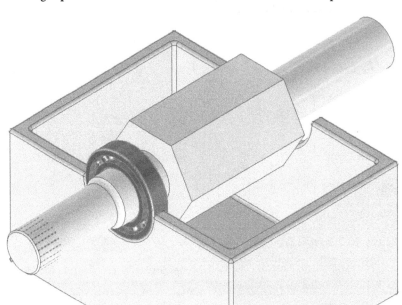

Figure 8.81: Completed shaft and bearing in the housing

The **Shaft Component Generator** and **Bearing Design Accelerator** tools have been used to define, specify, and place a custom shaft and bearing into the assembly.

Applying automatic bolted connections

Bolted connections are often a prominent and common feature of most mechanical design assemblies. Components such as nuts, bolts, and washers can be loaded independently into the model via the Content Center one at a time, but in some cases, it is advantageous to automate this procedure and apply multiple Content Center components that make up a bolted connection in one operation.

In this recipe, we will apply several bolted connection sub-assemblies to an existing assembly of parts.

Getting ready

To begin this recipe, navigate to `Inventor Cookbook 2023`, then `Chapter 8`, followed by `Stepper Motor Brake >`, and open `Nema 34 Stepper Motor Brake.iam`.

How to do it...

With Nema 34 Stepper Motor Brake.iam, we will apply several bolted connections in one operation to the assembly:

1. Ensure that you have Nema 34 Stepper Motor Brake.iam open. Navigate to the **Top View**. There are three countersunk threaded holes in the assembly. It is here that we will place automatic bolted connections:

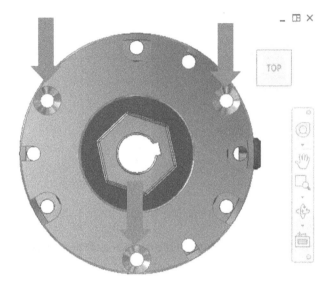

Figure 8.82: Locations on the Nema 34 stepper motor to place the bolted connections

2. Click on the **Design** tab and then select **Bolted Connection**:

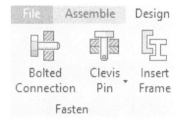

Figure 8.83: Bolted Connection

3. With the **Bolted Connection Design Accelerator** tool open, we can now define the placement of the connections and the types of content that will form this. Click on the drop-down menu and select **By hole**. This will allow you to define the placement of the bolted connection via an existing **Hole** feature:

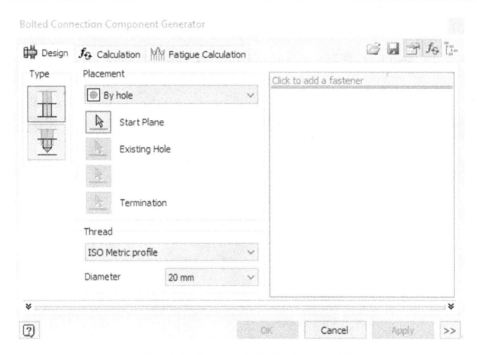

Figure 8.84: Selecting By hole for the Placement type

4. Select the red arrow next to **Start Plane** in the window and then select the face of the model, as shown in *Figure 8.85*:

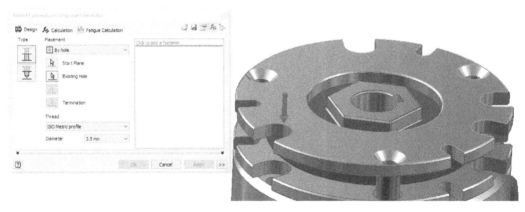

Figure 8.85: Reference plane to select

5. Select the red arrow next to **Existing Hole** in the window and then select one of the countersunk holes, as shown in *Figure 8.86*:

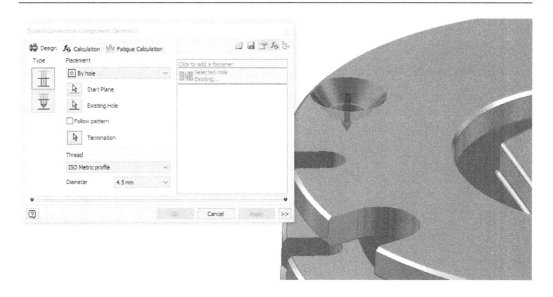

Figure 8.86: Reference Existing Hole selected

6. Select the red arrow next to **Termination** in the window. Navigate to the other side of the motor and select the face shown in *Figure 8.87*:

Figure 8.87: Termination face to select

7. Check the checkbox next to **Follow pattern**. This will ensure that bolted connections are placed on all instances of this **Hole** series in the model.

8. Select **Click to add a fastener**:

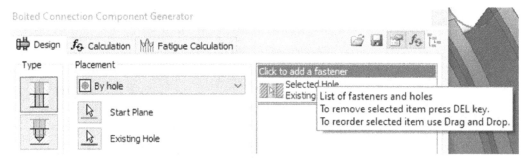

Figure 8.88: Click to add a fastener

9. Select the bolt shown in *Figure 8.89*. Note that because the hole has a countersink, Inventor only gives options for complimentary countersink bolts:

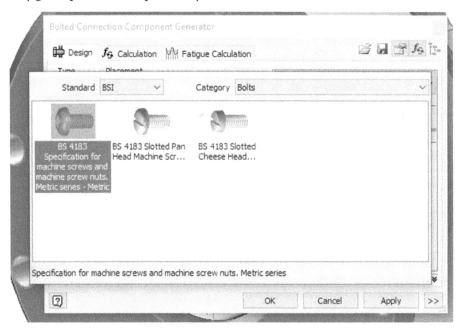

Figure 8.89: Countersunk head machine screw to select

10. The preview in the model will also update, showing how the bolted connection will be placed when completed. We do not require a washer on this size of the hole, but on the **Termination** side, we do. Select **Click to add a fastener**, as shown in *Figure 8.90*:

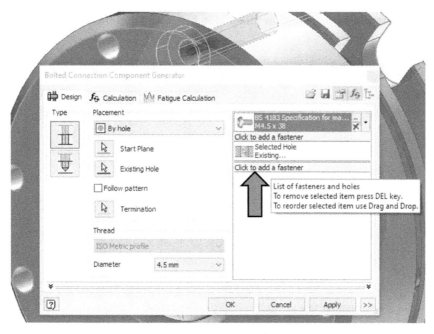

Figure 8.90: Click to add a fastener

11. Select the **IS 5556 B** washer. The preview will update:

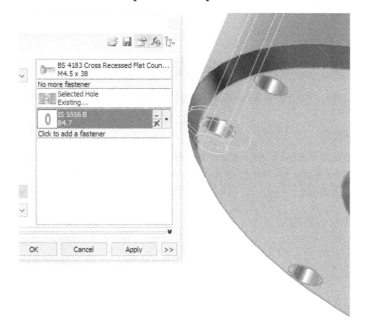

Figure 8.91: Washer added and preview updated

12. Select **Click to add a fastener** below the **IS 5556 B** washer. Select the highlighted **KS B** nut as shown in *Figure 8.92*:

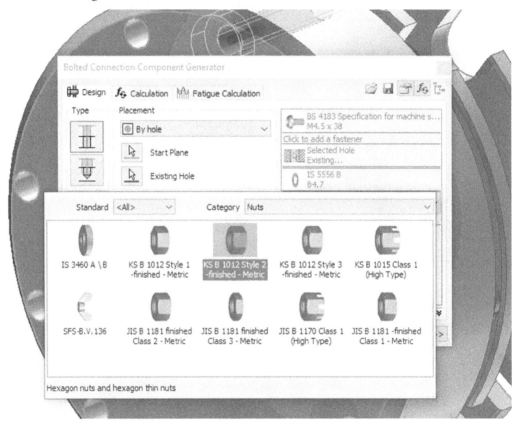

Figure 8.92: KS B 1012 Style 2-finished – Metric nut to add to the bolted connection

13. Select **OK** to apply.

The three separate bolted connections are placed and constrained in the correct position within the assembly and are visible in the **Model** browser. At any point, these connections can be edited and changed:

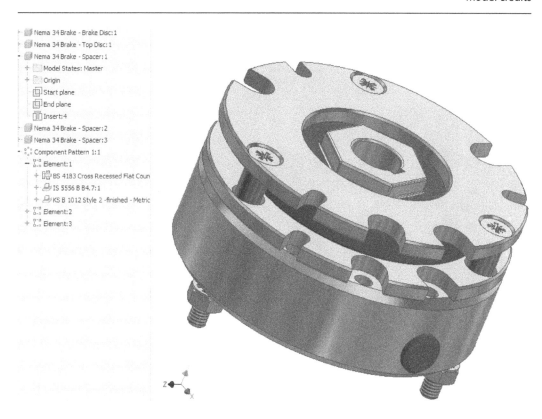

Figure 8.93: Completed bolted connections within the assembly

Using the **Bolted Connection Design Accelerator** tool, you have automatically specified and placed three different bolted connections within the context of an assembly.

Model credits

The model credits for this chapter are as follows:

Nema 34 DC 24V Electromagnetic Brake 4.0Nm (by Roy Pridham): `https://grabcad.com/library/nema-34-dc-24v-electromagnetic-brake-4-0nm-1/details?folder_id=10617223`

9

Design Communication – Inventor Studio, Animation, Rendering, and Presentation Files

A successful design project can be measured by how well the benefits and the solution are communicated and understood by the stakeholders. Therefore, being able to visualize designs in 3D within Inventor is highly important. Designers need to be able to render, animate, and sometimes quickly annotate parts and assemblies to communicate key design changes, updates, and showcase product features.

This chapter focuses on how a designer can implement renderings, explode views with **presentation files**, and create animations. The chapter also focuses on some of the basic areas of Inventor rendering, animation, and design change documentation. This will aid the designer in all aspects of product and concept presentation. Although Inventor can produce good rendered images/animations, it is not a specific tailored rendering application. In this respect, there are limitations, but for most mechanical design engineers, the functionality within Inventor is more than enough for what is required.

If more advanced rendering and animation are required as part of your workflow on a regular basis, I highly recommend *Autodesk 3Ds Studio Max* (interoperable with Inventor models). 3ds Max is a visualization tool, used to model, animate, and render.

In this chapter, you will learn the following:

- Creating an image render with Inventor Studio
- Animating components of an assembly in Inventor
- Presentation files – creating exploded views and animations of assemblies
- Creating and adding decals and custom appearances

This chapter also aims to give insight and advice on setting up renders for success and the technical and hardware requirements that may limit your rendering capabilities.

Technical requirements

To complete this chapter, you will need to access the practice files in the Chapter 9 folder within the Inventor Cookbook 2023 folder.

Creating an image render with Inventor Studio

Inventor Studio is a rendering and animation environment within the Autodesk Inventor part and assembly environment. With Inventor Studio, you can do the following:

- Create still and animated renderings of parts and assemblies
- Create still and animated renderings of products to show visual changes and/or movement within an assembly
- Create video animations using multiple cameras
- Reuse constraints or parameters between animations in one assembly file to drive and show movement, and much more

Inventor Studio works with model states and design representations. Before entering the Inventor Studio environment, it is good practice to have the model state you wish to animate or render active in the model.

With the rendering command, you can run this as standard with the part or assembly that is open, with no enhancements to the geometry, or you have the option to assign additional lighting styles, scenes, camera viewpoints, and materials. For quick design reviews or presentations, a full-scale render with cameras and lighting enhancements is not always required.

Various factors affect the end quality of the render; these include but are not limited to **anti-aliasing**, output resolution, material and bitmap selection, and geometry of the model. The end-specified image size and machine performance are also factors in render quality, as a larger image requires more machine resources and time (these factors also apply to rendered video output).

> **Anti-aliasing**
> Anti-aliasing is the process of smoothing jagged edges in digital images.

When a render is executed, the rendering progress is displayed in a separate window from the model. Inventor renders images locally on your machine, but other Autodesk solutions such as *Fusion 360* can render on the cloud.

You can save rendered output to standard formats: `.bmp`, `.gif`, `.jpg`, `.jpeg`, `.tif`, `.tiff`, and `.png`.

Custom lighting and materials can be created and saved into your **Template** and **Libraries** folders for future use.

In this recipe, you will render a realistic image of a model that has had a custom appearance, lighting, and camera customization assigned. We will start by applying **appearances** and **materials** to the geometry model. **Lighting styles** will then be defined, followed by shadows, cameras, and rendering options.

Getting ready

To begin this recipe, you will need to navigate to `Inventor Cookbook 2023|Chapter 9|Assy Mini Drone`, and then open `Assy Mini Drone.iam`.

How to do it...

To begin, ensure that you have the `Assy Mini Drone.iam` file open in Inventor. The first step will be to apply the relevant appearances to the model and override the material display:

1. Open `Assy Mini Drone.iam`.

2. In **Model Browser**, select **Frame**.

    Once selected, note that in the **Materials** toolbar, **PBT Plastic** is displayed. This is the material that has been specified previously to the part. Even though a material has been specified, you can override the appearance of that material with the **Appearance** browser.

3. To do that, in the **Quick Access** toolbar, select the **Appearance Browser** command.

Figure 9.1: Appearance Browser

4. Within the **Appearance** Browser, you can search and specify what appearance to apply to the highlighted part. These are libraries within Inventor that contain a list of pre-defined appearances (there is also an equivalent for materials). In the **Search** field, type `Metal 1600F Hot`.

5. The list populates results, and the specific appearance is available for selection. Hovering the cursor over this provides a preview in the **Graphics** window of the appearance applied.

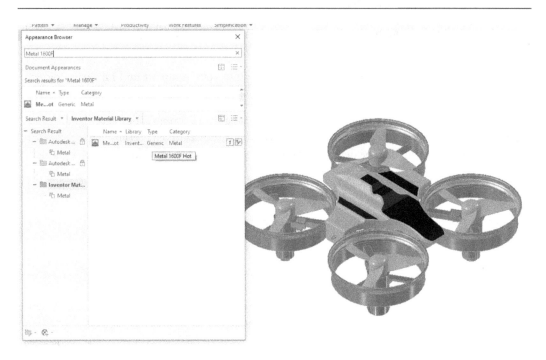

Figure 9.2: The Metal 1600F Hot appearance applied

6. Right-click on **Metal 1600F Hot** in the **Appearance** browser and select **Assign to Selection**. The frame changes its appearance to this material.

7. Close the **Appearance** browser.

8. Then select **Casing** from **Model Browser** and open the **Appearance** browser again.

9. Type `Orange` in the **Search** field and select **Smooth – Light Orange**. Right-click on **Smooth – Light Orange** and select **Assign to Selection**.

10. With the **Appearance** browser still open, hold *Ctrl* and select the **Casing** from the **Model** browser to deselect this from the current selection. Then, still holding *Ctrl*, select all four instances of **Propeller** from **Model Browser.**

11. Assign a **Smooth Red** material to the propellers.

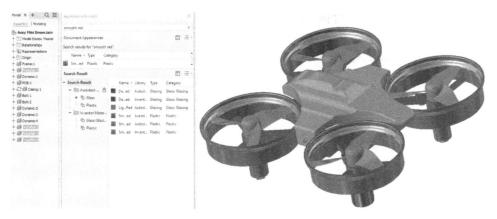

Figure 9.3: Smooth Red applied to the propellers

12. **Ground Plane**, **Reflections**, and **Shadows** now need to be defined. Click on the **View** tab. In the **Appearance** panel, select the dropdown next to **Ground Plane**. Select the checkbox to activate **Ground Plane**.

Figure 9.4: Ground Plane

13. Select **LEFT** on the view cube; the model should be level on the ground plane, as per *Figure 9.5*. It is important that this is correct; otherwise, shadows and reflections will not render accurately. If your model does not display this way, go to the **Ground Plane** settings and select **Automatic adjustment to model**, and then **Apply**.

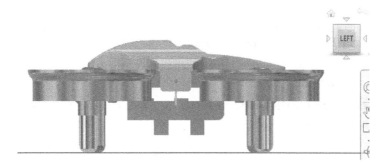

Figure 9.5: The ground plane activated, with the model set correctly

14. From the **View** tab, select **Appearance**, and then select **Reflections** to enable the reflections. The **Graphics** window will display this.

15. In the **View** tab, select **Appearance**, then select **Orthographic**, and change it to **Perspective**.

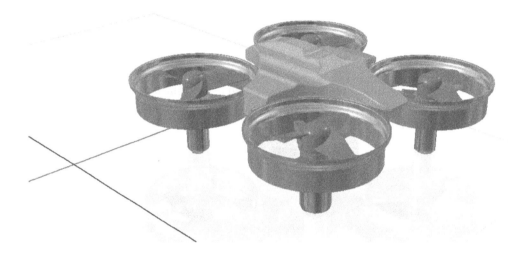

Figure 9.6: The Perspective view activated and turned on

16. Select **Default Light** and then **Warm Light** from the pre-defined drop-down list. This list contains a standard library of default lighting systems that can be applied; custom lighting styles can also be created and stored here.

17. Select the **Environments** tab, and then **Inventor Studio**.

18. From the rendering tools, we can further specify cameras, lighting, and components to animate. Select **Studio Lighting Styles** in the **Scene** panel of the ribbon.

19. Browse the lighting styles and environments, and select **Warm Light**. Right-click on **Warm Light** and select **Activate**.

20. Tick the checkbox next to **Display Scene Image** to include the scene with the lighting style in the rendering environment.

Figure 9.7: The Warm Light lighting style and the scene active

21. Select **Done**, followed by **Yes**, when prompted to save.

22. Select **Render Image** from the ribbon.

23. Maintain the settings in **General** as default, and select the **Output** tab. Select **Save Rendered Image** and browse to the `Mini Drone` folder in `Chapter 9`. Select **Save**.

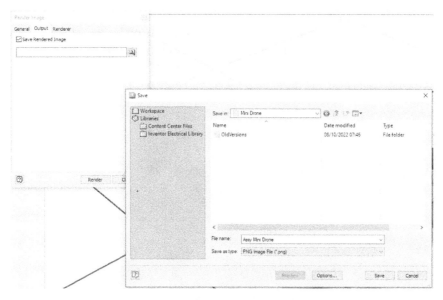

Figure 9.8: The location to save the completed rendered image

24. In the **Render Image** dialog box, select the **Renderer** tab. Then, select **Until Satisfactory** as the duration.

Figure 9.9: The Renderer settings to apply

25. Select **Render**. The render output window will open, and the image will begin to render. The longer the rendering process is given, the better the quality of the image. You can pause the renderer at any time if you deem that the image is satisfactory. After around 7 minutes, your rendered image should resemble *Figure 9.10*. Select **Stop Render** and save the image. Close the Inventor assembly and do not save.

Figure 9.10: The completed render

A successful still image render of `Mini Drone.iam` with lighting styles, environments, and appearance override has been created. Renders are often quite an iterative process, and to get the right image, it may take several renders and tweaks to lighting styles, cameras, and appearance choices to get the desired outcome or hero image of the product that is required. This recipe has shown that even with minimal input, a reasonable design visualization can be generated.

Animating components of an assembly in Inventor

In this recipe, you will create an animation of a drone assembly that shows the quad propellers spinning as it vertically takes off. Camera animations, lighting, and scenes will be applied, and then the end output exported through **Video Producer** as a `.avi` video.

Getting ready

To begin this recipe, you will need to navigate to `Inventor Cookbook 2023 | Chapter 9 | Drone Animation` and open `Assy Mini Drone.iam`.

How to do it...

To begin, ensure that you have the `Assy Mini Drone.iam` file open in Inventor. All appearances and relationship constraints required to drive and create the animation have already been applied:

1. Click on the **Environments** tab from the ribbon and select **Inventor Studio**.

2. We want to establish our scene and lighting. In this example, we will use one of the existing lighting and scene styles. Select **Studio Lighting Styles** from the **Scene** panel.

3. Right-click on **Empty Lab** from the list and select **Active**.

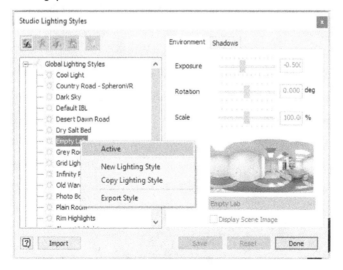

Figure 9.11: Activating the lighting and scene styles

4. Select the checkbox next to **Display Scene Image**. Then, zoom out using the mouse wheel. Displaying the scene with this lighting style shows an empty laboratory with the drone assembly on a table; zoom out using your curser wheel to see this.

5. In the **Environment** tab of the **Studio Lighting Styles** menu, change the **Scale %** value from **100.000**% to 30.000%. This makes the scene smaller, and the drone appears much larger.

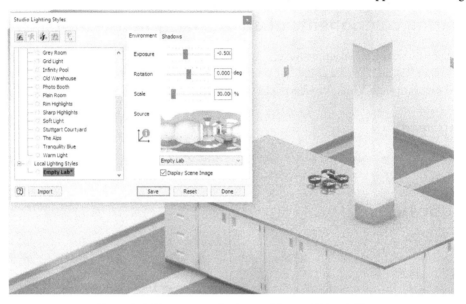

Figure 9.12: The scene scale changed from 100% to 30%

6. Next, we will adjust the scene rotation so that it is more suitable for our animation. In the **Rotation** section, change the value from **0.000** degrees to 180.000 degrees to rotate the scene.

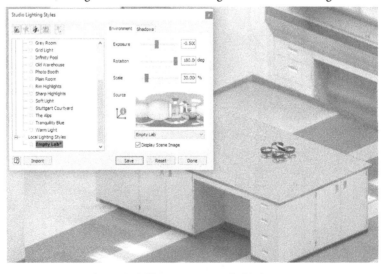

Figure 9.13: The scene rotated 180 degrees

7. Select **Done**, followed by **Yes**.

8. Now, select the **View** tab. In the **Appearance** panel, change the **Orthographic** view to **Perspective**.

9. Orientate your view so that it resembles *Figure 9.14*.

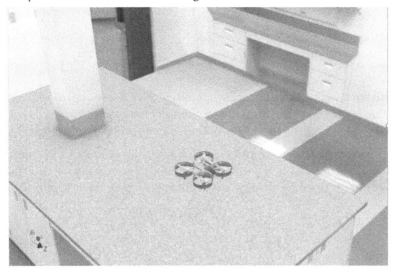

Figure 9.14: The view to orientate

10. We can now make this **Current** view the **Camera** view for the animation. Select the **Render** tab, right-click **Cameras** in **Model Browser**, and select **Create Camera from View**. This creates a new camera view from which a render or video can be taken, based on the current view in the model space. Cameras can also be created manually with the **Camera** command.

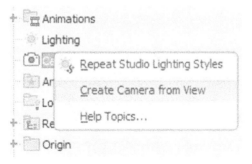

Figure 9.15: Create Camera from View

11. Now that lighting and cameras are placed, we can specify how the animation will flow. Select **Animation Timeline** from the ribbon, followed by **OK** if prompted.

12. The **Animation** timeline appears at the bottom of the screen. This is where your actions will populate once they have been created. Select the **Animations Options** command from the bottom right of the screen (highlighted in *Figure 9.16*).

Figure 9.16: Animations Options

13. In the **Animations** options, we can define specifics about how fast actions will run throughout the duration of the animation. In the **Seconds** area of the window, change the value from **30** seconds to 5 seconds. This is the length of time the animation will run. Select **OK** to complete.

14. Now, we will animate the four propellers to spin. Select the **Animate Constraints** command, as shown in *Figure 9.17*. As there are directed angle mate constraints in place on the propellors already, we can use this directed angle constraint to drive and rotate the propellers in place.

Figure 9.17: The Animate Constraints command

15. For the action, expand the first instance of the propellors in **Model Browser**, and select the angle constraint.

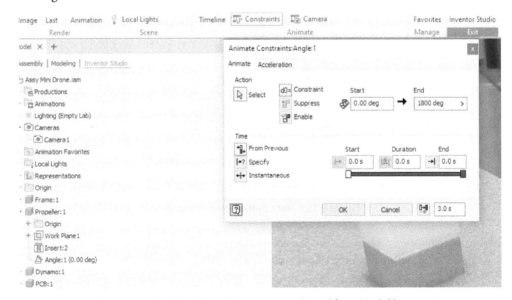

Figure 9.18: The directed angle constraint selected from Model browser

16. For the **End** degrees, change the value from **0.00** to 1800.

17. Under **Time**, select **Specify**. Then, change **Start** to 0.3 seconds and **End** to 5 seconds. Then, select **OK**.

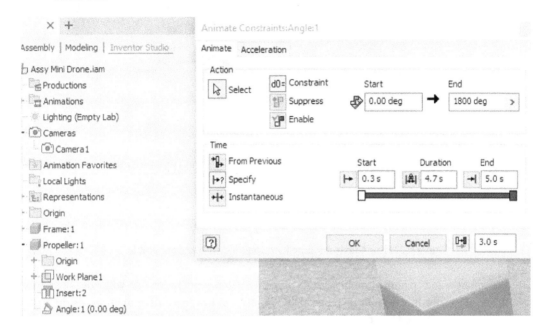

Figure 9.19: Time specifications for the animation of the propellers

18. Select **Expand Action Editor**, as shown in *Figure 9.20*. This expands the **Animation** timeline. You will see that the animate constraints for the directed angle constraint are visible on the timeline. Select the **Play** button from the timeline to preview and play the animation.

Figure 9.20: Expand the Animation timeline

19. Only one of the propellors will spin at this stage, as we have only applied the animation to the constraint located in **Propellor 1**. For the other three propellors to spin at the same rate and speed, repeat *steps 14–19* for each of the three remaining instances of **Propellor**.

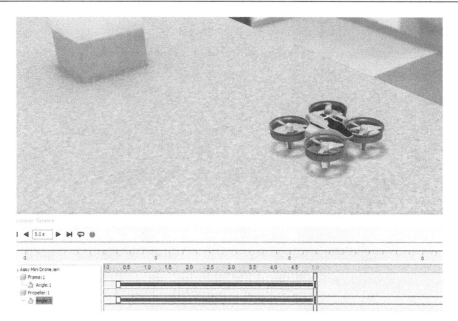

Figure 9.21: The Animation timeline expanded with actions visible

20. We will now add an animation to several components in one operation that will move the mini drone vertically off the table as the propellers spin from the previous animation. As the first step, select **Animate Constraints**.

21. Select **Mate:3** under **Frame** in **Model Browser** as the constraint to animate.

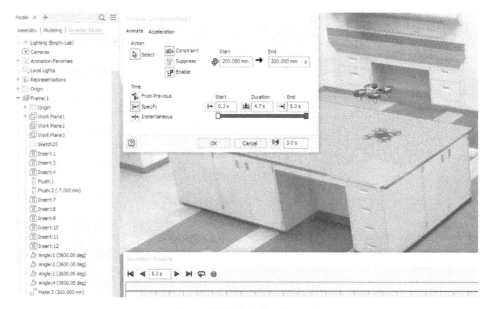

Figure 9.22: Mate 3 animated

22. Ensure the **Start** value is set to 100.000 mm and type 200.00 mm as the **End** value. Then, select **Specify** under **Time** and set **Start** as 0.3 seconds and **End** as 5.0 seconds. Then click **OK**.

23. Upon pressing **Play** in the **Animation** timeline, the mini drone will vertically take off 100 mm from the tabletop as the propellors spin. Select **Play** to view the animation.

24. The camera that we previously created has not yet been used in this animation. Now, we will animate the camera so that it partially revolves around the mini drone as it takes off. To do so, select **Animate Camera**.

Figure 9.23: Animate Camera

25. Once the **Animate Camera** dialog opens, ensure that under **Camera**, **Camera 1** is selected as the active camera. Then, select the **Turntable** tab within the **Animate Camera** window.

26. Select the checkbox next to **Turntable**. In the drop-down menu under **Axis**, select **Y Origin**.

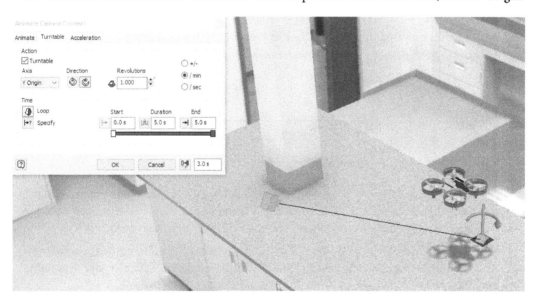

Figure 9.24: Turntable and Y Origin selected

27. Select +/- and change **Revolutions** from **1.000** to **.65**. Then, select **OK** to complete.

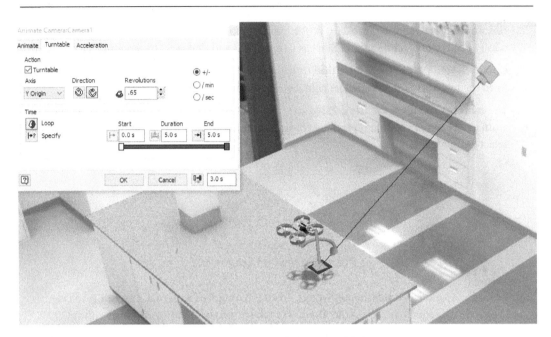

Figure 9.25: .65 revolutions specified

28. In the **Animation** timeline, rewind back to the start. In the dropdown on the right of the **Animation** timeline, ensure that **Camera 1** is selected.

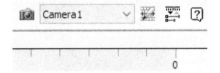

Figure 9.26: Camera 1 selected

29. Play the animation. The mini drone's propellors should spin, and it should rise above the table as **Camera 1** pans around it at .65 of a revolution in the Y axis. The preview may appear to lag; once the animation is rendered properly, this will be a much smoother transition.

30. There is much more that can be done to tweak an animation, but in our case, we will now proceed to **Render Animation** to export and render the animation. Select **Render Animation** from the ribbon.

31. Select the **Output** tab. Select **Browse** and save the file as a `.avi` file in the `Drone Animation` folder. Then, change **Time Range** to 0–5 seconds and increase **Frame Rate** to 25 frames per second.

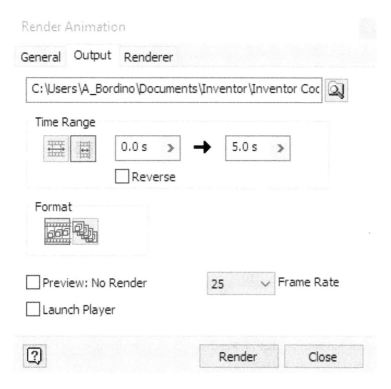

Figure 9.27: Render Animation output

32. Click on the **Renderer** tab. Change the iteration from **400** to **15** and the accuracy from **High** to **Low**. If you need a more polished render, increasing the iterations and quality will achieve this. For this recipe, low quality will suffice, as that will reduce the processing time.

Figure 9.28: Renderer options

33. This is optional, but you can click on **Renderer**. In the dropdown, select **Uncompressed** as the compressor, and then select **OK**.

Figure 9.29: Video Compression options

34. The render will take a few minutes to complete each frame. Once finished, play the `.avi` video file created. Anytime in the **Inventor Studio** environment, you can navigate back to the **Inventor Assembly** environment and continue modeling or making changes. All settings and animations applied in the **Inventor Studio** environment will be contained within the part or assembly file.

Using **Inventor Studio**, you have successfully created an animation of a mini drone taking off vertically in an Inventor environment, with custom camera animations and lighting. The result has been exported as a `.avi` video format.

Presentation files – creating exploded views and animations of assemblies

Engineers and designers must frequently produce exploded views of products or assemblies to communicate design intent. Often, it is a requirement to produce an animation of this. Inventor has specific tools and file formats to enable the efficient creation of such animations. These are known as **presentations**.

Presentations (`.ipn` files) can be created from any assembly with two or more parts contained within it. In this recipe, you will create a **presentation file** of a toy figure, showing all its parts exploding outward. The result will be exported as a `.avi` video. The recipe will also show how you can use **snapshot views** within the **Presentation** environment to create an exploded view for a drawing file.

Getting ready

To begin this recipe, we will have to create a new `.ipn` file and then import `LMan.iam` into the presentation file. Once this is complete, we can start the process of creating an exploded animation of the product. You will need access to the `LMan.iam` file located in the `Inventor 2023 Cookbook` folder, the `Chapter 9` folder, and then select `Lman.iam`.

How to do it...

To begin, we will need to create a new `.ipn` presentation file:

1. From the Inventor **Home** screen, select **New**. Browse for `Standard (mm).ipn` and select **Standard (mm).ipn**, and finally, click on **Create**.

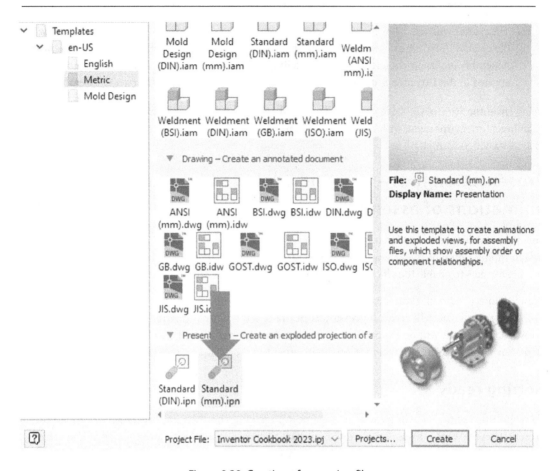

Figure 9.30: Creation of a new .ipn file

2. Browse the following location when prompted, Inventor 2023 Cookbook| Chapter 9|
    LMan, and then open LMan.iam. This imports the assembly file (.iam) into the presentation
    file (.ipn).

3. LMan.iam opens within the **Presentation** environment. From here, we can now tweak
    components using a timeline to show parts exploding from the assembly. The opacity of parts can
    also be controlled. To begin the exploded view, we will start to tweak some of the components.

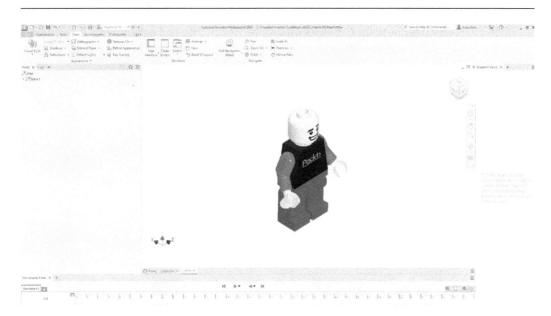

Figure 9.31: LMan.iam open in the presentation file

4. Along the timeline, you will notice **Storyboard 1** is active; this is the active storyboard we are working within. You can create multiple storyboards if required, but for this recipe, the default will be fine. At the top of the **Presentation** ribbon, select **Tweak Components**.

5. Select `LMan Head.ipt` in the **Graphics** window.

6. The following appears, as shown in *Figure 9.32*. Using the **X** and **Y** arrows, you can pull and push the component and control how you would like to animate it during the presentation. You can also type values and specify exactly the required movement. During a presentation file tweak, a part can either be moved or rotated.

Figure 9.32: Tweak components active on Head.ipt

7. Select the vertical arrow in the **Graphics** window on Head.ipt and pull upward. Type 40mm in **Distance**, and change **Duration** to 2 seconds. By keeping **Full Trail** active, a trail line will be left from where the part originated in the assembly. Then select the green tick.

Figure 9.33: A tweak of Head.ipt

8. The first tweak has been created and is populated in the storyboard. Within the storyboard area, tweaks can be edited and moved in the animation.

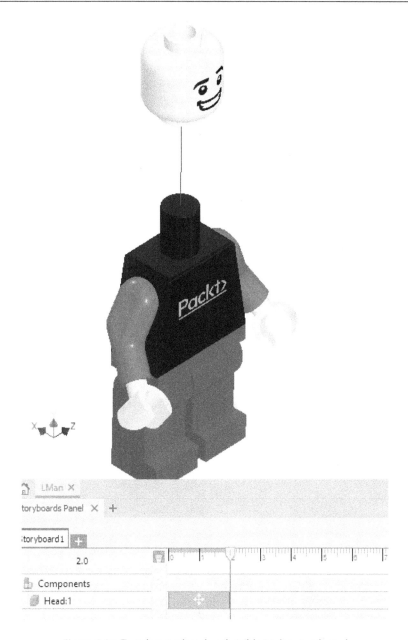

Figure 9.34: Tweak completed and visible in the storyboard

9.   Select **Play** in the timeline, and the animation will commence, showing the head moving from the body.

10. Now, select **Tweak Components**. Then, select Arm 1 and Hand 2 from **Model Browser**.

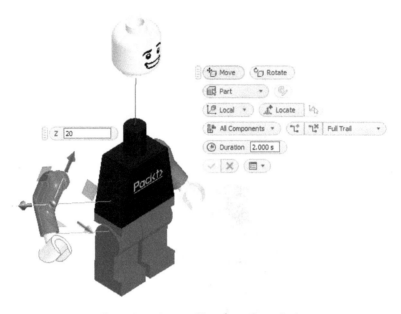

Figure 9.35: Arm and hand tweak applied

11. Pull the horizontal arrow for the arm and hand, and type 20mm in **Distance**. Ensure **Duration** is kept at 2.000 seconds. Select the green tick to complete.

12. Repeat *steps 10–11* with the other arm so that the result resembles *Figure 9.36*.

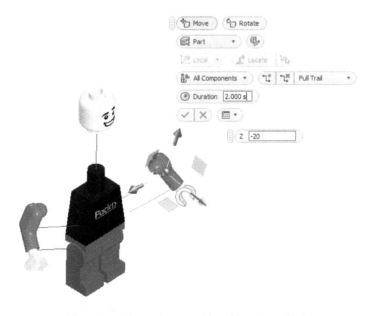

Figure 9.36: Second arm and hand tweak applied

13. By default, each **Tweak** animation happens sequentially. In the storyboard, click and drag the animation blocks that represent the arm and hand tweaks, and place them so that they have the same start and end times. This will make each arm break off from the body concurrently in the animation. *Figure 9.37* shows before and after examples. Click on **Play** to view the updated animation.

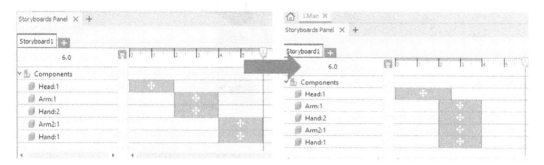

Figure 9.37: Before and after the tweaks' manual adjustment within the storyboard

14. Create two more tweaks of the Leg components so that this resembles *Figure 9.38*.

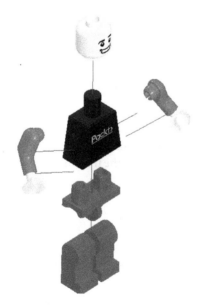

Figure 9.38: Two additional tweaks showing the leg components moving away from the body

15. Now that the additional tweaks have been created, in the timeline, edit the animation blocks so that the legs move away from the body at the same time as the head and arms. Ensure that **Duration** for all movements is set to 2.000 seconds.

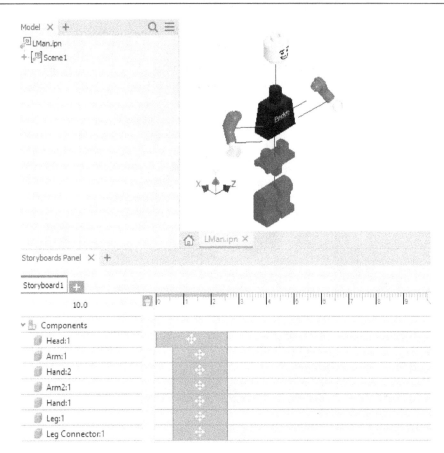

Figure 9.39: Timeline actions edited so that movement is all at the same time

16. The animation at present is only **2.6** seconds long. Drag the animation blocks for all components in the timeline to **4.5** seconds to increase the time that these parts move.

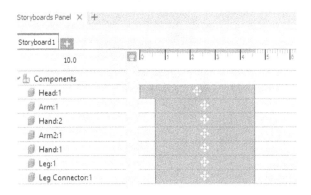

Figure 9.40: Edits to the animation blocks

17. Press **Play** to view the animation.

If edits are required to the tweaks, they can be made in various ways; one of the easiest ways is to expand **Model Browser**, as shown in *Figure 9.41*.

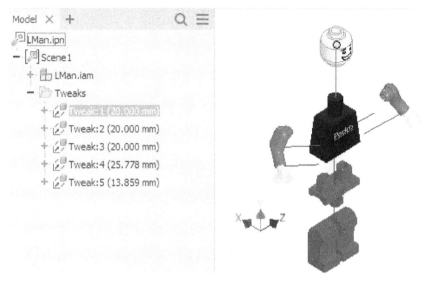

Figure 9.41: Tweaks created in Model Browser

18. Now that the animation is complete, we can further refine the display options to get the best results. Center and change your view of the model so that it resembles that of *Figure 9.42*.

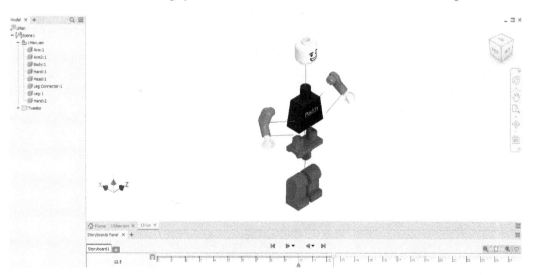

Figure 9.42: Model centered in the Graphics window

19. Navigate to and select the **View** tab. Choose **Orthographic** and select **Perspective**.

20. Select **Shadows** as **Active**. Then, select the drop-down menu and untick the **Ground Shadows** option.

21. Select **Visual Style**, and then choose **Realistic**.

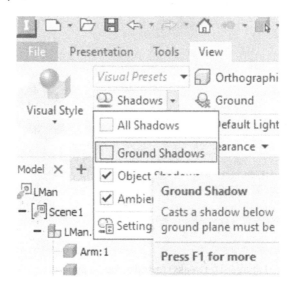

Figure 9.43: Visual display options to change

22. Currently, a white background is visible; this can be changed if desired. Select the **Tools** tab in the ribbon.

23. First, select **Application Options** and then **Colors**.

24. Within the **Colors** options, set **In-canvas Color Scheme** to **Light**, and select **Background** as **Gradient**, as shown in *Figure 9.44*.

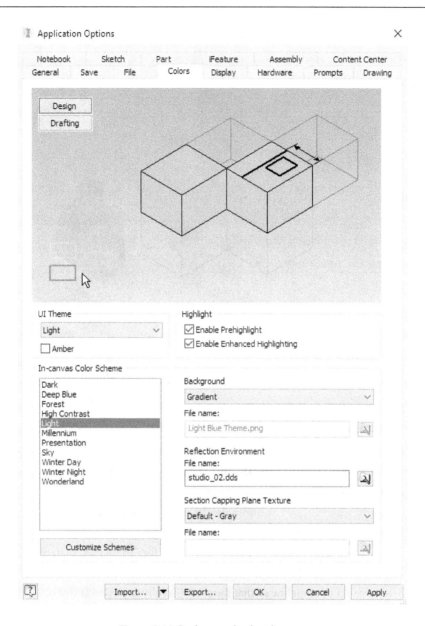

Figure 9.44: Background color changes

25. Click on **Apply**, followed by **Close**.

26. Select the **Presentation** tab in the ribbon. Select **Publish to Video**.

27. In the **Publish to Video** window, select **Current Storyboard**. Select and change **Video Resolution** to **1920 x 1080 (16:9)**. Browse the location to save the video in the Chapter 9 folder.

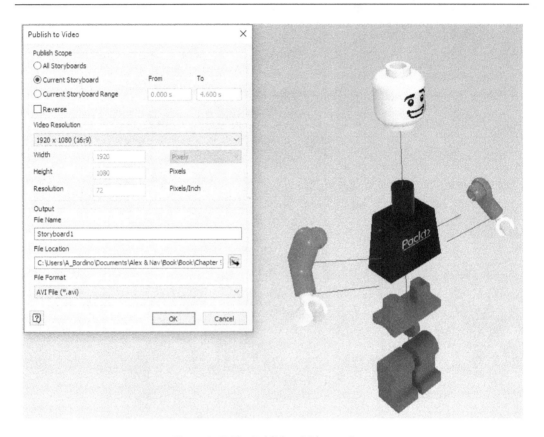

Figure 9.45: The Publish to Video options

Change **File Format** to `.avi`. Select **OK**. Then, select **Uncompressed** as the compressor. Select **OK** to continue. The video will begin to publish.

28. Open the `Storyboard1.avi` file from the `Chapter 9` folder. Play the file. The video will play and show the model exploding into separate parts.

29. Drawing views can also be created from a presentation file. To do that, first, navigate back to the `LMan.ipn` file in Inventor.

30. Select **4.7** seconds on the timeline so that the model is fully exploded at this point. Select **New Snapshot View** from the ribbon.

31. The new snapshot view is populated on the right of the screen. Right-click this and select **Create Drawing View**.

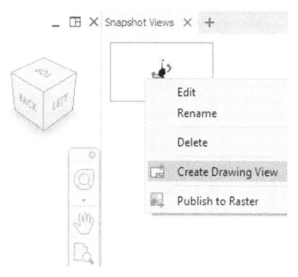

Figure 9.46: Create Drawing View from the snapshot

32. Select ISO.idw from the options. Then, select **Create**.

33. Change **Scale** to **3 : 1**, and **Style** to **Shaded**. Then, select **OK**.

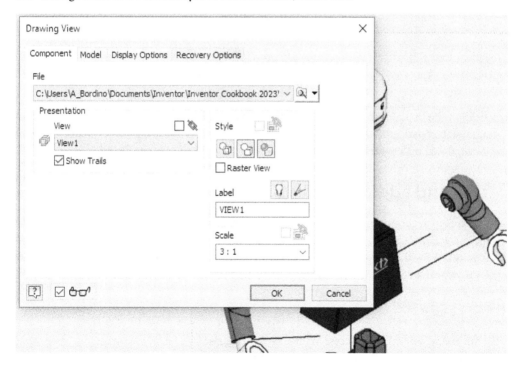

Figure 9.47: Scale and Style changed

34. The snapshot view created in the presentation file is placed into the ISO.idw drawing.

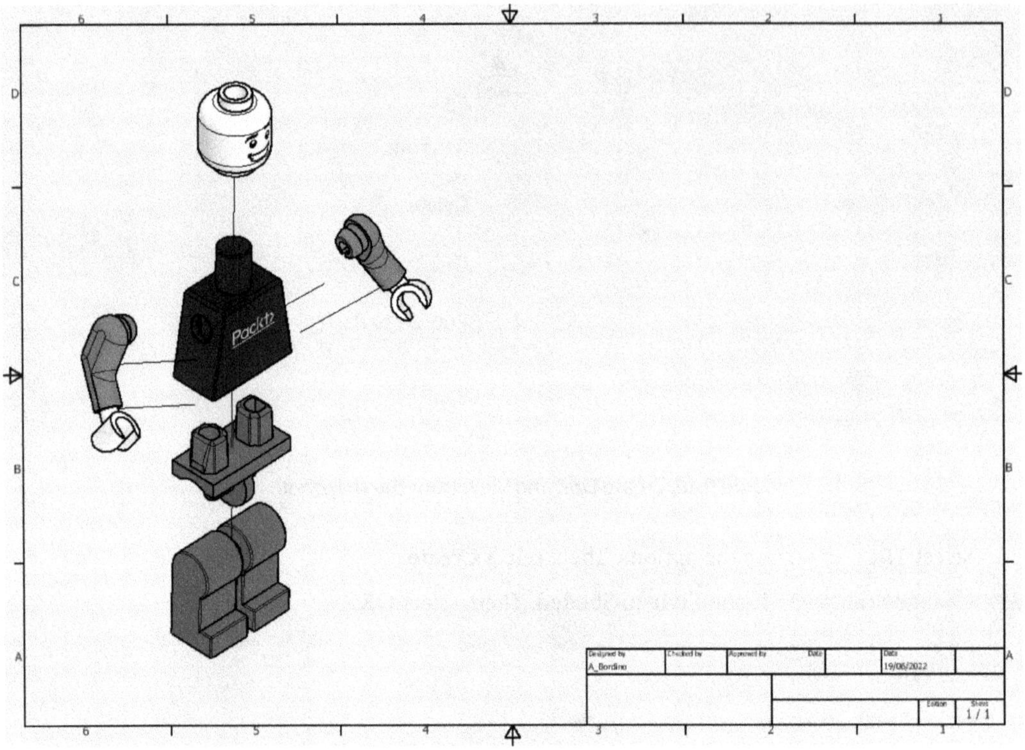

Figure 9.48: The snapshot view from the presentation file placed on the drawing sheet

A presentation file of an existing assembly has been successfully completed, showing a product exploding into its separate parts. The output has been generated as a video, and an exploded view of this has been placed on a separate drawing sheet.

Creating and adding decals and custom materials

Decals of company logos or other images are sometimes required to be featured on a CAD model. In this recipe, we will place a logo decal onto a model, placing it on various geometry types, including curved features. We will also specify and create a custom material for the model and save this in the material and appearance libraries (custom appearances can also be created in the same way).

Getting ready

To begin this recipe, you will need access to the Bearing Bracket.ipt file. Select the Inventor 2023 Cookbook folder and then select the Chapter 9 folder to find the part file.

How to do it...

To begin, we will start to create the custom material, define its properties, and also show how it will display in the model:

1. Ensure that you have `Bearing Bracket.ipt` open in Inventor.

2. Navigate to the **Quick Access** toolbar, and you will notice that the active material at present is **Iron Cast**. Select the **Material Browser** command.

Figure 9.49: The Material Browser command

3. In **Material Browser**, select **Create a New Material**.

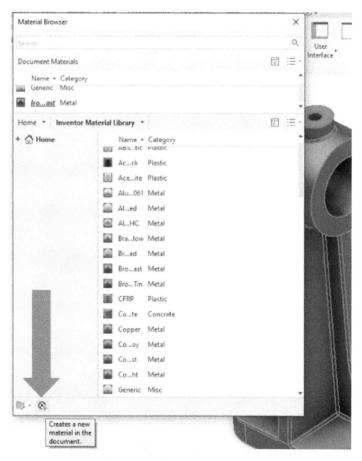

Figure 9.50: Create a New Material

4. **Material Editor** is now open. Now, we can define the properties and physical attributes of our material as well as how the appearance will appear in the model. You do not have to enter all fields; you can just change the name of the material to `Steel 101` and set **Description** as `A Generic Steel`.

5. For **Type**, select **Metal** from the drop-down menu. For **Keywords**, enter `Steel` and `101`.

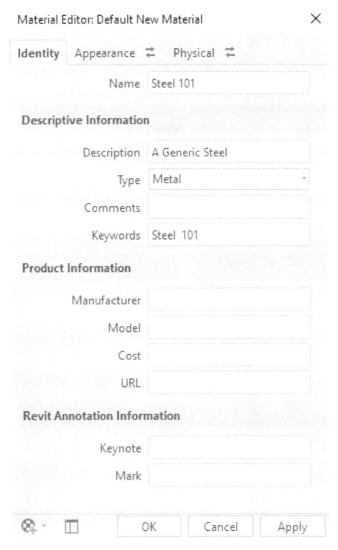

Figure 9.51: Identity fields to enter

6. Select the **Appearance** tab in **Material Editor**. Select **RGB 80 80 80**. Then, select the light gray option and select **OK**.

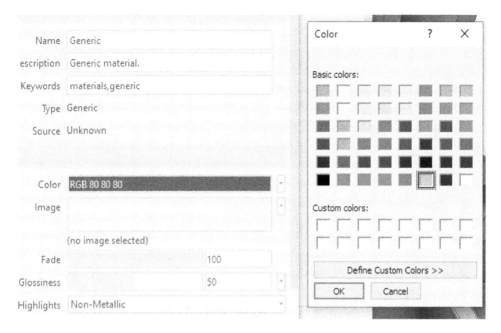

Figure 9.52: Light gray RGB selected

7. Change **Glossiness** to **75**, and for **Highlights**, select **Metallic** from the options.

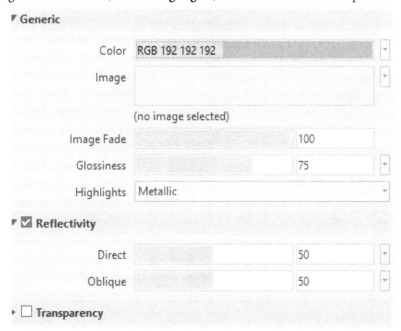

Figure: 9.53: The Glossiness, Highlights, and Reflectivity options to change

8. Tick the **Reflectivity** checkbox.

9. Select the **Physical** tab. Here, you can define the material-specific information for the **Mechanical**, **Thermal**, and **Strength** properties. In this instance, leave these as default.

10. The **Steel 101** custom material is now visible to select and apply in **Material Browser**. Select **Bearing Bracket** from **Model Browser**. Right-click on **Steel 101** in **Material Browser** and select **Assign to Selection**. The custom material is added to the part.

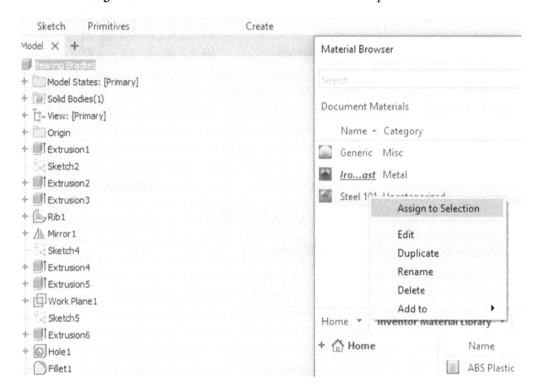

Figure 9.54: Steel 101 is assigned to Bearing Bracket.ipt

11. Close **Material Browser**.

12. We will now add two decals to the model. The first will be to a flat face, and the second we will wrap around curved geometry. Select the face of the model shown in *Figure 4.55* and select **Create Sketch**.

Figure 9.55: Create a sketch on this face

13. In the **Sketch** environment, select **Image** from the **Insert** panel in the ribbon.

Figure 9.56: The Insert Image command

14. Browse the `Chapter 9` folder, and locate and select `Packt-Logo.png`. This has already been created for you and has a transparent background. Select **Open**.

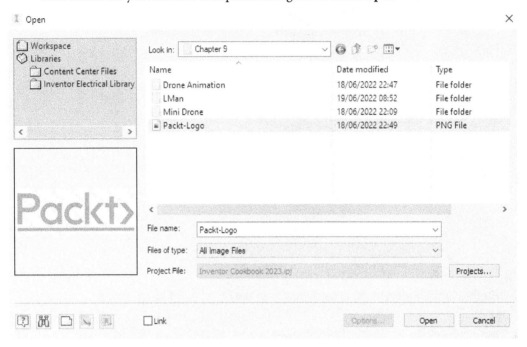

Figure 9.57: Packt-Logo selection

15. Left-click to place `Packt-Logo.png` on the face.

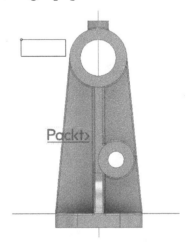

Figure 9.58: Packt-Logo.png placed on the face of the model

16. The logo now needs to be orientated and scaled correctly. This can either be done by clicking and dragging the bounding box of the image in the **Graphics** window, or by using the **Move**, **Copy**, or **Rotate** commands in the **Modify** section of the ribbon.

Figure 9.59: The Move, Copy, and Rotate commands

Scale and orientate the image using the methods outlined in *step 17* so that it resembles *Figure 4.60*.

Figure 9.60: The image correctly orientated and scaled

17. Select **Finish Sketch**. We will now convert this into a decal.

18. Select the **3D Model** tab and then **Decal**.

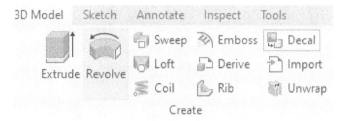

Figure 9.61: The Decal command

19. The image should be selected by default. Select the face as shown in *Figure 4.62* as the face to wrap the decal.

Figure 9.62: The image converted to a decal

20. Select **OK** to complete. The decal is completed and is also visible in **Model Browser** as an editable item.

21. Now, we will create a decal that's wrapped around a curved face. Select **Plane**, then, **Tangent to Surface Parallel to Plane**. For the **Plane** references, select the curved face highlighted in *Figure 4.63*, followed by the **XYP** plane, and create a plane as shown in *Figure 4.63* – that is, tangent to the curved surface highlighted.

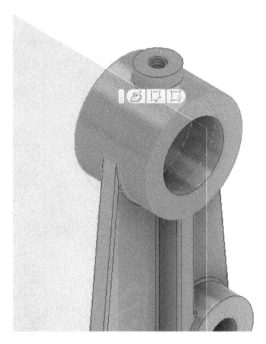

Figure 9.63: Tangent to Surface Parallel to Plane on the surface highlighted

22. The image must first be placed on a flat plane before being wrapped to a curved geometry surface. Create a new sketch on the new workplane.

23. Place an instance of Packt-Logo.png on the sketch. Orient and scale the image so that it resembles *Figure 4.64*.

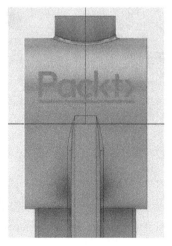

Figure 9.64: Packt-Logo.png placed on the new sketch on the newly
created Tangent to Surface Parallel to Plane workplane

24. Select **Finish Sketch**.

25. Select the **3D Model** tab, and then select **Decal**. Select the curved face shown in *Figure 4.65*. Then, select the checkbox next to **Wrap to Face**. Select **OK**.

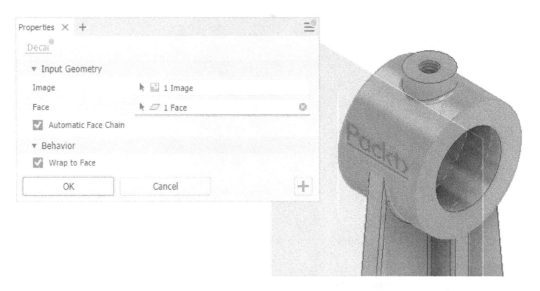

Figure 9.65: The image converted to a decal and wrapped to the face

26. The decal is successfully wrapped to the curved face of the geometry.

Figure 9.66: The decal wrapped to the curved face of the model

In this recipe, you have created a custom material with a unique appearance, applied this to the model, and created and applied decals to various types of geometry faces.

Model credits

The model credits for this chapter are as follows:

- *Drone Mini* (by MBardan): `https://grabcad.com/library/drone-mini-2`

- *Lego Man* (by Safira A): `https://grabcad.com/library/lego-man-49/details?folder_id=10688431`

- *3D CAD Basic exercise/Bearing bracket/Autodesk Inventor Pro* (by Alejandra Cervantes Tetrika México): `https://grabcad.com/library/3d-cad-basic-exercise-bearing-bracket-autodesk-inventor-pro-1`

10

Inventor iLogic Fundamentals — Creating Process Automation and Configurations

The **iLogic** functionality within Autodesk Inventor enables designers and engineers to create rules that drive model geometry, parameters, and behavior, all from a user-friendly interface utilizing **VB.NET code**.

iLogic rules are created and organized using built-in snippets and other code statements to run functionality within a part's file, assembly, or drawing. The scope of what can be achieved with iLogic is immense: complete model configurations can be created using rules that define the shape, size, suppression or unsuppression of features, material type, and more. iLogic can also be used for drawing layouts and assembly nomenclature. iLogic rules can be run internally and manually inside a file, externally from the file, or configurations can be driven from iLogic **forms** or **triggers** in the file.

iLogic's core purpose and functionality is to provide a level of automation and intelligence to digital CAD models, enabling engineers to readily generate configurations. It resides within the core Inventor product, out of the box, and does not require additional purchases or installation.

In terms of incorporating intelligence into designs, there is a progression of solutions in terms of levels of complexity:

- Parameters and equations
- iFeatures, iParts, iAssemblies, and spreadsheet-driven models
- iLogic
- VBA Macros and apps developed in the Inventor API

In this chapter, we will focus on the third level, iLogic. You must have a good understanding of the functionalities outlined in *levels 1 and 2* before proceeding with this chapter, as knowledge and experience in practical terms with these functionalities is essential for understanding and utilizing iLogic.

iLogic may seem a little intimidating to readers who have no experience in coding, but this need not be the case. The iLogic environment and use of **snippets** (ready-made pieces of code) make using iLogic out-of-the-box simple, and no previous coding experience is required.

In this chapter, we will cover the following topics:

- Understanding the fundamentals and usage of iLogic
- Creating a configurable part with iLogic rules
- Creating a configurable assembly with iLogic
- Creating an iLogic form and Event Triggers

Technical requirements

To complete this chapter, you will need access to the practice files in the `Chapter 10` folder within the `Inventor Cookbook 2023` folder. It is important that you have a good understanding of equations, parameters, iFeatures, iParts, and iAssemblies before proceeding. You can find information on these topics in *Chapter 3, Driving Automation and Parametric Modeling in Inventor*.

Understanding the fundamentals and usage of iLogic

In this recipe, you will learn how to operate and utilize an existing iLogic assembly of a ladder. You will become familiar with the iLogic browser and how you can access and edit rules to construct models. You will also make changes to the assembly using the predefined iLogic form.

iLogic itself is a large topic, but this chapter aims to introduce readers to the core functionalities. Here are some common areas where iLogic is utilized:

- Suppress or unsuppress parts or features
- Control assembly constraints
- Run multiple operations on a single input
- Make configurations
- Update iProperties of models
- Work with iFeature, iParts, and iAssemblies
- Control drawing borders and title block information

As with most features and functionality in Inventor, there is a recommended workflow. For iLogic, this is as follows:

1. Create and define the model.

2. Create the iLogic rules.

3. Set rule triggers (if applicable).

4. Edit rules.

5. Create an iLogic form (if applicable).

As iLogic rules mainly run from existing equations and parameters defined during the creation of the model, it is very important that from the outset, the model has been properly defined at all stages, and there are no unresolved parts, mates, or missing geometry/undefined sketches. The base model must be solid before progressing with the implementation of iLogic rules.

It is also advisable to have set goals and objectives of what you want your end part or configuration to be. Mapping out the options prior to starting will make the process of applying iLogic code much easier.

Creating the rules is done in the iLogic environment of Inventor. A single rule or multiple rules can be defined, and there is no limit to the number of rules that can be applied. A single rule can in fact run multiple operations. Rules can be contained within the relevant file or externally, and at any point, rules can also be edited. Conditional statements such as the following form the basics of most iLogic rules:

```
If (Something) = True Then
  (Do Something)
Else
   (Do Something Else)
End If
```

What makes iLogic easy to configure is that Inventor already has a library of snippets – pre-defined VB.NET code statements that can be applied and edited to define the rules. You can of course create your own from scratch.

When editing the rules, the model may have to be flexed and tested with the rule to ensure that it is working to the desired outcome.

Within Inventor, there are two types of iLogic rules, internal and external. Both are created within the context of Inventor, inside the iLogic browser. **Internal rules** are rules that are created and stored within the context of a file. **External rules** are not stored within Inventor files, but in a local or networked directory.

Internal rules are accessible to users that have permission to access those files. External rules have an added level of security in that they are stored in a directory either locally on a user system or centrally on a server. It is up to you to decide which type of rule you wish to create, depending on the objective.

Getting ready

Navigate to Inventor Cookbook 2023 | Chapter 10|iLogic Access Ladder.

How to do it...

To start the recipe, we will open the model and observe the predefined form that appears:

1. Open the Access Ladder.iam file in Inventor.

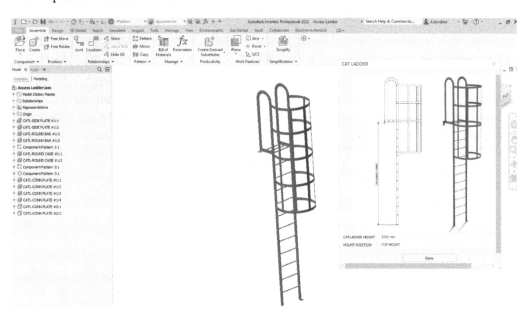

Figure 10.1: Access Ladder assembly open with the iLogic form to the right

Upon opening the file in Inventor, you will see the model displayed in the default state at the Origin. You will also notice that an iLogic form is visible and open. This iLogic form has been created to control the two internal iLogic rules within the file so that you can configure the Access Ladder.iam file efficiently without editing the VB.NET code of the rules.

In this example, you can either adjust the **HEIGHT** or **MOUNT POSITION** of the access ladder.

2. Select the dropdown next to **MOUNT POSITION** and select **SIDE MOUNT**.

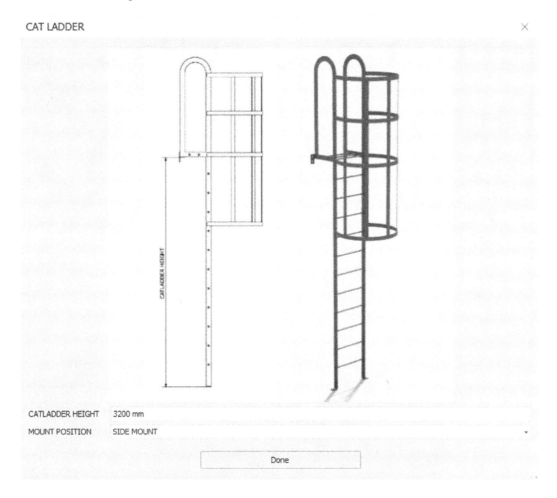

Figure 10.2: SIDE MOUNT selected on the iLogic form

Once SIDE MOUNT is selected, the assembly updates automatically to display a side mounting plate.

3. Select **LADDER HEIGHT** and type 1500mm. The model updates as shown in *Figure 10.3*. Access Ladder.iam is shortened to 1500mm in length and the guard is removed as dictated by the iLogic rules. To close the iLogic form, select **Done**.

Figure 10.3: Updated iLogic form and assembly

4.   We will now access the iLogic rules themselves and understand how they operate. From the Model Browser, select the **iLogic** tab to open the iLogic browser.

Figure 10.4: iLogic browser selected

5. Clicking on **Rules** displays active rules within this file. From the list, we have two rules that are active: **CAT LADDER RULE #1** and **OPEN**. Right-click on **OPEN** and select **Run Rule**. The iLogic form will appear on the screen. This rule is used to open the iLogic form as the Access Ladder.iam file is opened so that the user can begin configurations immediately.

6. Right-click on **CAT LADDER RULE #1** and select **Edit Rule**. The **Edit Rule** menu now opens:

Figure 10.5: Edit Rule menu open for CAT LADDER RULE #1

7. This environment is where you can create, edit, and define the iLogic rule functionality. As *Figure 10.5* shows, it's quite a busy menu, so let's break it down by looking at *Figure 10.6*:

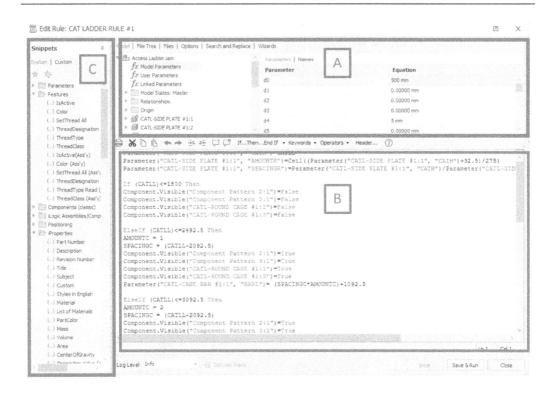

Figure 10.6: Edit Rule menu areas

This is what each menu area in *Figure 10.6* means:

A. This section is where we can access features, parts, parameters, and equations that make up the parts in our assembly. You can actively select and reference parameters, equations, dimensions, or other model features for use in your iLogic rule. Ultimately, it is parameters and equations that will provide the building blocks to most rules.

B. This is where the iLogic VB.NET code is written and displayed. This is where most of the work to create the iLogic rule takes place. There are tools to enable you to cut, paste, comment, and easily apply `If Then` and `End If` statements, operators, and keywords to help you build code and functionality quickly.

C. The **Snippets** area is where we can browse an existing library of readymade snippets of VB.NET code that you can paste, edit, and use within *section B* of *Figure 10.6* to develop the iLogic rules.

Look closely at the start of the code in *Figure 10.7*. We will now break down what the code means and how it is driving the model.

```
Parameter("CATL-SIDE PLATE #1:1", "CATH")=CATLL
Parameter("CATL-SIDE PLATE #1:1", "AMOUNTR")=Ceil((Parameter("CATL-SIDE PLATE #1:1", "CATH")+32.5)/275)
Parameter("CATL-SIDE PLATE #1:1", "SPACINGR")=Parameter("CATL-SIDE PLATE #1:1", "CATH")/Parameter("CATL-SID
|
If (CATLL)<=1800 Then
Component.Visible("Component Pattern 2:1")=False
Component.Visible("Component Pattern 3:1")=False
Component.Visible("CATL-ROUND CAGE #1:1")=False
Component.Visible("CATL-ROUND CAGE #1:3")=False

ElseIf (CATLL)<=2492.5 Then
AMOUNTC = 1
SPACINGC = (CATLL-2092.5)
Component.Visible("Component Pattern 2:1")=True
Component.Visible("Component Pattern 3:1")=True
Component.Visible("CATL-ROUND CAGE #1:1")=True
Component.Visible("CATL-ROUND CAGE #1:3")=True
Parameter("CATL-CAGE BAR #1:1", "BARL")= (SPACINGC*AMOUNTC)+1092.5

ElseIf (CATLL)<=3092.5 Then
AMOUNTC = 2
SPACINGC = (CATLL-2092.5)
Component.Visible("Component Pattern 2:1")=True
Component.Visible("Component Pattern 3:1")=True
```

Figure 10.7: Close-up of the code we will examine

The first three lines of code define how certain parameters will work:

```
Parameter("CATL-SIDE PLATE #1:1", "CATH")=CATLL
Parameter("CATL-SIDE PLATE #1:1",
    "AMOUNTR")=Ceil((Parameter("CATL-SIDE PLATE #1:1",
    "CATH")+32.5)/275)
Parameter("CATL-SIDE PLATE #1:1",
    "SPACINGR")=Parameter("CATL-SIDE PLATE #1:1",
    "CATH")/Parameter("CATL-SIDE PLATE #1:1", "AMOUNTR")
```

These sections define how the ladder rungs are placed in terms of the number and spacing.

Next are End If/Else If statements, which define how these parameters should behave, depending on the critical CATLL value (which is the ladder's height) entered in the iLogic form:

```
If (CATLL)<=1800 Then
Component.Visible("Component Pattern 2:1")=False
Component.Visible("Component Pattern 3:1")=False
Component.Visible("CATL-ROUND CAGE #1:1")=False
Component.Visible("CATL-ROUND CAGE #1:3")=False
```

This states that if the value of CATLL (the ladder height) is less than 1800mm, then the following features are to be suppressed, and this includes the round protective cage. This is what we demonstrated in *step 3* when we defined **LADDER HEIGHT** as 1500mm in the iLogic form.

The next portion of the code states that if the value of **LADDER HEIGHT** is set to <2492.5, then create the cage but only with the values and spacings shown here:

```
ElseIf (CATLL)<=2492.5 Then
AMOUNTC = 1
SPACINGC = (CATLL-2092.5)
Component.Visible("Component Pattern 2:1")=True
Component.Visible("Component Pattern 3:1")=True
Component.Visible("CATL-ROUND CAGE #1:1")=True
Component.Visible("CATL-ROUND CAGE #1:3")=True
Parameter("CATL-CAGE BAR #1:1", "BARL")=
   (SPACINGC*AMOUNTC)+1092.5
```

The next parts of the preceding code starting with `Component.Visible` are to calculate the size of the round cage section as the height of the ladder increases.

This next section of code defines how the side or top mounting plates are visible; it relates to the drop-down menu in the iLogic form, defining which components are visible or suppressed:

```
If BASE1 = "TOP MOUNT " Then
Component.Visible("CATL-CONN PLATE #1:1")= True
Component.Visible("CATL-CONN PLATE #1:2")= True
Component.Visible("CATL-CONN PLATE #2:1")= False
Component.Visible("CATL-CONN PLATE #2:2")= False

End If


If BASE1 = "SIDE MOUNT" Then
Component.Visible("CATL-CONN PLATE #1:1")=False
Component.Visible("CATL-CONN PLATE #1:2")=False
Component.Visible("CATL-CONN PLATE #2:1")=True
Component.Visible("CATL-CONN PLATE #2:2")=True
```

The final section of code sets a limit to the configuration in that if a value for **LADDER HEIGHT** in the iLogic form is set to <18,000, then the model displays a working error message "TOO BIG, LOWER LADDER HEIGHT", "ERROR":

```
If (CATLL)>18000 Then
MessageBox.Show("TOO BIG, LOWER LADDER HEIGHT",
```

```
    "ERROR")
  End If
```

This communicates to the user that a lower value must be entered to proceed and compute.

1. We will now test the final line of code. Select **Close** in the **Edit Rule** menu.

2. From the iLogic browser, select **Forms**, then **CATLADDER**.

3. Choose **CATLADDER HEIGHT** and enter a value of 20,000. Hit *Enter*.

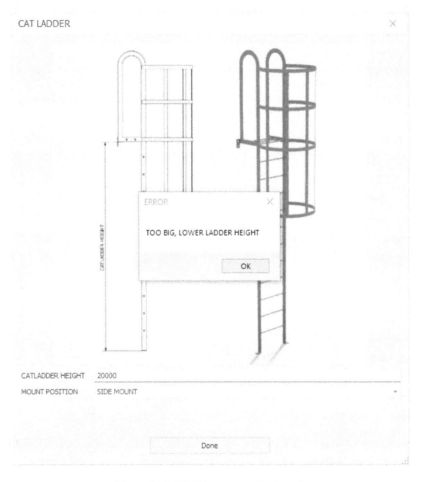

Figure 10.8: ERROR message displayed

The rule is executed as soon as the value is higher than 18,000, displaying the warning message.

4. Now we will add an additional element to the rule that changes the color of all components of the configuration when it is greater than 3092.5 in length. Select the iLogic browser, right-click on **CAT LADDER RULE #1**, and select **Edit Rule**.

Scroll down the code until you reach the section shown in *Figure 10.9* and select where the arrow is pointing. This is where the cursor will then deploy, and we will write new functionality within the rule.

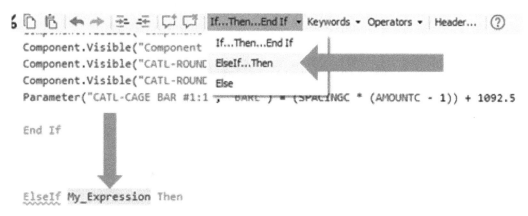

```
             If...Then...End If ▾ Keywords ▾ Operators ▾ Header...    (?)
ElseIf (CATLL)>3092.5 Then
inc = 1
AMOUNTC = Ceil(((CATLL-2092.5)/1000)+1)
SPACINGC=(CATLL-2092.5)/Ceil((CATLL-2092.5)/1000)
Component.Visible("Component Pattern 2:1")=True
Component.Visible("Component Pattern 3:1")=True
Component.Visible("CATL-ROUND CAGE #1:1")=True
Component.Visible("CATL-ROUND CAGE #1:3")=True
Parameter("CATL-CAGE BAR #1:1", "BARL") = (SPACINGC * (AMOUNTC - 1)) + 1092.5

End If
```

Figure 10.9: Section in the rule where we will write new code

5. Hit *Enter* a few times to create some space between the sections of code. It is important to keep your code organized neatly, as this makes understanding it and future edits much easier.

6. Select **If…Then…End If**, from the toolbar in Rule Editor. This adds the conditional statement to your code without you having to type anything. This is the basis of our new function within the rule.

```
               If...Then...End If ▾ Keywords ▾ Operators ▾ Header...    (?)
Component.Visible("Component    If...Then...End If
Component.Visible("CATL-ROUND   ElseIf...Then
Component.Visible("CATL-ROUND   Else
Parameter("CATL-CAGE BAR #1:1                (SPACINGC * (AMOUNTC - 1)) + 1092.5

End If



ElseIf My_Expression Then
```

Figure 10.10: If…Then…End If statement added to the code

7. We now need to change `My-Expression` to a parameter and a value. Select `My_Expression` and type `If (CATLL) >3092.5 Then`. This states that if the CATLL parameter value entered is greater than `3092.5` then do something.

8. Now we must define what it is exactly we want the code to do when the value of the CATLL parameter reaches the threshold. After Then in the code, hit *Enter* to start a new line.

9. In the **Snippets** area of the **Edit Rule** window, select the Components folder, and select **Color** by double-clicking. The default snippet is inserted into the code. This snippet allows us to define what color the component should change to when the rule is executed.

Figure 10.11: Component.Color snippet added to the code

10. Change Component.Color("PartA:1") to Component.Color("CATL-SIDE PLATE #1:1") = "Cyan". Now we have defined the component that we want to change the color of when the rule is run, and we have specified a color found in the Appearance Library.

We now need to repeat this for all other components we want to follow this rule. Add additional lines of code as follows. You can view the components in the model in the **Model** and **File Tree** area of the **Edit Rule** menu, shown as *section A* in *Figure 10.6*.

Type the following code into the iLogic Rule, or use copy and paste from `CAT LADDER #1 iLOGIC RULE COMPLETE.txt` in the `iLogic AccessLadder` folder:

```
If (CATLL) >3092.5 Then
Component.Color("CATL-SIDE PLATE #1:1") = "Cyan"
Component.Color("CATL-SIDE PLATE #1:2") = "Cyan"
Component.Color("CATL-ROUND BAR #1:1") = "Cyan"
Component.Color("CATL-ROUND BAR #1:2") = "Cyan"


Component.Color("Component Pattern 1:1") = "Cyan"


Component.Color("CATL-ROUND BAR #1:1") = "Cyan"
Component.Color("CATL-ROUND BAR #1:2") = "Cyan"


Component.Color("CATL-ROUND CAGE #1:1") = "Cyan"
Component.Color("CATL-ROUND CAGE #1:3") = "Cyan"
Component.Color("Component Pattern 2:1") = "Cyan"
Component.Color("Component Pattern 3:1") = "Cyan"


Component.Color("CATL-CONN PLATE #1:1") = "Cyan"
Component.Color("CATL-CONN PLATE #1:2") = "Cyan"
Component.Color("CATL-CONN PLATE #1:3") = "Cyan"
Component.Color("CATL-CONN PLATE #1:4") = "Cyan"
Component.Color("CATL-CONN PLATE #2:1") = "Cyan"
Component.Color("CATL-CONN PLATE #2:2") = "Cyan"


End If
```

11. With all the components that we want to change the color of defined for this rule, select **Save & Run**. Inventor will check your code for errors and syntax before completing and will warn you if there are any errors. Be mindful as spelling is key here. Close the **Edit Rule** menu. We will now test the new piece of code.

12. Open the iLogic form for the `Access Ladder.iam` file. Set **CATLADDER HEIGHT** to 3500. Hit *Enter*. The model updates as per the code and all components turn cyan in color. The rule has run successfully with the edits we made.

    In the iLogic form, change **CATLADDER HEIGHT** to 2200 and hit *Enter*. The model updates correctly in terms of geometry, but the color for all components is left as cyan. Our rule stated that this should only occur if the value of **CATLADDER HEIGHT** is greater than 3092.5. This is because we are lacking an `Else If` statement in our code. We need to define all possible scenarios so that if the value is not greater than 3092.5, then the color should be **Canary**.

13. In the iLogic Browser, right-click **CAT LADDER RULE #1** and select **Edit Rule**.

14. Add the following after `Component.Color("CATL-CONN PLATE #2:2") = "Cyan"`:

```
Else
Component.Color("CATL-SIDE PLATE #1:1") = "Canary"
Component.Color("CATL-SIDE PLATE #1:2") = "Canary"
Component.Color("CATL-ROUND BAR #1:1") = "Canary"
Component.Color("CATL-ROUND BAR #1:2") = "Canary"

Component.Color("Component Pattern 1:1") = "Canary"

Component.Color("CATL-ROUND BAR #1:1") = "Canary"
Component.Color("CATL-ROUND BAR #1:2") = "Canary"

Component.Color("CATL-ROUND CAGE #1:1") = "Canary"
Component.Color("CATL-ROUND CAGE #1:3") = "Canary"
Component.Color("Component Pattern 2:1") = "Canary"
Component.Color("Component Pattern 3:1") = "Canary"

Component.Color("CATL-CONN PLATE #1:1") = "Canary"
Component.Color("CATL-CONN PLATE #1:2") = "Canary"
Component.Color("CATL-CONN PLATE #1:3") = "Canary"
Component.Color("CATL-CONN PLATE #1:4") = "Canary"
Component.Color("CATL-CONN PLATE #2:1") = "Canary"
Component.Color("CATL-CONN PLATE #2:2") = "Canary"
End If
```

This informs Inventor that if the value is less than `3092.5` in the iLogic form, then the **Component Color** of all components must be set to `Canary`.

15. Now we need to check the entirety of the code. Ensure that your code matches the completed code in `Inventor Cookbook 2023 | Chapter 10 | iLogic Access Ladder | CAT LADDER #1 iLOGIC RULE COMPLETE.txt`.

    Once your code matches that of the preceding `.txt` file, select **Save & Run**. Close the **Edit Rule** menu. The assembly should update with all components canary in color. In the iLogic form, set **CATLADDER HEIGHT** to `3100` and hit *Enter*. The model geometry updates and the color is changed back to cyan.

You have configured an iLogic model using an iLogic form to drive an Inventor assembly with more than one iLogic rule. You have also edited one of the rules to add additional functionality using snippets, and `If Then/End If` conditional statements, to drive a global color change across all components in the assembly, depending on user input of one of the parameters through the iLogic form.

Creating a configurable part with iLogic rules

In this recipe, the objective is to take an existing support bracket design with existing parameters and equations defined and create a new iLogic rule with multiple functions that can generate three different design variants: Heavy, Regular, and Light. The rule will be run through a custom user-defined multi-value parameter.

Getting ready

To start the recipe, we will open the model and examine the existing equations and parameters that have been set. We will also establish the design criteria for the finished parts and what the final configurations will look like.

How to do it...

To begin this recipe, you will need to navigate to Inventor Cookbook 2023 | Chapter 10 | iLogic Support Bracket and open Support Bracket.ipt:

1. Open Support Bracket.ipt.

    The bracket is displayed in its default regular arrangement. 3D annotations are visible in the **Graphics** window to illustrate the changes in dimensions. In *Figure 10.12*, you can see the three design variants (Light, Regular, and Heavy) that we will generate from the iLogic rule that we will create.

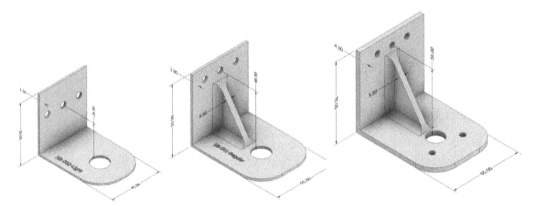

Figure 10.12: Final configurations of each support bracket generated
by the iLogic Rule: Light, Regular, and Heavy

2. There are currently no iLogic rules defined in this model. Select the **Manage** tab, and select **Parameters**. We will now examine the existing equations and parameters that are present and how they control the behavior of the model.

    Looking at the existing equations, you can see that the height value is of great importance and is featured throughout many of the equations. It is this primary dimension that drives most of the parameters and thus the behavior in the model.

Parameters   ✕

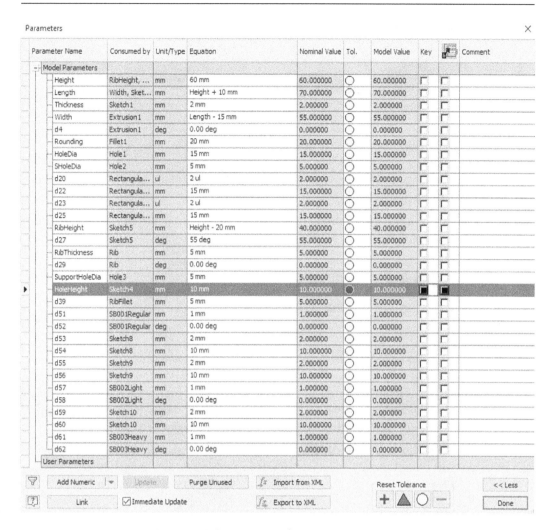

Parameter Name	Consumed by	Unit/Type	Equation	Nominal Value	Tol.	Model Value	Key	🖼	Comment
−┤Model Parameters									
Height	RibHeight, ...	mm	60 mm	60.000000	○	60.000000	☐	☐	
Length	Width, Sket...	mm	Height + 10 mm	70.000000	○	70.000000	☐	☐	
Thickness	Sketch1	mm	2 mm	2.000000	○	2.000000	☐	☐	
Width	Extrusion1	mm	Length - 15 mm	55.000000	○	55.000000	☐	☐	
d4	Extrusion1	deg	0.00 deg	0.000000	○	0.000000	☐	☐	
Rounding	Fillet1	mm	20 mm	20.000000	○	20.000000	☐	☐	
HoleDia	Hole1	mm	15 mm	15.000000	○	15.000000	☐	☐	
SHoleDia	Hole2	mm	5 mm	5.000000	○	5.000000	☐	☐	
d20	Rectangula...	ul	2 ul	2.000000	○	2.000000	☐	☐	
d22	Rectangula...	mm	15 mm	15.000000	○	15.000000	☐	☐	
d23	Rectangula...	ul	2 ul	2.000000	○	2.000000	☐	☐	
d25	Rectangula...	mm	15 mm	15.000000	○	15.000000	☐	☐	
RibHeight	Sketch5	mm	Height - 20 mm	40.000000	○	40.000000	☐	☐	
d27	Sketch5	deg	55 deg	55.000000	○	55.000000	☐	☐	
RibThickness	Rib	mm	5 mm	5.000000	○	5.000000	☐	☐	
d29	Rib	deg	0.00 deg	0.000000	○	0.000000	☐	☐	
SupportHoleDia	Hole3	mm	5 mm	5.000000	○	5.000000	☐	☐	
HoleHeight	Sketch4	mm	10 mm	10.000000	○	10.000000	■	■	
d39	RibFillet	mm	5 mm	5.000000	○	5.000000	☐	☐	
d51	SB001Regular	mm	1 mm	1.000000	○	1.000000	☐	☐	
d52	SB001Regular	deg	0.00 deg	0.000000	○	0.000000	☐	☐	
d53	Sketch8	mm	2 mm	2.000000	○	2.000000	☐	☐	
d54	Sketch8	mm	10 mm	10.000000	○	10.000000	☐	☐	
d55	Sketch9	mm	2 mm	2.000000	○	2.000000	☐	☐	
d56	Sketch9	mm	10 mm	10.000000	○	10.000000	☐	☐	
d57	SB002Light	mm	1 mm	1.000000	○	1.000000	☐	☐	
d58	SB002Light	deg	0.00 deg	0.000000	○	0.000000	☐	☐	
d59	Sketch10	mm	2 mm	2.000000	○	2.000000	☐	☐	
d60	Sketch10	mm	10 mm	10.000000	○	10.000000	☐	☐	
d61	SB003Heavy	mm	1 mm	1.000000	○	1.000000	☐	☐	
d62	SB003Heavy	deg	0.00 deg	0.000000	○	0.000000	☐	☐	
└User Parameters									

▽	Add Numeric ▾	Update	Purge Unused	ƒx Import from XML	Reset Tolerance		<< Less
[?]	Link	☑ Immediate Update		ƒx Export to XML	＋ ▲ ○ ─		Done

Figure 10.13: Parameters and equations in the model

Now we will create a user-defined parameter that will enable us to choose between Light, Regular, and Heavy variants of the bracket. This user-defined parameter is what we will reference in the iLogic rule that we will create.

3. In the **Parameters** window, click the drop-down arrow next to **Add Numeric** and select **Add Text**.

4. Rename **User Parameter** BracketType. Right-click on **Text** in the **Unit/Type** column for **BracketType** and select **Make Multi-Value** from the options.

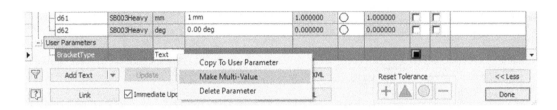

Figure 10.14: User-defined parameter BracketType created and Make Multi-Value selected

5. **Value List Editor** opens. Type Light, Regular, and Heavy into the **Add New Items** section and select **Add**. Then select **OK**.

    The multi-value options defined in the preceding step are added to the user-defined custom parameter: BracketType.

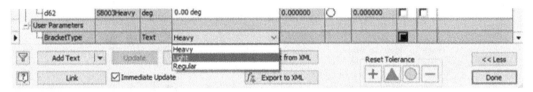

Figure 10.15: Multi-value options for BracketType added

6. Cycling through the options currently will have no effect on the model as an iLogic rule driving the changes has not yet been specified. Change the value to **Regular** as this model is currently in the state of the Regular bracket. Select **Done** to close the **Parameter** window.

7. Select the **iLogic** tab next to the **Model** tab in the Model Browser. Right-click on the empty space and select **Add Rule**. We will now define the rule to create the configurations.

Figure 10.16: Add rule

8. Name the rule Bracket Config and select **OK**.

Before starting the rule, we will outline our objective and design requirements as **comments** first. Comments are non-actionable pieces of code that enable you to communicate information in iLogic rules. It is good practice to do this because in the future, if further edits are required, you will be able to get a good understanding of the purpose of the rule without needing to read through the whole code. Comments are also helpful for structuring your iLogic code. Neat and tidy code is good code!

9. Type the following into the code writing area (you can also copy and paste this section from the `support bracket.dox` file into the iLogic Support Bracket folder):

```
'An iLogic Rule To define 3 Variants of the Support
  Bracket .ipt
'Light, Regular and Heavy BracketType

'Regular
'Light
'Heavy

'Regular =
'Height Of 60
'Rib Active
'Material Thickness Of 2 mm
'Additional Support Holes Inactive On rounded plate
'SB-001-Regualar Serial Number Stamped

'Light =
'Height Of 50
'Rib Inactive
'Material Thickness Of 1.5 mm
'Additional Support Holes Inactive On rounded plate
'SB-002-Light Serial Number Stamped
'Hole Height On back plate lowered 10mm


'Heavy =
'Height Of 70
'Rib Active
'Rib Fillet Active
```

```
'Rib Thickness 5mm
'Material Thickness Of 4 mm
'Additional Support Holes Active On rounded plate
'SB-003-Heavy Serial Number Stamped
'Hole Height On back plate lowered 10mm
```

10. This now defines what we want to achieve in the iLogic rule. Select all the code and select **Comment out the selected lines** to turn the text into a comment within the rule. This is shown in *Figure 10.17*.

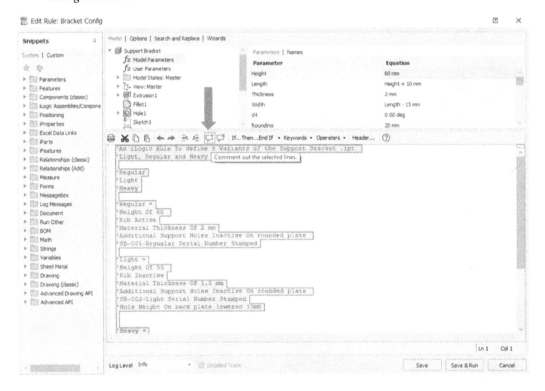

Figure 10.17: Comment out the selected lines

11. We will now start to write functionality into the iLogic rule, starting with a conditional statement. The first section of the code will define the Regular support bracket configuration. Place the cursor after the comment section and hit *Enter* twice.

12. Type the following commands into the code area:

```
'Regular Bracket Config
If BracketType = "Regular" Then
```

```
        Height = 60
        Thickness = 2
'       Rib and Hole feature controls:
        Feature.IsActive("Rib") = True
        Feature.IsActive("Hole3") = False
        Feature.IsActive("Mirror1") = False
        Feature.IsActive("Hole2") = True
        Feature.IsActive("Rectangular Pattern1") = True
        Feature.IsActive("Rectangular Pattern2") = True
        HoleHeight = 10
'       Serial Number:
        Feature.IsActive("SB001Regular")=True
```

Let's break down the code. The following segment of code states that if the part is set to Regular, then the Height must be 60 and the Thickness of the material must be 2 mm:

```
If BracketType = "Regular" Then
        Height = 60
        Thickness = 2
```

The following segment controls what features are to remain active or inactive at the selection of the Regular variant:

```
Rib and Hole feature controls:
        Feature.IsActive("Rib") = True
        Feature.IsActive("Hole3") = False
        Feature.IsActive("Mirror1") = False
        Feature.IsActive("Hole2") = True
        Feature.IsActive("Rectangular Pattern1") = True
        Feature.IsActive("Rectangular Pattern2") = True
        HoleHeight = 10
```

This mainly controls the size and positioning of the central rib and hole features.

The final section enables the correct serial number to be displayed on the bracket:

```
Serial Number:
        Feature.IsActive("SB001Regular")=True
```

13. Now highlight the section of iLogic code that you created in *step 12*. Copy and paste a new instance of this below the previous code added. Then change the values and statements of the copied code to detail the Light variant of the configuration; it should look like the following code:

```
Light Bracket Config:
Else If BracketType = "Light" Then
    Height = 50
    Thickness = 1.5
'    Rib and Hole feature controls:
    Feature.IsActive("Rib") = False
    Feature.IsActive("Hole3") = False
    Feature.IsActive("Mirror1") = False
    Feature.IsActive("Hole2") = True
    Feature.IsActive("Rectangular Pattern1") = True
    Feature.IsActive("Rectangular Pattern2") = True
    HoleHeight = 20
'    Serial Number:
    Feature.IsActive("SB001Regular") = False
    Feature.IsActive("SB002Light")=True
```

14. Copy the preceding snippet of code for `Heavy Bracket Config` and paste below the previous code as before. Then change the values to `Heavy` and change `Height`, `Thickness`, and all other values as per the following code:

```
'Heavy Bracket Config
Else If BracketType = "Heavy" Then
    Height = 70
    Thickness = 4
'    Rib and Hole feature controls:
    Feature.IsActive("Rib") = True
    Feature.IsActive("Hole3") = True
    Feature.IsActive("Mirror1") = True
    Feature.IsActive("Hole2") = True
    Feature.IsActive("Rectangular Pattern1") = True
    Feature.IsActive("Rectangular Pattern2") = True
    Feature.IsActive("RibFillet") = True
    HoleHeight = 10
    d39 = 5
```

```
'     Rib Thickness change:
     RibThickness = 5
'     Serial Number:
     Feature.IsActive("SB001Regular") = False
     Feature.IsActive("SB002Light") = False
     Feature.IsActive("SB003Heavy") = True


End If
```

We have now defined all the code required for the iLogic rule and created the custom user-defined parameter to create the three configurations of the bracket automatically. Check that the iLogic code you have written is identical to that found in the following document: support bracket.docx located in the iLogic Support Bracket folder.

15. Select **Save & Run**. The **Bracket Config** rule is created, which drives multiple functions to drive the Support Bracket.iam configuration.

16. Select the **Manage** tab and then **Parameters**. Navigate to the **BracketType** user parameter and select **Light**. The rule will run, and the model will update to the Light configuration. Select **Heavy**, and the same again will happen, only this time the Heavy variant is produced. All three design variants can be cycled using the BracketType parameter, which executes the *Bracket Config* iLogic rule.

You have successfully created a new iLogic rule with multiple functions that is driven by a user-defined, multi-value parameter within the model and creates three different design configurations of a support bracket.

Creating a configurable assembly with iLogic

iLogic rules can also be referenced in a top-level assembly or new rules can be created at this level. In this recipe, you will both reference existing part-level iLogic rules and create new iLogic rules in a top-level assembly to control the configurations of a lamp assembly.

The rule you will create will allow the user to automatically configure the lamp in Inventor in terms of the following aspects: lamp height, lamp length, sleeve color, base shape type, and part numbering.

An example of one of the configurations is shown in *Figure 10.18*, with labels that correlate to the features and references we will be using as part of the iLogic rule.

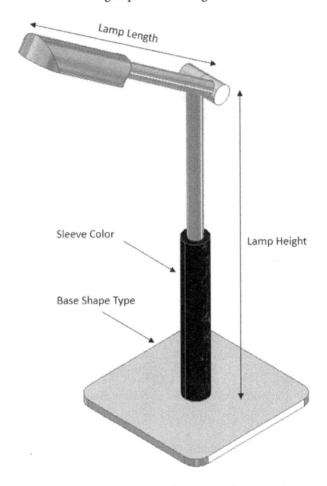

Figure 10.18: Lamp with references to things that will change in the rule to be created in the assembly

Getting ready

To begin this recipe, you will need to navigate to Inventor Cookbook 2023, then select the Chapter 10 folder, followed by the iLogic Lamp folder. Then open iLogic Lamp.iam.

How to do it...

To start the recipe, we will open the model and understand the iLogic rules that have been created and how these influence the model. You will also need to familiarize yourself with the table shown in *Figure 10.19*, as this details all the configurations of the lamp that our new iLogic rule will control, from the assembly.

Lamp Configuration Name	Lamp Height /mm	Lamp Length /mm	Sleeve Color	Base Shape Type	iProperties- Part Number
Long Circle Black	280	120	Black	Circle	LongCircleBlack
Long Circle Blue	280	120	Blue	Circle	LongCircleBlue
Long Circle Cork	280	120	Cork	Circle	LongCircleCork
Medium Circle Black	240	70	Black	Circle	MediumCircleBlack
Medium Circle Blue	240	70	Blue	Circle	MediumCircleBlue
Medium Circle Cork	240	70	Cork	Circle	MediumCircleCork
Small Circle Black	220	50	Black	Circle	SmallCircleBlack
Small Circle Blue	220	50	Blue	Circle	SmallCircleBlue
Small Circle Cork	220	50	Cork	Circle	SmallCircleCork
Long Square Black	280	120	Black	Square	LongSquareBlack
Long Square Blue	280	120	Blue	Square	LongSquareBlue
Long Square Cork	280	120	Cork	Square	LongSquareCork
Medium Square Black	240	70	Black	Square	MediumSquareBlack
Medium Square Blue	240	70	Blue	Square	MediumSquareBlue
Medium Square Cork	240	70	Cork	Square	MediumSquareCork
Small Square Black	220	50	Black	Square	SmallSquareBlack
Small Square Blue	220	50	Blue	Square	SmallSquareBlue
Small Square Cork	220	50	Cork	Square	SmallSquareCork

Figure 10.19: Configurations of the lamp that the iLogic rule will generate

Now that we understand all the configuration requirements, we can now start to create the iLogic code to generate these by following these steps:

1. Open iLogic Lamp .iam. The assembly consists of three parts. Select the **iLogic** tab in the Model Browser.

2. There are no iLogic rules at this top level of the assembly. Instead, we will need to create a new rule here. First, we will look at each part individually and examine any iLogic rules that are already present. Select the **Model** tab in the Model Browser.

3. Right-click on the **lamp:1** part and select **Edit**. Then select the **iLogic** tab in the Model Browser. There are three iLogic rules that control this part.

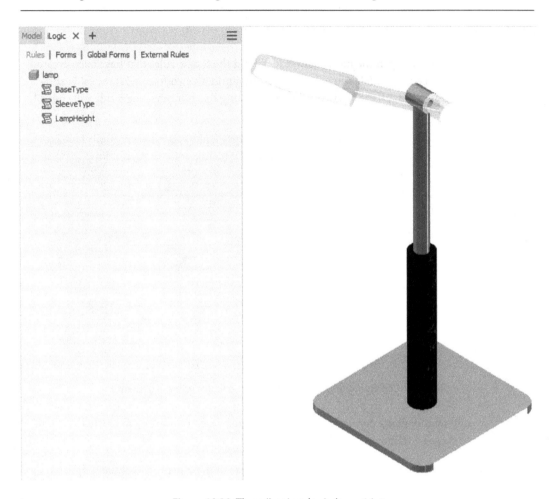

Figure 10.20: Three iLogic rules in lamp:1.ipt

4.   We will now discess how each iLogic rule functions. Right-click on the **BaseType** iLogic rule and select **Edit Rule**.

5.   Select **User Parameters** to view the user-defined parameters in the model.

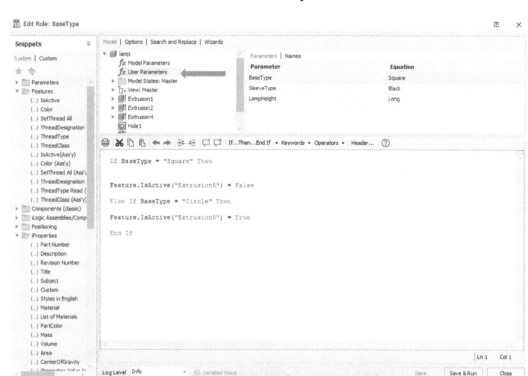

Figure 10.21: Edit Rule window for the BaseType rule open, with User Parameters selected

Looking at the iLogic code, this rule controls the shape of the base of the lamp, by either activating or deactivating an extruded cut that is part of **Extrusion5**. A conditional If Then, End If statement controls this with a True or False option.

6.   Select **Close**.

7.   Right-click on the **SleeveType** iLogic rule and select **Edit Rule**.

8.   Select **User Parameters** to view the user-defined parameters in the model.

The **SleeveType** iLogic rule controls the color of the sleeve that partially covers the vertical stand of the lamp. If the parameter in the model is changed to Black, for example, then the color of the **Sleeve** is also set to black. There are three options within this rule: Black, Blue, and Cork.

9.   Select **Close**.

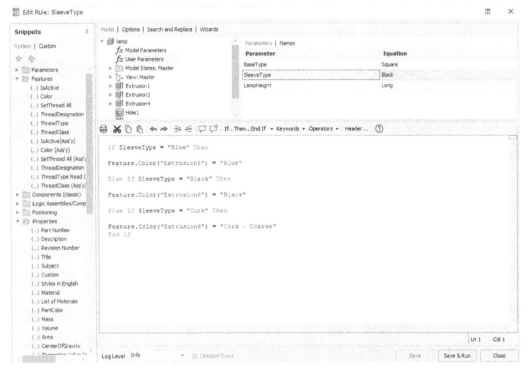

Figure 10.22: Edit Rule for SleeveType open, with User Parameters selected

10.   Right-click on the **LampHeight** iLogic rule and select **Edit Rule**.

11.   Select **User Parameters** to view the user-defined parameters in the model.

The **LampHeight** iLogic rule controls the height of the extrusion, which defines the lamp's height (the dimension name that controls lamp height is d6) and sets the exact value this must be when either **Long**, **Medium**, or **Small** is selected.

12. Select **Close**.

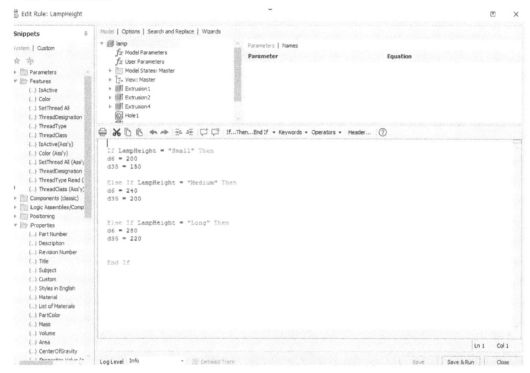

Figure 10.23: Edit Rule window for LampHeight open, with User Parameters selected

13. In the ribbon, select **Return** to return to the assembly.

14. Select the **Model** tab in the Model Browser; right-click on **lamp_prt2:1** and select **Edit**.

15. Open the **iLogic** tab in the Model Browser.

16. Right-click on **LampLength** and select **Edit Rule**.

17. Select **User Parameters** to view the user-defined parameters in the model.

The **LampLength** iLogic rule controls the length of the extrusion, which defines the lamp's length (this is dimension *d4*) and sets the exact value this must be when either **Long**, **Medium**, or **Small** is selected.

18. Select **Close**.

Figure 10.24: Edit Rule window for LampLength open, with User Parameters selected

19. In the ribbon, select **Return** to return to the assembly.

20. We now need to create an iLogic rule in the top level of iLogic Lamp.iam, which references all the iLogic rules that are at the part level, so that we can create the configurations outlined in *Figure 10.19*. Select the **Manage** tab, then select **Parameters**.

21. In the **Parameters** window, we will create a user-defined multi-value parameter that will be used to select the configuration by name. The iLogic rule we will create will reference this. Under **User Parameters**, select **Add Text Parameter**. Add the name LampConfig to the parameter.

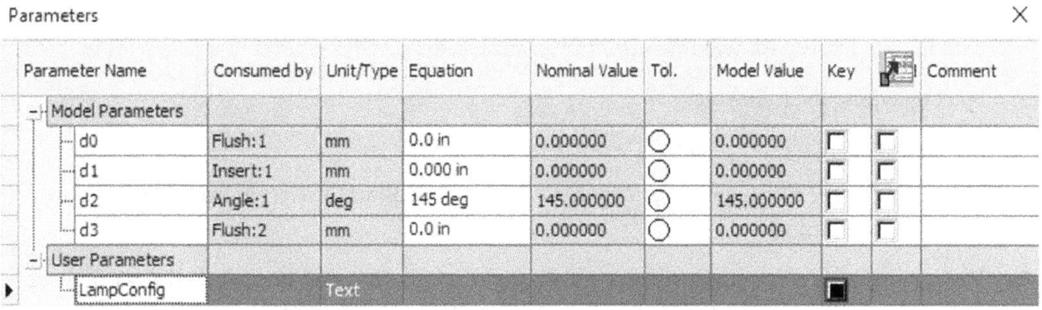

Parameter Name	Consumed by	Unit/Type	Equation	Nominal Value	Tol.	Model Value	Key		Comment
— Model Parameters									
d0	Flush:1	mm	0.0 in	0.000000	○	0.000000	☐	☐	
d1	Insert:1	mm	0.000 in	0.000000	○	0.000000	☐	☐	
d2	Angle:1	deg	145 deg	145.000000	○	145.000000	☐	☐	
d3	Flush:2	mm	0.0 in	0.000000	○	0.000000	☐	☐	
— User Parameters									
▶ LampConfig		Text						■	

Figure 10.25: LampConfig user parameter created

22. Right-click on **LampConfig** and select **Make Multi-Value**. In the **Value List Editor** window, type the following:

```
Long Circle Black
Long Circle Blue
Long Circle Cork
Medium Circle Black
Medium Circle Blue
Medium Circle Cork
Small Circle Black
Small Circle Blue
Small Circle Cork
Long Square Black
Long Square Blue
Long Square Cork
Medium Square Black
Medium Square Blue
Medium Square Cork
Small Square Black
Small Square Blue
Small Square Cork
```

23. Then select **Add**, followed by **OK**. This adds the configuration names into the user parameter dropdown for selection. Now you can click **Done**.

    We now need to write a rule to reference this parameter.

24. Select the **iLogic** tab. Right-click and select **Create New Rule**. Name the iLogic rule `LampConfig` and select **OK**. The **Rule Editor** window opens.

We now have to write iLogic code for every configuration we wish to generate from the pre-defined user parameter list. We will reference the existing rules at the part level of this assembly.

25. Start by typing the following:

```
If LampConfig = "Long Circle Black" Then
```

This identifies the rule in our assembly that if **Long Circle Black** is selected from the dropdown, then... do something.

26. We now need to reference a specific existing parameter. Expand **lamp:1** in the model tree within the **Edit Rule** window. Select **User Parameters**. With your cursor on the next line of code, double-click the **LampHeight** parameter to import it into your code. Type = "Long" after this snippet.

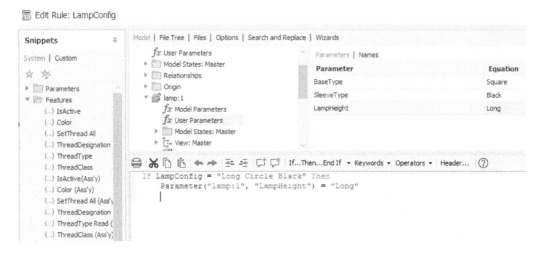

Figure 10.26: Parameter("lamp:1", "LampHeight") = "Long" added

This rule now states that if LampConfig is set to Long Circle Black, change LampHeight to the value stated in the **LampHeight** rule, that is, = Long.

27. We must now repeat this process for all other aspects of the configuration. Add the following code below Parameter("lamp:1", "LampHeight") = "Long":

```
Parameter("lamp:1", "SleeveType") = "Black"
Parameter("lamp:1", "BaseType") = "Circle"
Parameter("lamp_pt2:1", "LampLength") = "Long"
iProperties.Value("Project", "Part Number") =
   "LongCircleBlack"
```

The result of this is displayed in *Figure 10.27.*

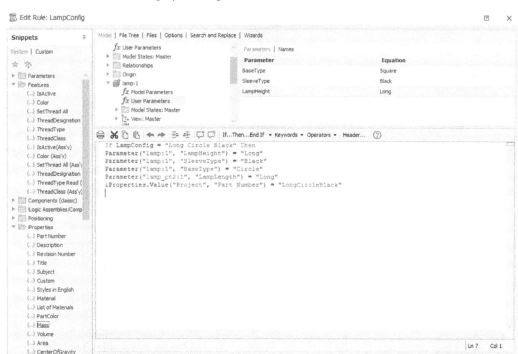

Figure 10.27: Completed code for LampConfig= Long Circle Black

The last line of code added changes Part Number in the iProperties of the assembly to the text shown automatically when the rule runs.

This code must now be duplicated and edited to reflect all the other possible configurations in the **LampConfig** dropdown.

28. The following code needs to now be added below the current paragraph of code shown in *Figure 10.27*. You can also copy and paste this from iLogic Lamp Code.docx located in the Inventor Cookbook 2023|Chapter 10 folder.

29. Once you have added the code in *step 29*, select **Save & Run**.

30. Select **Parameters** and cycle through the **LampConfig** drop-down options. As you select each configuration, the iLogic rule in the assembly runs and references existing rules in the part files to generate the desired configuration of the lamp automatically.

Figure 10.28: Completed iLogic assembly; LampConfig options selected for 'Medium Square Cork'

Using iLogic rules defined in the top level of an assembly, you have created a rule that references multiple existing iLogic rules at the part level to produce automatic configurations of a lamp that enable the user to define the base type, height, length, and color, and set automatic part numbers in the iProperties.

Creating an iLogic form and Event Triggers

iLogic forms represent the simplest way for users to generate configurations of models, and they provide an easy-to-use interface to do so. In this recipe, you will create an iLogic form that opens upon the action of an **iTrigger**. The iLogic form will allow the user to customize a mounting plate, based on predefined iLogic rules and parameters in the model.

Getting ready

To begin this recipe, you will need to navigate to Inventor Cookbook 2023 folder, then select Chapter 10. Select the iLogic Form folder and then Mounting Plate.ipt.

How to do it...

The model has already been created, and there are already two iLogic rules, which are as follows:

- *Material Selection* allows the material of `Mounting Plate.ipt` to be set either as steel alloy or as titanium

- *Plate Sizing* controls the size of the length and width parameters so that only three variants can be picked

You will now start the configuration of an iLogic form to make the configuration of material and plate sizing of this much easier:

1. Select the **iLogic** tab and select **Forms**.

2. Right-click on **Forms** and select **Add Form**. **Form Editor** opens. Here we can define and create a live preview of what the final iLogic form will look like.

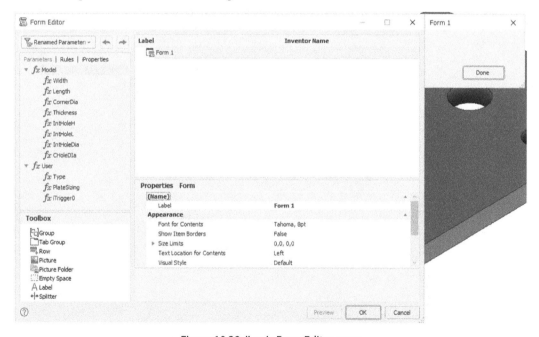

Figure 10.29: iLogic Form Editor open

3. Under **Label**, select **Form 1** and type a new form name: `Mounting Plate Configuration`.

Figure 10.30: Label changed to Mounting Plate Configuration

4. From the **Toolbox** area, click and drag the **Picture** element over to **Form Workspace**, as shown in *Figure 10.31*.

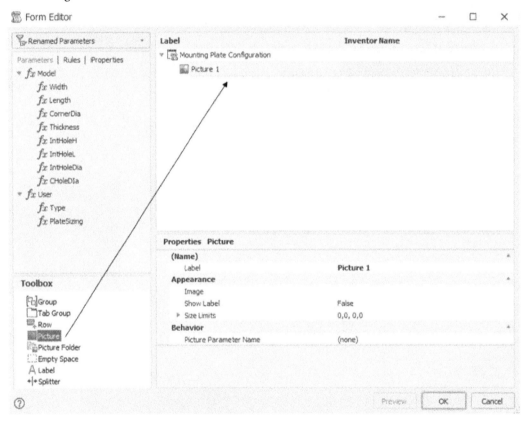

Figure 10.31: Picture element dragged to Form Workspace

5.   Select the browse button, ..., as shown in *Figure 10.32*.

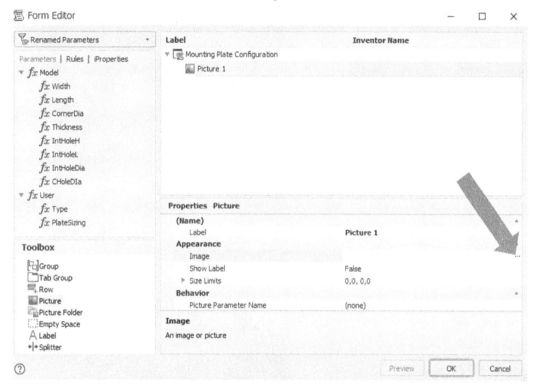

Figure 10.32: The iLogic Form Editor with the Image Browser command show.

6.   Then browse the `Chapter 10` folder, and select `Mounting Plate.png`. Select **Open**. The image is incorporated into **iLogic Form Preview**.

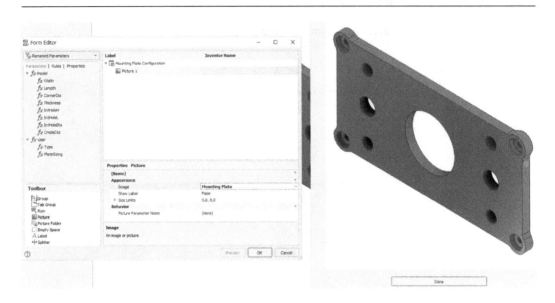

Figure 10.33: Mounting Plate image now incorporated into the iLogic Form Preview

7. Now we will start to add functionality so that the mounting plate can be configured in the form. Click and drag three instances of **Tab Group** from the **Toolbox** into **Form Workspace**.

8. Rename the *Tab Groups* as shown in *Figure 10.34.*

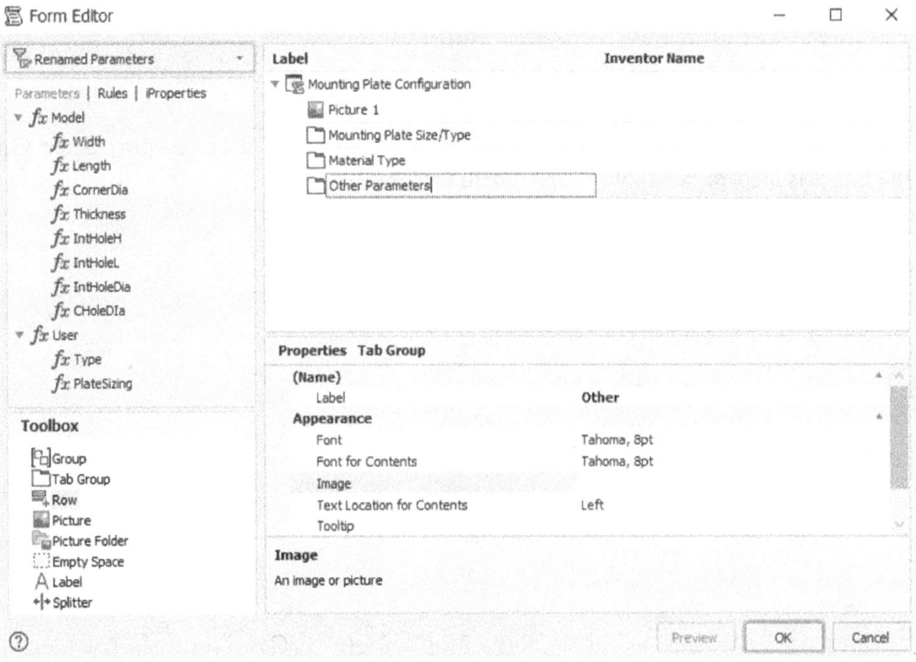

Figure 10.34: Tab Groups added and renamed

9. Click and drag the following parameters to the renamed **Tab Group** folders in **Form Editor** as shown in *Figure 10.35*.

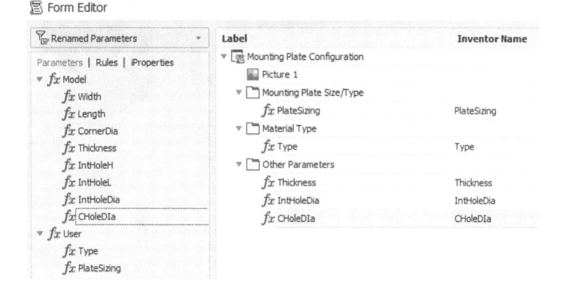

Figure 10.35: Parameters dragged into the correct Tab Group folder

The preview of the form will have updated, reflecting these changes.

10. We will now make the Inventor names for the parameters more meaningful for the purpose of the form. Select **PlateSizing**, then type `Mounting Plate Type & Sizing` under **Label**, as shown in *Figure 10.36*.

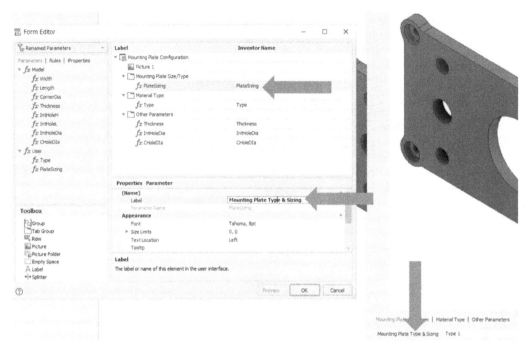

Figure 10.36: PlateSizing renamed to Mounting Plate Type & Sizing in the iLogic form

Form Preview will automatically update to reflect the change.

11. Repeat *step 10* with the other parameters:

- **Type** = Material Selection
- **Thickness** = Thickness of Plate
- **IntHoleDia** = Internal Hole Dia
- **CholeDia** = Central Hole Dia

12. Additional buttons will now be added to the form. Select the top level in **Form Editor**: Mounting Plate Configuration. Scroll down the **Properties** form to **Behavior**. Select **Done**. Under **Predefined Buttons**, select **OK Cancel Apply**.

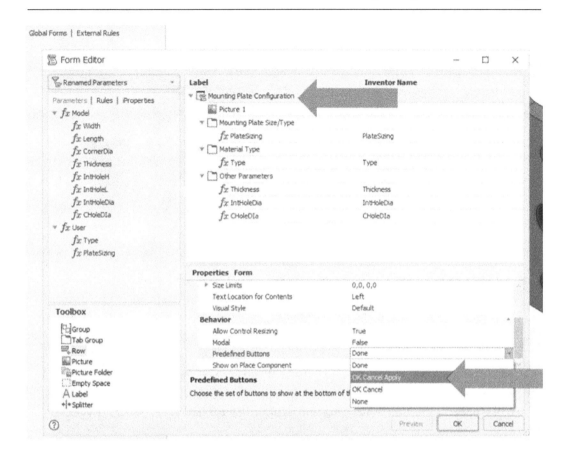

Figure 10.37: OK Cancel Apply selected as additional buttons in the iLogic form

13. Select **OK** to complete the form.

14. On the **iLogic** tab, select **Form**, and select **Mounting Plate Configuration**.

15. The iLogic form is launched; test the form by editing the drop-down menus and the allowable values. The part file will update automatically.

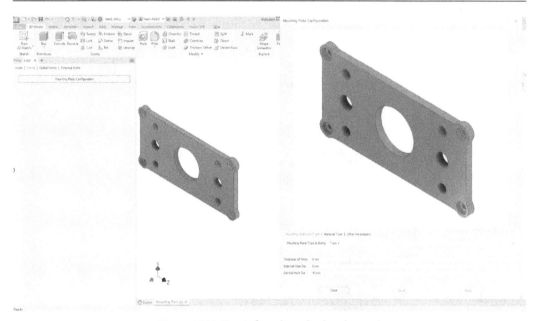

Figure 10.38: iLogic form launched and tested

16. Close the iLogic form. Currently, this iLogic form requires the user to manually open it.

17. We will now create an iTrigger to automatically open the form as soon as Mounting Plate. ipt is opened. Select the **Manage** tab and navigate to the **iLogic** panel. Select **iTrigger**.

Figure 10.39: iLogic iTrigger command

18. Select **Parameters**. An **iTrigger0** user parameter has been created in *step 17*. Select **Done**.

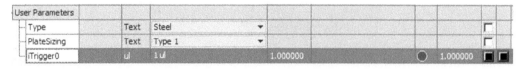

Figure 10.40: iTrigger0 added to User Parameters

19. Close **Parameters**. Navigate to the **Manage** tab, and in the **iLogic** panel, select **Add Rule**.

20. Name the rule iLogic Form Launch. Select **OK**.

21. Next, type trigger=iTrigger0 as the first line of code. Then navigate to the **Forms** folder in the **Snippets** area and double-click **Show Form**. Then adjust the code so that Mounting Plate Configuration is entered as the specific form to show. This will launch the form automatically as the document is opened. Select **Save & Run**. Close the form.

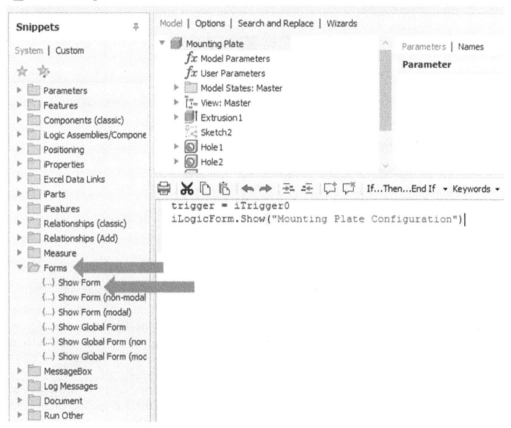

Figure 10.41: Adding code for the iTrigger

22. Now we will create an Event Trigger to execute a rule automatically. This will be performed on the *Material Selection* rule so that if the material is manually changed in the model by the user, it will revert to the only choices available in the form: steel alloy or titanium. In the **Manage** tab, find the **iLogic** panel and select **Event Triggers**.

23. Click and drag the **Material Selection** rule into the **Material Change** event category.

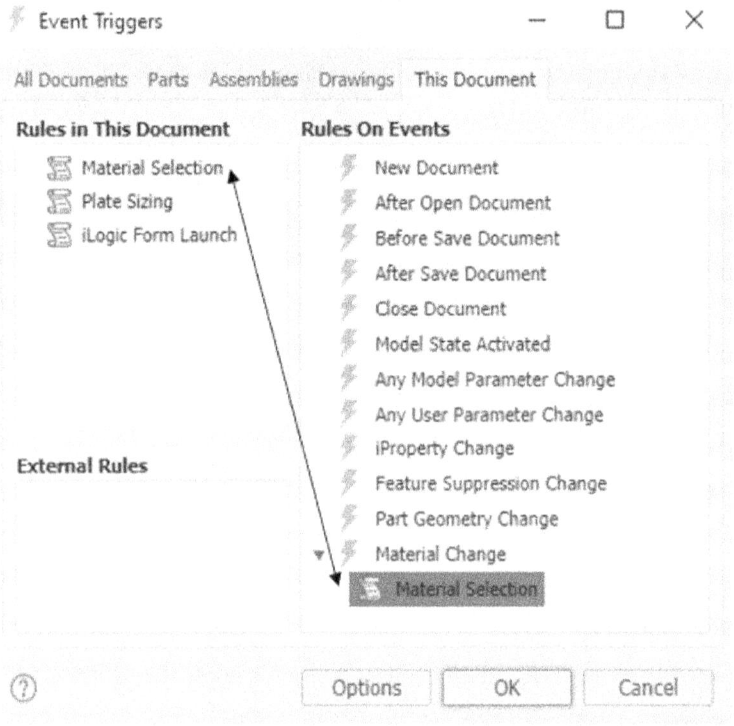

Figure 10.42: Material Selection rule added to Material Change Event Trigger

24. Select **OK**. If the material is changed in the Material Editor and not the iLogic form, the material will revert to steel alloy automatically as the Event Trigger is triggered and runs the Material Selection rule.

You have created an iLogic form with multiple functions and rules that enables the configuration of a mounting plate part. iTriggers have been set to launch the form upon the document opening, and Event Triggers have been configured to maintain the correct material selection.

Model credits

The model credits for this chapter are as follows:

- **iLOGIC CAT LADDER** (by Jacobus Willlemse): `https://grabcad.com/library/ilogic-cat-ladder-1`

- **Modern Lamp** (by Kyle Canalichio): `https://grabcad.com/library/modern-lamp-1`

11

Inventor Stress and Simulation – Workflow and Techniques

Autodesk Inventor can conduct a variety of accurate **Finite Element Analysis** (FEA) studies on models and assemblies. The Inventor environment used to facilitate this is known as the **Stress Analysis** environment, located in the **Environments** tab in both the **Part** and **Assembly** ribbons. FEA is a computerized method for simulating and predicting how parts and products react to real-world forces, vibration, heat, fluid flow, and other physical effects. FEA can be used to simulate whether a product will fail, work, or when the life cycle of the product will end. The advantage of using FEA in your design process is that products can be tested and optimized before expensive, real-world prototypes are made. FEA enables you to fail early and fast in the design process, which results in a more optimal solution that is validated and brought to market more quickly.

FEA is a complex and large topic that requires very specialist analysis knowledge in some cases. Not all aspects of Inventor FEA can be explained and delivered in a single chapter. The aim is to bring to you the very basics of what Inventor's Stress Analysis Environment and Frame Analysis Environment are capable of. Not all FEA functionalities will be covered and if FEA studies are something you will be doing frequently, then it is recommended you seek out additional resources. This chapter will give a basic overview and highlight key and common areas that most designers will use within the context of FEA in Inventor.

Before embarking on the recipes, it is important to understand the following key fundamental theory behind FEA, how it works in Inventor, and its limitations.

How does FEA work?

FEA involves taking your existing CAD model and breaking it down into thousands of finite elements. Then, mathematical equations, material data forces, and constraints are applied to create a virtual test. During a simulation study, Inventor combines all the individual behaviors of the defined finite elements and then uses this with the mathematical equations and forces to predict the real-world behavior of the part or assembly. FEA is never 100% accurate and is always subject to error, be that in the study setup, design, or even the interpretation of results. It is always recommended to run a series of studies, seek professional analysis expertise, and verify it with real-world testing. The place of FEA is to validate designs quickly and efficiently so that fewer real-world prototypes are required.

Although FEA can be used to predict a wide variety of behaviors in models, in Inventor, this is limited only to **linear analysis**. For **nonlinear analysis** and additional FEA capability, this can be achieved with the **Inventor Nastran** product. This is an additional product found within the **Autodesk Product Design & Manufacturing Collection**. Users that only have Inventor Professional will not have the Nastran in-CAD functionality.

With the standard Stress Analysis Environment in Inventor Professional, you can run FEA studies to determine the following:

- Mechanical stress
- Mechanical vibration
- Motion

Why FEA is important to the design process?

- Explores more design options and iterations earlier in the design process
- Limits the reliance and usage of building expensive real-world prototypes
- Predicts performance and improves safety by understanding where products will fail
- Modeling decisions are driven by specific data

Limitations to Inventor FEA

- The primary skill for success with FEA is engineering judgment. With this, all the simulation inputs can be properly quantified.
- Inventor FEA can only perform linear analysis.
- Inventor FEA is suited to components with small deformations and operational loading conditions.
- Inventor FEA assumes material properties' parts remain linear after the yield limit.
- Inventor FEA with linear analysis assumes that all materials are ridged and homogenous.
- Inventor FEA with linear analysis assumes all material properties are isotropic.

Inventor Stress Analysis utilizes a variety of different first- and second-order tetrahedrons and thin elements to compute simulations. These tetrahedrons are what Inventor uses to create a mesh around a model and break the surface down into smaller "finite" elements. The accuracy of the study is increased due to the number of tetrahedrons and the density of the mesh. By increasing the mesh size and density, the computing power required is also increased, as there are more tetrahedrons to compute. The mesh size and density can either be controlled on the model or locally on specific areas of a model. Dependent on the model type, these are the different elements that can be used:

- 4-noded tetrahedron: Used for linear elements

- 10-noded tetrahedron: Used for quadratic or curved elements

- 2-noded beam: Linear elements, often used to simulate structural beams

- 4-noded shell: Used for linear elements

- 8-noded shell: Used for quadratic or curved elements

Figure 11.1 shows the different finite elements that Inventor utilizes:

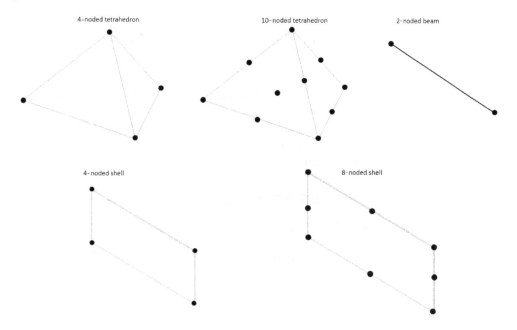

Figure 11.1: Types of finite element method elements

The FEA process in Inventor Stress Analysis

These are the typical steps followed in a standard workflow to conduct a linear stress analysis study on a model:

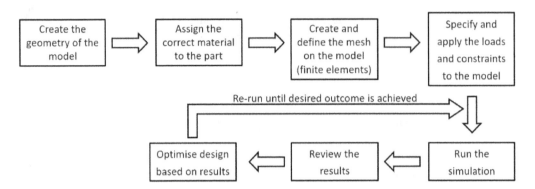

Figure 11.2: The FEA process in Inventor Stress Analysis

In this chapter, you will learn about the following:

- Introduction to the Stress Analysis Environment – how to conduct a simple analysis study, workflow, and interface
- Performing stress analysis on bolted connections
- Calculating wind loads and using midsurfaces
- Calculating weld sizes using stress analysis and the Weld Calculator

Technical requirements

To complete this chapter, you will need access to the practice files in the Chapter 11 folder within the Inventor Cookbook 2023 folder.

Introduction to the Stress Analysis Environment – how to conduct a simple analysis study, workflow, and interface

In this recipe, you will learn how to conduct a simple **Inventor Stress Analysis** study on a part file. You will learn about the basic process and workflow of creating a Stress Analysis study, how to assign materials, apply a mesh, and generate a comprehensive report of results from the study. In the study, you will apply a **Load** of 500N to one end of AE_Bracket and examine the **Displacement** and **Safety Factor** settings of the part.

Getting ready

To begin this recipe, you will need to navigate to the Inventor Cookbook 2023 folder and then open Chapter 11. Select and open AE_Bracket.ipt.

How to do it...

With `AE_Bracket.ipt` now open, we can begin to set up the model so that it is ready to run the simulation. In this example, we will fix `AE_Bracket.ipt` at one end and apply a force of `500N` to the other as static analysis and observe the results. This is one of the simplest applications of the Stress Analysis tools in Inventor:

1. Select the **Environments** tab and then select **Stress Analysis**. This environment can also be accessed in the assembly environment:

Figure 11.3: The Stress Analysis command

2. The **Analysis** tab now becomes active, and we can begin to set up a study, as shown here:

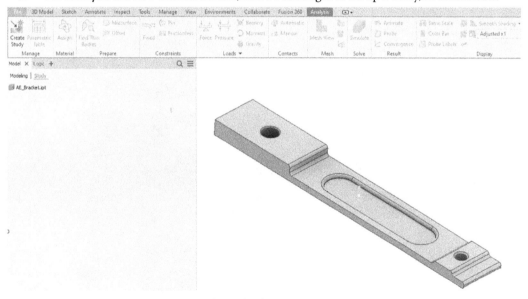

Figure 11.4: Stress Analysis Environment active

At any point, you can navigate back to the modeling commands. This is so that, after a study, you can go back and optimize the design.

3. From the ribbon, select **Create Study**. We will now define the type of analysis that we will simulate.

4. The **Create New Study** window appears. Rename the study Study 1. Ensure **Static Analysis** is selected. Change **Tolerance** to 0.010mm. Ensure that **Type** is set to **Bonded**, as this is a single machined piece of steel, so there is no movement between components. Change **Shell Connector Tolerance** to 1:

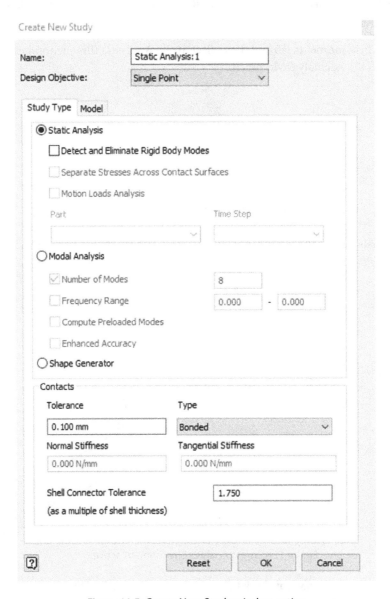

Figure 11.5: Create New Study window active

> **Static and Modal Analysis**
>
> **Static Analysis** is the most common type of structural analysis using FEA. It is used to determine the stress, strain, and deformation of components.
>
> **Modal Analysis** is used to find the natural frequencies and their associated modes (as in, the deformation shapes) of a structure.
>
> **Shape Generator** enables Inventor to recommend design solutions in relation to the FEA study.

5.   Select **OK**.

6.   In the **Material** tab of the **Analysis** ribbon, select **Assign**. We will now define the material of the part, as this will influence how it behaves under the defined stress. In the **Assign Materials** window, navigate to the **Override Material** dropdown and select **Steel, Carbon** from the list. Then, select **OK**:

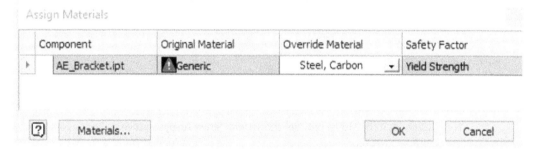

Figure 11.6: Assigning materials to the part

7.   The material for the study is defined. Notice that on the left of the screen, the **Model** browser has changed to show stress analysis-specific details. Expand the **Material** node in the **Model** browser and you will see that **Steel, Carbon** is shown in *Figure 11.6*. You can edit any of these aspects of the study from the **Model** browser:

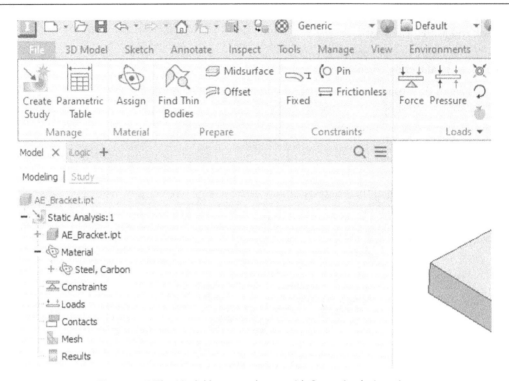

Figure 11.7: The Model browser shown with Stress Analysis active

8.   In the **Constraints** section of the ribbon, select **Fixed**.

9.   Then, select the face of the part shown in *Figure 11.7*. Select **Apply** and then close the dialog:

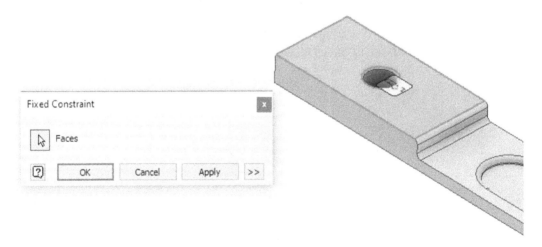

Figure 11.8: Fixed constraint applied to the part

10. Select the **Force** command from the ribbon.

11. Select the face of the part shown to apply the **Force** feature. Then, ensure **Direction** is as shown in *Figure 11.8*. Enter a **Magnitude** value of 500N and then select **OK**:

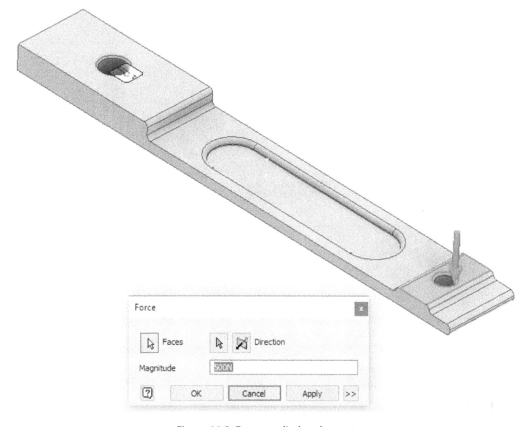

Figure 11.9: Force applied to the part

12. We will now define **Gravity** in the study. Select the **Gravity** command from the **Force** area of the ribbon:

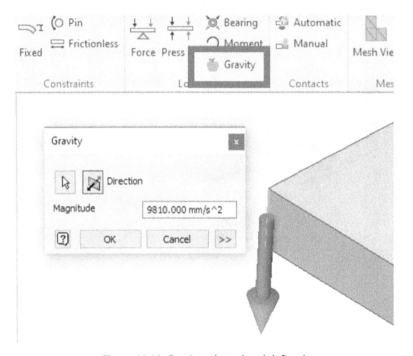

Figure 11.10: Gravity selected and defined

13. Select the edge of the part shown in *Figure 11.9* to apply a **Direction** setting to **Gravity**. Flip the **Direction** setting if required using the **Direction** button in the **Gravity** dialog. Then, select **OK**.

14. Select **Mesh View**. The default mesh settings are applied to the model, which will be approximately 3,433 nodes and 1,829 elements:

Figure 11.11: Mesh View applied to the part

15. To have a more accurate study, we will edit **Mesh Settings** and apply a much denser mesh to the model. This will increase the number and density of elements. Next to the **Mesh View** command, select **Mesh Settings**:

Mesh View
Simulate
Mesh
Solve

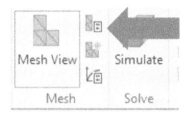

Figure 11.12: Mesh Settings

16. Change **Mesh Settings** as follows – **Average Element Size** as `0.010`, **Minimum Element Size** as `0.020`, and **Grading Factor** as `1` – and then select **OK**:

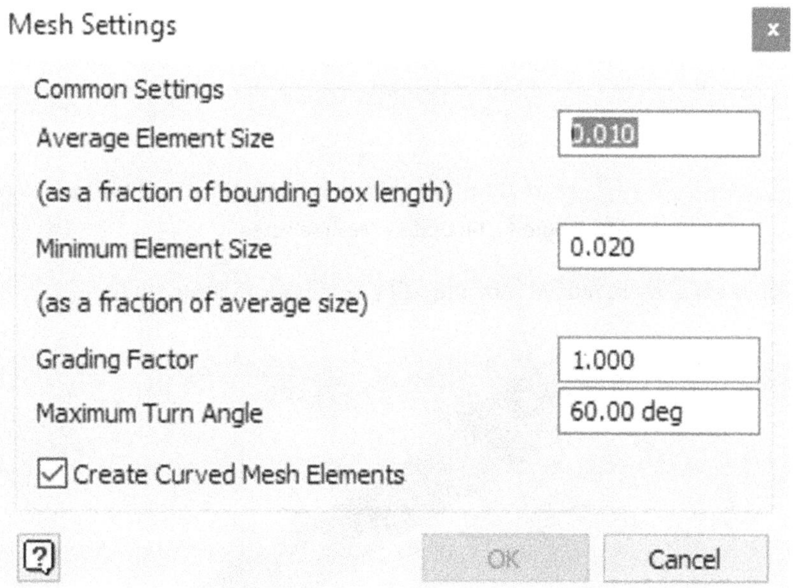

Mesh Settings
Common Settings
Average Element Size
0.010
(as a fraction of bounding box length)
Minimum Element Size
0.020
(as a fraction of average size)
Grading Factor
1.000
Maximum Turn Angle
60.00 deg
Create Curved Mesh Elements
OK
Cancel

Figure 11.13: Mesh Settings changed

17. The mesh will not automatically update upon selecting **OK** in **Mesh Settings**. Navigate to the **Model** browser, right-click on the mesh icon, and select **Update Mesh**. Inventor will calculate and then reapply the mesh view to the model:

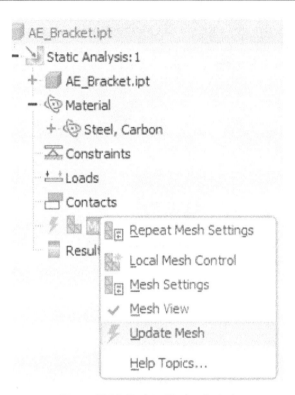

Figure 11.14: Update Mesh selected

The mesh is much more refined now and will give a more accurate study:

Figure 11.15: Refined mesh view

18. All parameters for this simple Static Analysis study have now been defined. Select **Simulate** and then select **Run.**

19. We can now interpret and interrogate the results of the FEA study. The **Von Mises Stress** value is used to determine whether the materials will yield or fracture:

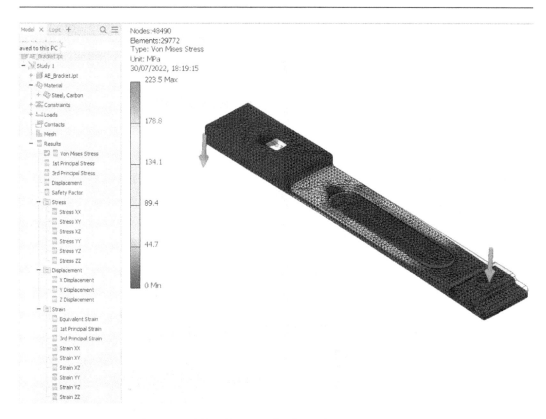

Figure 11.16: Von Mises results shown from the FEA study

Within the **Model** browser, all aspects and results of the study can be accessed. Expand the folders for **Stress**, **Displacement**, and **Strain** (as shown in *Figure 11.16*). Each study listed can be accessed and the results can be viewed by double-clicking on it. As the studies are selected, the graphics window will update to show the results.

20. We can interrogate the results of each study further. In this example, we will look at the **Von Mises** results. Within the **Result** area of the ribbon, we can use **Animate** on the results, conduct a **Convergence** plot, and use **Probe** on the model to view the study data on specific areas. We will select **Animate**:

Figure 11.17: Result and Display options in the Stress Analysis Environment

21. By selecting **Animate**, you can press **Play** and control the playback speed and steps of how the stress will act on the part during the loading of the 500N value in the location we specified.

22. Select **Probe** and then select a point on the model. The exact value at that point will be shown. Selecting **Probe Labels** in the **Display** area of the ribbon will hide or unhide any **Probe Labels** you have previously placed:

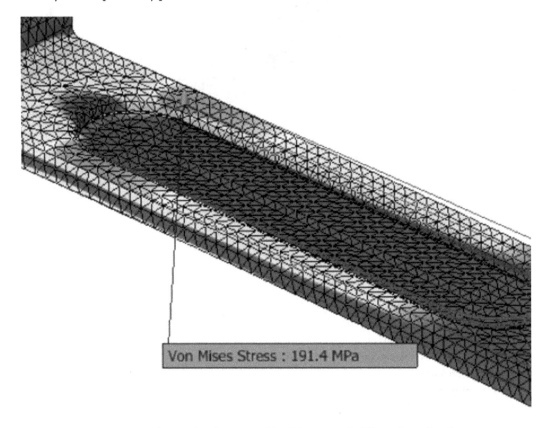

Figure 11.18: Probe used to show exact Von Mises stress in MPa at the point shown

23. By selecting **Color Bar**, you can define the **Minimum** and **Maximum** values shown on the model within the study:

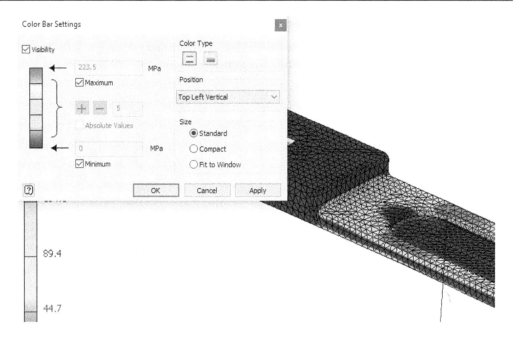

Figure 11.19: Color Bar Settings

24. The maximum and minimum values in the study can quickly be identified by selecting **Maximum** and **Minimum**. Once these values are selected, automatic **Probe labels** are placed at these points on the model in the graphics window. Select **Maximum** and **Minimum** to show these values:

Figure 11.20: Maximum and Minimum commands

25. The **Display** options, shown in *Figure 11.20*, enable you to define **Shading** and also the graphical representation of results. Change this from **Adjusted x1** to **Actual**:

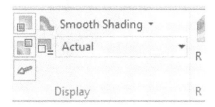

Figure 11.21: Actual selected

This will change the display so that the model shows the actual predicted changes.

26. Select **Displacement Study** in the **Model** browser with a double click.

27. Navigate the model to **Front View**. You can see that with this **Load** value applied, the part will displace by a maximum of **7.117mm** from its original position:

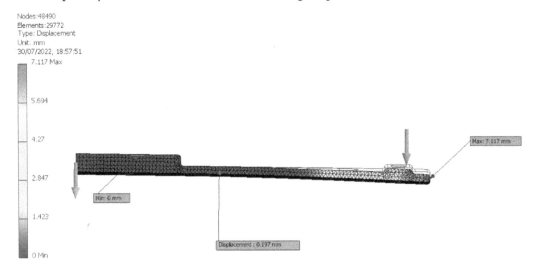

Figure 11.22: Displacement of the part shown

28. Navigate to the **Model** browser and select the **Safety Factor** study:

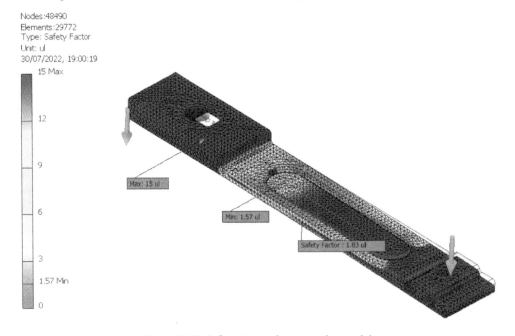

Figure 11.23: Safety Factor shown on the model

The **Safety Factor** values are all above the minimum value of **1**, meaning that the component will fail above the intended design load of 500N, and therefore, could support additional loading. In this instance, the component would probably not fail. However, it would be advisable to run further studies and consider design optimization to ensure that **Displacement** is reduced and the **Safety Factor** value is increased.

29. Now, we will generate a report of all the studies we have conducted for further analysis. Select **Report**:

Figure 11.24: Report options

30. In the **General** tab, change **Images Size** to **800 x 600**. In this example, we will generate a report of all studies conducted in the simulation. Navigate to the **Format** tab and select **PDF**.

31. Select **OK**. The Stress Analysis report is compiled and created as a .PDF file.

You have navigated and used the basic functionality of the Stress Analysis Environment to perform a simple static stress analysis on a bracket, to examine the effect on the **Displacement** and **Safety Factor** settings when 500N is applied in one direction, on one end of the bracket. A .PDF report of the results has then been generated.

Performing stress analysis on bolted connections

In this recipe, you will learn how to apply a stress analysis study on a typical bolted connection, to verify a bolted connection calculation for the given scenario.

The scenario is that we have two plates of 20 mm-thick 405 stainless steel, joining with a 90 mm contact overlap. A force of 15 Kn is applied in each direction away from the bolt. Using **Bolted Connection Strength Calculator**, we will first determine the current results with one bolt and determine what the present safety factor is.

Then, we will use **Bolted Connection Strength Calculator** to recommend several bolted connections to apply, to achieve a factor of safety of above 2.5 with an allowable bolt thread pressure of 20 MPa. Once complete, we will update the assembly to contain the new additional bolted connections, and then test this calculation with an FEA study.

This will involve incorporating a new concept of stress analysis connections within an Inventor assembly.

Figure 11.25 provides a visual for the scenario just explained:

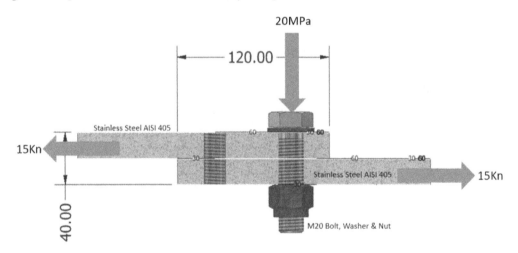

Figure 11.25: Half-section view of the assembly and scenario

These are the conditions for the study:

- Two overlapped mechanically joined plates of stainless steel AISI 405 subject to a bidirectional force of 15 Kn on each side

- 20 MPa of allowable thread pressure
- **Factor of safety (FOS)** of >2.5 required

The following are the objectives for the purpose of the study:

- What is the current FOS?
- Does the current arrangement fail?
- How many bolts are required to meet the desired FOS?
- Test and validate the bolted connection calculation with an FEA stress analysis study

Getting ready

To begin this recipe, you will need to navigate to Inventor Cookbook 2023|Chapter 11 | Bolted Connection and open Bolted Connection.iam.

How to do it...

The first step is to open **Bolted Connection Design Accelerator**, determine the current conditions of the assembly, and whether the current design fails based on our requirements. Once we have done this, we will test the calculations and then modify the design to the required FOS:

1. The assembly contains two plates of stainless steel AISI 405 that are lapped and bonded together with a single bolted connection. The specification of the bolted connection is shown in *Figure 11.26*:

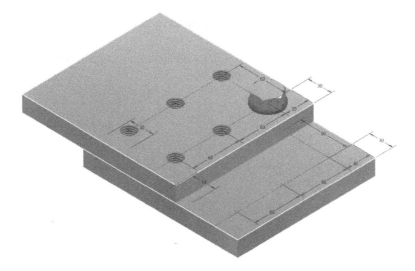

Figure 11.26: Specification of the bolted connection

Select the **Design** tab from the ribbon and select **Bolted Connection**.

2. Navigate and select the **Calculation** tab from the **Bolted Connection Component Generator** window.

3. The **Calculation** tab allows various calculations to be performed in relation to **Bolted Connections** to ensure that the right components are selected for the design requirements and structural loads.

   For **Type of Strength Calculation**, select **Check calculation**:

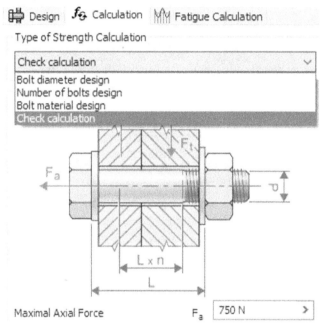

Figure 11.27: Check calculation selected

4. We will now enter the scenario criteria to work out whether the current bolted connection can withstand the forces acting on the two plates.

   Enter the following data into the fields:

   I.   **Maximal Tangent Force (Ft)** – Enter 15,000 N. This is the bidirectional force applied to each plate.

   II. **Required Safety Factor (Ks)** – Enter 2.5ul. This is the required **Safety Factor** we need to meet for this application.

   III. **Functional Width (L)** – Enter 40mm. This is the combined depth of the two steel plates.

IV.   **Number of bolts (z)** – Enter `1ul`. Only one bolted connection is present.

V.   **Allowable Thread Pressure (Pa)** – Enter `20MPa`.

Leave all other fields as default, as shown here:

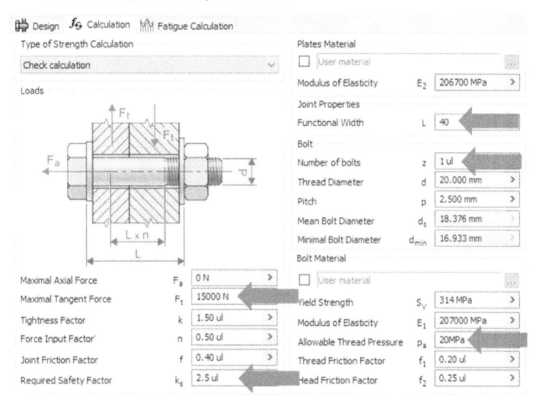

Figure 11.28: Values to change in the strength calculation

5.   Select **Calculate**. The results are generated in the preview – see the text highlighted in *Figure 11.29*:

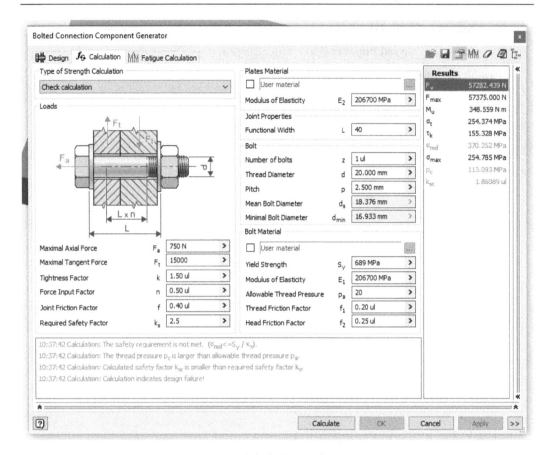

Figure 11.29: Calculation result summary

The given **Calculation** has indicated a design failure in the maximum force and thread pressure and critically, a FOS of 1.9 has been indicated. This presents a design failure. We will now verify this with a stress analysis study. Your values calculated may be slightly different.

6.   Close **Bolted Connection Component Generator**.

7.   Select the **Environments** tab in the ribbon, and then select **Stress Analysis**, followed by **Create Study**.

8.   Name the study Bolt Study – 1. Leave all other settings as default and select **OK.**

9.   Materials have already been assigned in the model, so we can proceed to define forces and pressure. Select **Force** from the **Constraint** area of the ribbon. Apply two instances of a 15,000N force on each side of the two plates, as shown in *Figure 11.30*. Ensure the direction of the force applied is the same as the arrows that are shown in *Figure 11.30*:

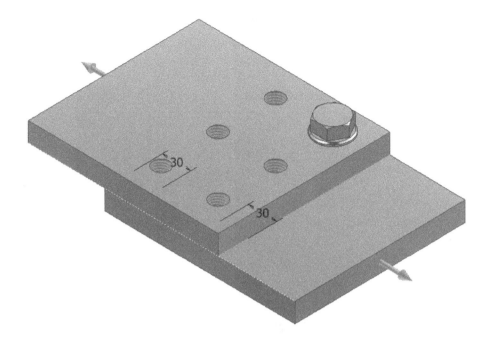

Figure 11.30: 15,000 N force applied bi-directionally to the plates

10. Select **Gravity** and apply this to the edge and **Direction** shown in *Figure 11.31*:

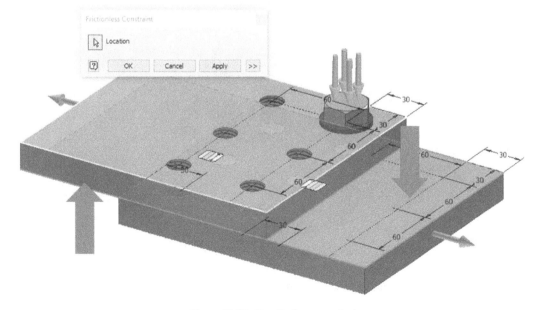

Figure 11.31: Gravity force applied

11. Select **Pressure** and apply 20 MPa to the bolt head:

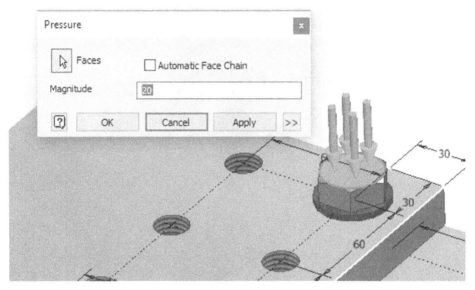

Figure 11.32: Pressure applied to the bolt head

12. Select **Frictionless Constraint** and select the two faces shown to apply a **Frictionless Constraint** between the plates:

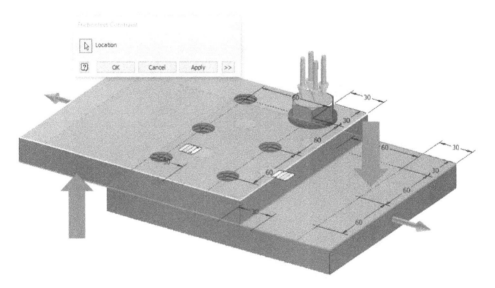

Figure 11.33: Frictionless constraint applied

This tells Inventor that the two faces are separate and can slide.

13. We now need to define contacts between each individual component. Contacts are a set of rules that communicate to Inventor how parts will behave with one another as the simulation is calculated. As we have several parts already, we can select **Automatic** from the ribbon to auto-calculate the contacts.

Select the **Automatic** contacts:

Figure 11.34: Automatic contacts

14. Default bonded contacts are applied to all components. On expanding the **Contacts** area of the **Model** browser, you will see the contacts are automatically applied. We need to change a number of these to **Sliding / No Separation**.

This is because it will allow sliding contact but will not allow the separation of parts. Having lots of separation only in contacts can result in a much longer analysis run, as these contacts are non-linear and take longer to process convergence results.

Figure 11.35 shows the **Bonded** automatic contacts created:

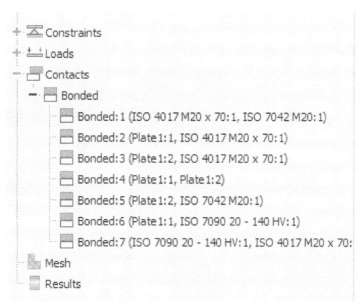

Figure 11.35: Bonded contacts applied by default

Select all contacts, as shown in *Figure 11.36,* and then right-click and select **Edit Contact:**

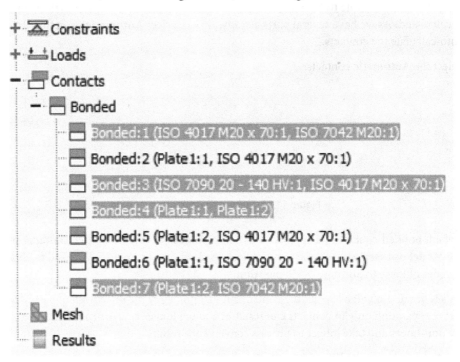

Figure 11.36: Bonded contacts selected

15. Select **Sliding / No Separation** and select **OK.**

16. Select **Mesh View**. Automatically, a denser mesh has been applied around the bolt, so there is no need to adjust this to a finer resolution.

17. Select **Simulate** and then **Run**. Note that the threads on the model are displayed as Cosmetic Threads only, which is to speed up analysis. If real-world threads were modeled, the analysis would be overly complex and take longer to run due to the extra geometry.

18. Select **Safety Factor** from **Results** in the **Model** browser.

19. Right-click on **Plate1:1** from the **Study** browser and select **Visibility**. This will hide the **Plate** components from the results temporarily. Repeat this process for **Plate2:2** to temporarily hide this too:

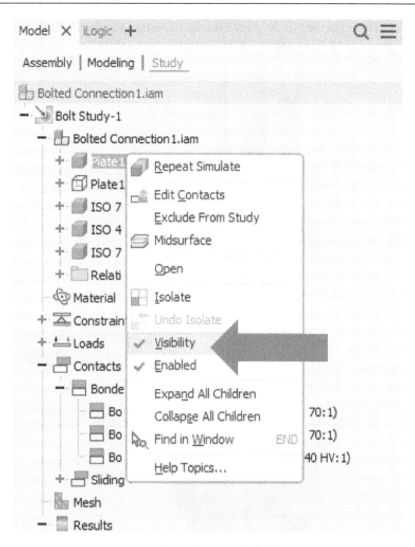

Figure 11.37: Isolate selected on bolted connection

20. The bolt is displaying stress wherever expected due to the forces acting upon it. From the **Safety Factor Results** bar, we can see that clearly, the mid-section of the bolted connection will fail with these forces acting upon it. The minimum value detected at a single point was 0.52 and using the probe in this region of the bolt, you will get values of around 1 to 1.7. This is accurate to our previous calculation in *step* 5 of 1.9.

The stress analysis has verified the given **Bolted Connection Strength Calculation**:

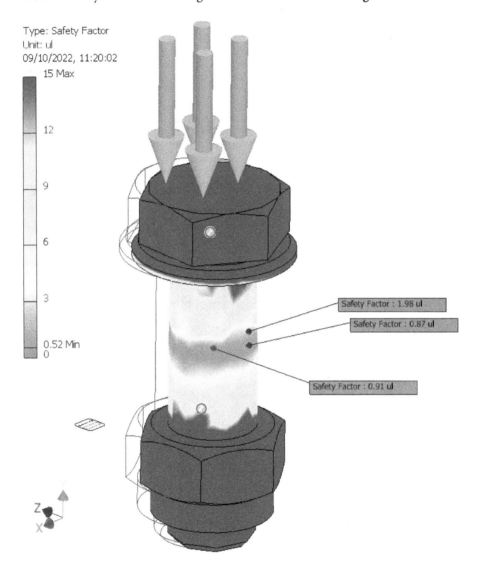

Figure 11.38: Bolted connection failing under loads with a FOS of 0.52

21. Select **Finish Analysis** from the ribbon.

22. Back in the assembly environment, right-click on **Plate1** and **Plate2** in the **Model** browser and select **Visibility** to unhide the other components.

23. We will now use **Bolted Connection Strength Calculator** to recommend how many additional bolted connections we need to solve this design problem.

24. Select the **Design** tab and then select **Bolted Connection**.

25. Select the **Calculation** tab in the **Bolted Connection** window.

26. Change **Type of Strength Calculation** from **Check calculation** to **Number of bolts design**.

27. Ensure that the values match those of *Figure 11.39*:

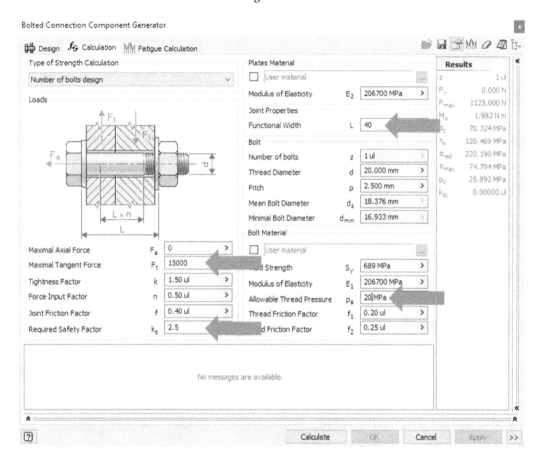

Figure 11.39: Calculation values must match this

28. Select **Calculate**. The number of bolts required to attain a FOS of <2 . 5 is 6. We will now adapt the assembly to incorporate six bolted connections and then rerun the simulation to verify. More detailed results can be found by selecting the **Results** button (the notepad icon) as shown here in *Figure 11.40*, which can be found in the top-right corner of *Figure 11.39*:

Figure 11.40: Results button

29. Close **Bolted Connection Strength Calculator**.

30. Select the **Assemble** tab and then select **Pattern**.

31. Select all elements of the **Bolted Connection** subassembly as the component to pattern. Then, select the rectangular pattern from the icon:

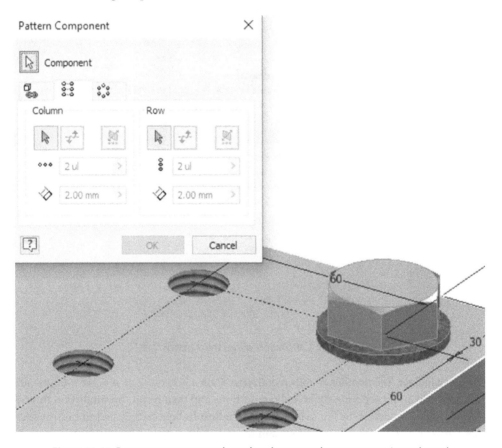

Figure 11.41: Pattern component selected and rectangular pattern option selected

32. Select the red arrow for **Column**. Then, select the edge of the plate shown in *Figure 11.42*. Ensure the **Direction** setting of the pattern is the same as in *Figure 11.42*:

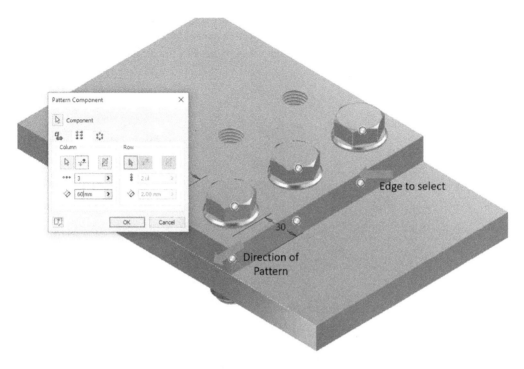

Figure 11.42: Edge and direction of the pattern

33. Change the **Pattern Component** values to 3 ul and 60 mm:

Figure 11.43: Pattern component values to change

34. For **Row**, select the edge shown in *Figure 11.44*. Ensure the **Direction** setting of the pattern is the same as in *Figure 11.44*. Change the number of instances to 2 and the distance to 60 mm. The six bolted connections will populate the holes. Select **OK** to complete:

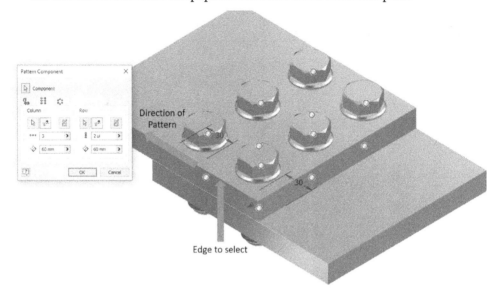

Figure 11.44: Row pattern values to add

35. Select the **Environments** tab and select **Stress Analysis**.

36. Right-click on **Bolt Study – 1** from the **Model** browser and select **Copy Study**:

Figure 11.45: Copy Study

37. Rename the study `Bolt Study - 6`.

38. Right-click on **Contacts** in the **Model** browser and select **Update Automatic Contacts**.

39. With the help of the *Ctrl* key, select the bonded contacts (**1**, **3**, **4**, **7**, **8**, **14**, **15**, **16**, **17**, **18**, **29**, **30**, **31**, **32**, and **33**), as shown in *Figure 11.46*. Select **Edit Contact** and change these to **Sliding / No Separation**:

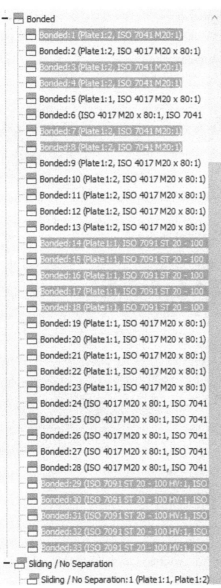

Figure 11.46: Bonded contacts to change to Sliding / No Separation

40. Select **Pressure** and apply 20 MPa on each of the five remaining bolt heads:

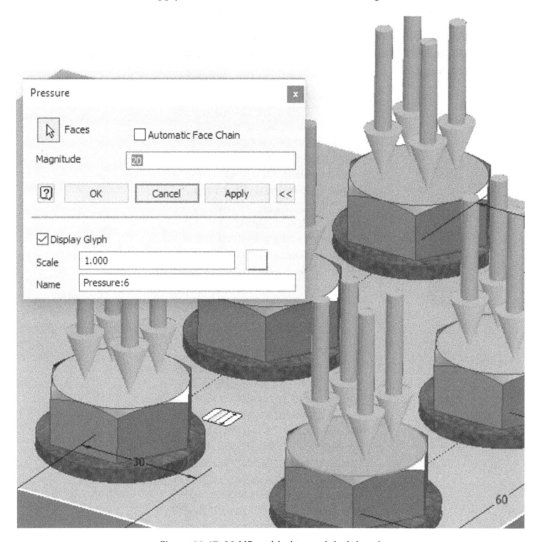

Figure 11.47: 20 MPa added to each bolt head

41. Select **Simulate** and then **Run.** Both studies will compute.

42. Hide the visibility of the plates from the **Model** browser. Then, select the safety factor results from **Bolt Study – 6**. The minimum value detected for the FOS is 4.53, which is similar to the calculation value we got from **Bolted Connection Strength Calculator**. The five additional bolts have solved the design problem, as shown here:

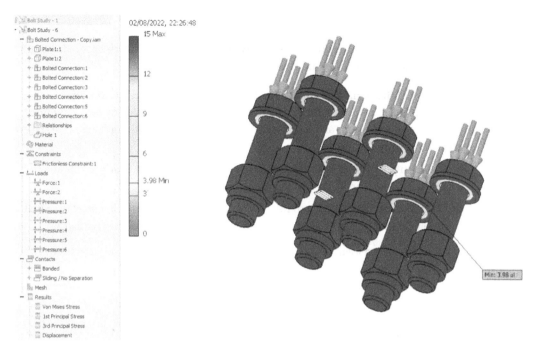

Figure 11.48: Results of second study, FOS shown

You have used bolted connection calculations and stress analysis of bolted connections to determine the number of bolts required to meet the required factor of safety with the stress and loads previously outlined. You have also successfully completed a stress analysis of multiple components and parts with the use of contacts.

Calculating wind loads and using midsurfaces

This recipe will involve calculating the wind load acting on a free-standing solar panel frame assembly. The model itself is made up of several different bodies and solids of varying thicknesses. To create a more accurate result, we will further simplify the framework, and then convert the assembly to **midsurfaces** prior to the analysis. In some instances, when conducting a stress analysis study, using solid elements for a part or assembly, where the length and width of the bodies are too large compared to the thickness (such as with thin parts), will lead to performance and accuracy problems. An example of this would be sheet metal components. As the mesh element size reduces, it leads to a large number of mesh elements and decreased mesh quality. Using midsurfaces in the analysis enables you to further simplify and shell the components so that the results are more accurate.

Getting ready

To begin this recipe, you will need to navigate to the `Inventor Cookbook 2023|Chapter 11` folder and open `Solar Panel Frame.iam`.

How to do it...

The first step will be to review the model and simplify it. We will then start the analysis procedure and create the midsurface before running the study and calculating the wind loads:

1. Review the model of the frame and ensure that **Steel, Mild** is selected as the **Material** setting for the frame and **Aluminum 6061** for **Fixing Bracket**. The model has already been simplified somewhat as bolted connections and fixings have been removed and the solar panel cells have also been excluded from the model.

2. Select **Environments** and then select **Stress Analysis**. Select **Create a Study** and name the study `Solar Panel Wind Load`. Select **OK** to complete.

3. Select **Find Thin Bodies**. This will search the model and pick out elements that meet the **Shell Features** criteria. Select **OK** on the next window, indicating that some elements qualify.

    Fixing Brackets have been identified as meeting the criteria for a midsurface. Select **OK**:

Figure 11.49: Fixing brackets identified as meeting shell feature criteria for a midsurface

4. The frame section that bears the solar cells has been simplified and converted into a midsurface for analysis. At any point from the **Model** browser in the Stress Analysis environment, the midsurface can be edited within the **Midsurface** folder:

Figure 11.50: Midsurfaces created from the model

5. To calculate the wind load of the solar panel, we will use the generic formula F = A x P x Cd, where the following applies:

    - F is the force or wind load – unknown at this stage
    - A is the projected area of the object – 3.36m2
    - P is the wind pressure – P = 0.02256 x V2
    - Cd is the drag coefficient – 2 for rectangular shapes

    For a real-world application, there are far many more factors that would have to be considered in this instance, such as design wind speed based on zip code number, exposure rating, risk category, elevation above sea level, and panel positioning.

    In this example, we will calculate for 60 mph winds:

    F = A x P X Cd

    P = 0.02256 x 602 = 81.252

    Force = 3.36 x 81.252 x 2 = 546 N

6. Select **Fixed Constraint** and select the faces shown in *Figure 11.51*. This is where the frame would be welded to the other parts of the frame that have been excluded from the study:

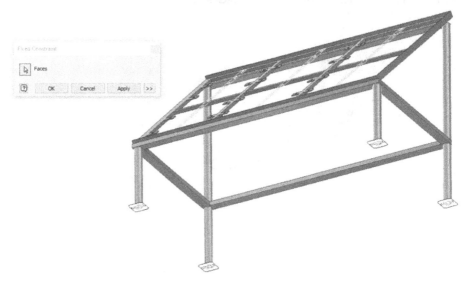

Figure 11.51: Fixed elements in the study

7. Select **Gravity**, followed by the face shown in *Figure 11.52* to add **Gravity** at the default value. Select **OK** to complete:

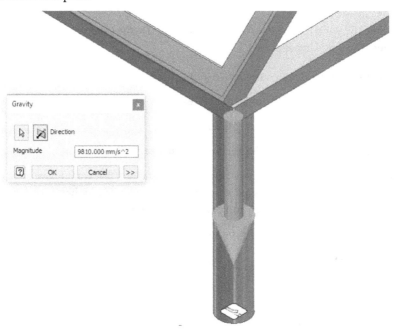

Figure 11.52: Gravity added to the model

8. Select **Force** and enter the value of **Magnitude** – 546 N. Apply the force to all areas shown in *Figure 11.53*:

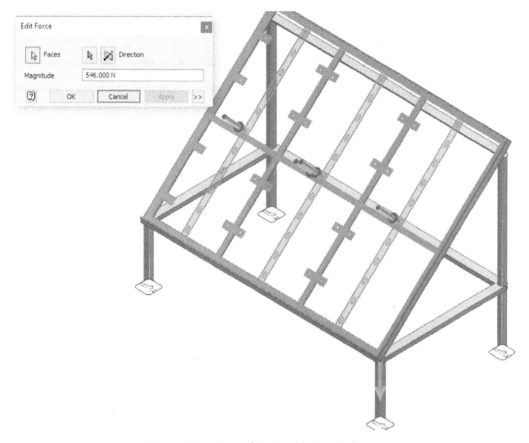

Figure 11.53: Force of 546 N added to the frame

9. We now need to allocate the weight of the solar panel cells (not shown in the model). Select **Force** again and apply 400 N (40 kg). As this is the weight of the solar panel, the direction of this force needs to be the same as gravity.

    Add this to the same faces as the Force in *step 8*, except with a direction the same as the gravity we specified earlier, shown in *Figure 11.53*.

10. Select **Simulate** and then **Run.**

Using a midsurface, you simplified a model, converted thin bodies to shells, and performed a wind load calculation in the Inventor stress analysis environment.

Calculating weld sizes using stress analysis and the Weld Calculator

Applying welds within a CAD model is sometimes required, not only to give a value for the true manufactured weight of a part (particularly in aerospace) but also to examine how and if the correct weld has been applied to the part, and whether it will exceed or meet the desired safety factors.

In this recipe, you will perform a weld calculation to work out the desired weld size for a two-part welded bracket in the following scenario:

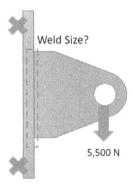

Figure 11.54: Two-part mild steel bracket with 5,500 N force acting on the lug

Once the calculation has validated the results, we will further validate with a stress analysis study, including the weldments. Using **Weld Calculator** will eliminate guesswork and enables us to validate a weld size prior to setting up a stress analysis study.

Things to note about welds in Inventor stress analysis:

- You should never expect accurate stresses in or near welds in a finite element model
- More accurate modeling of the weld beads will not create a more accurate analysis result
- Real-world testing and prototyping are always advised
- Welds in production will bear little or no resemblance to welds in a CAD model

This is because welds vary from part to part and can vary within the weld itself. FEA does not consider the following:

- Weld chemistry
- Weld temperature
- Microcracking
- Residual stress, warpage, alignment, surface finish, and likewise

Getting ready

To begin this recipe, you will need to navigate to Inventor Cookbook 2023, select the Chapter 11 folder, then select the Weldment folder, and open Weldment Bracket.iam.

How to do it...

We will start by converting the assembly to a weldment and then using **Weld Calculator** to input values and generate a minimum **Weld Height**:

1. Open Weldment Bracket.iam. At present, the welds have not been applied to the model. From the ribbon, select the **Environments** tab, and then select **Convert to Weldment**:

Figure 11.55: Convert to Weldment

2. Select **Yes**, followed by **OK.**

3. From the newly created **Weld** tab, select **Weld Calculator**. Then, select **Fillet Weld Calculator (Spatial)**:

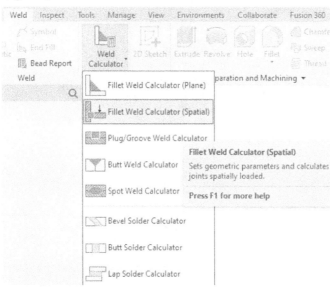

Figure 11.56: Weld Calculator

4. In **Weld Calculator**, we can now specify the load requirements and the size of the plates that the weld will form. This will enable us to examine whether our suggested **Weld Height** will meet the FOS requirements.

The values that we will consider are as follows:

- **Force** acting on the lug = 5,500N

- Initial **Weld Height** = 2mm

- **Beam Height** = 78mm (taken directly from the model)

- **Beam Width** = 10mm (thickness of the mild steel plate)

- **Safety Factor** goal = 2ul

Before entering these values, select **Weld Form** and select the box type weld, as this is what the actual weld will look like around the circumference of the lug from above:

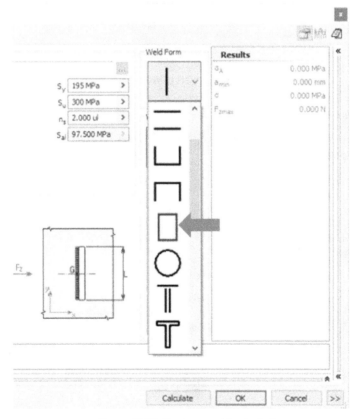

Figure 11.57: Box type weld selected

You can select **Front** on **View Cube** to see this on the model if required.

5.   Then, select the bending force parallel with the weld plane as the **Weld Loads** option:

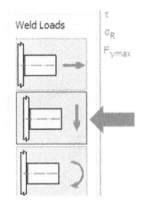

Figure 11.58: Bending force parallel with the weld plane

6.   Enter the values as indicated in *step 4*, as shown in *Figure 11.59*:

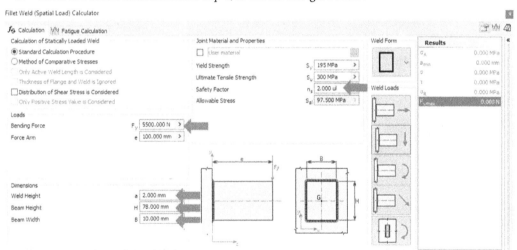

Figure 11.59: Values to add to the calculator

7.   Select **Calculate**. The calculator indicates that the weld will fail based on our defined requirements. Change **Weld Height** to 5mm.

8.   Select **Calculate**. The calculation has indicated that the requirements are met. The true FOS is max force/bending force, which means `13194.474N/5500 N = 2.4ul FOS`.

9.   Now, we can test this and validate it once more in the stress analysis study. Select **OK** to close **Fillet Weld (Spatial Load) Calculator**. The results are saved in the **Model** browser under **Fillet Weld.** Before testing, we now need to apply the weld geometry for the weld that we just calculated.

10. Select **Welds**, followed by **Fillet Weld.**

11. For **Reference 1**, select the face of the model shown in *Figure 11.60*:

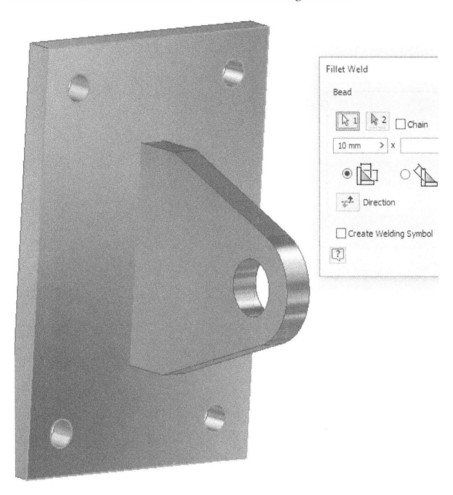

Figure 11.60: First reference for the weld

12. Select **2** in the **Fillet Weld** command and then select the four faces of the lug, as shown in *Figure 11.61*:

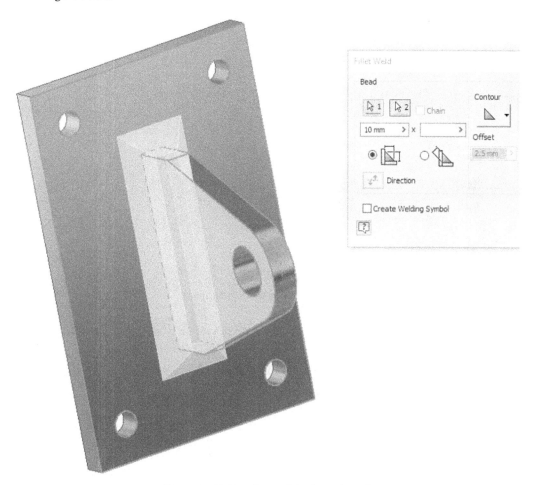

Figure 11.61: Four faces of the lug selected

13. Now, we need to define **Weld Height**. Enter 5mm as the values shown in *Figure 11.62*. Select **OK** to complete the **Fillet Weld** configuration:

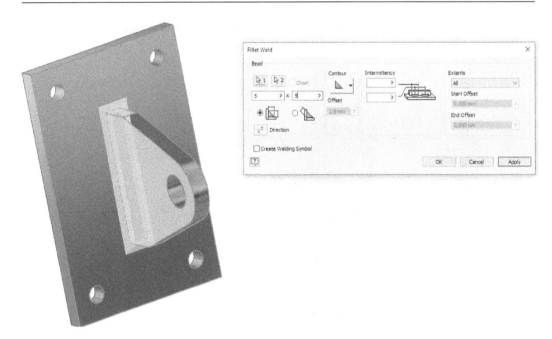

Figure 11.62: A 5 mm fillet weld applied

14. The weld is applied. We now need to define the stress analysis study. Select **Return** and then select the **Environments** tab.

15. Select **Stress Analysis**, followed by **Create Study**. Name the study Weldment and then select **OK**.

16. Apply fixed constraints on the four fixing holes, as per *Figure 11.63*:

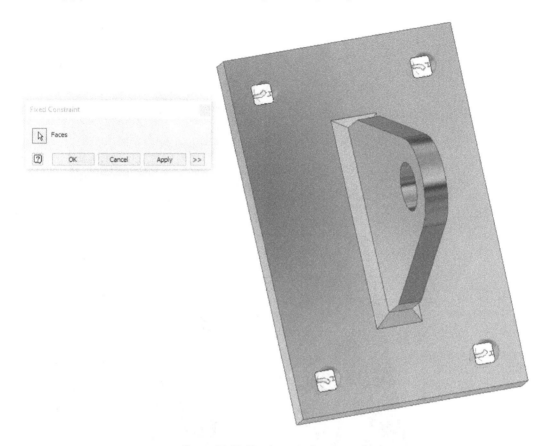

Figure 11.63: Fixed constraints to apply

17. Select **Gravity** and apply as per *Figure 11.64*:

Figure 11.64: Gravity applied to the model

18. Select **Force** and apply a **Magnitude** setting of 5,500 N to the inner circular face of the lug in the direction shown:

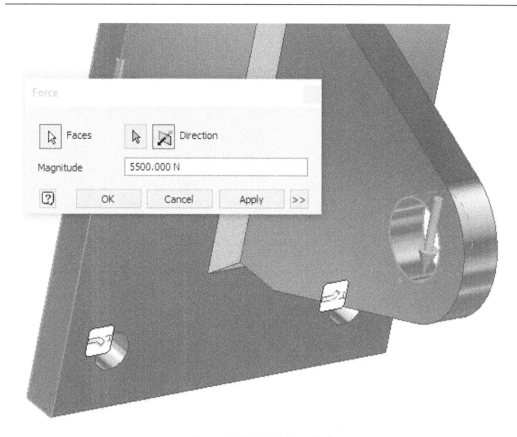

Figure 11.65: 5,500 N applied

19. Select **Automatic Contacts** and ignore the warning about weldments.

20. Select **Simulate** and then **Run.**

21. Select **Safety Factor** – the minimum value shown is **10.4**. This is much higher than the calculation.

Using **Weld Calculator**, you have calculated the minimum height value of the weld required to satisfy the design requirements of the bracket. The results have then been validated with a stress analysis study, which includes the weldments.

Model credits

The model credits for this chapter are as follows:

SOLAR PANEL STAND (by Suman pal): `https://grabcad.com/library/solar-panel-stand-3`

12

Sheet Metal Design – Comprehensive Methodologies to Create Sheet Metal Products

Autodesk Inventor has a specific design environment for the creation, editing, and detailing of **sheet metal** CAD components. Sheet metal parts are parts that are fabricated from a sheet of material with a uniform thickness. These sheet metal parts are subject to certain design constraints and sheet metal-specific manufacturing processes. Although specific to sheet metal parts, any parts that you create of uniform thickness in a standard part file can also use the sheet metal functionalities.

Within Inventor, sheet metal parts can be displayed as a folded model or a flat pattern. With the sheet metal functionality, you can work on sheet metal parts in both a folded and an unfolded state. Inventor enables you as the designer to set important **sheet metal defaults/rules** based on the equipment you have available for manufacture, for example, bend radii, K factors, material thickness, and corner seam type.

The ability to fold and refold the material, generate bend tables, and so on means the digital model created can greatly improve and enhance the physical manufacturing process.

A typical sheet metal design workflow is as follows.

The creation of a sheet metal part starts with a sheet metal template. The sheet metal template file is configured and stores the specific set of sheet metal rules. Adjusting and changing the rules enables the selection of different materials, thicknesses, and more.

Initially, a base feature is created. A sheet metal base feature is often a face of a shape. Additional sheet metal-specific features are then added. The sheet metal component does not have to start exclusively with a single flat face; alternatively, this could be a contour roll.

If a part is created with consistent thickness, with the standard modeling method, it can be converted into a sheet metal part in Inventor. This is also the case for non-native imported CAD models.

The typical sheet metal workflow is as follows:

1. Define sheet metal rules.

2. Start the creation of a base feature/face.

3. Apply secondary sheet metal features (holes, cuts, flanges, and hems).

4. Unfold and refold the model as appropriate.

5. Document the design in the drawing area of Inventor.

6. Export the folded and flat pattern design in the desired format, such as `.dxf` for manufacture.

Figure 12.1 shows an example of a sheet metal part created in Inventor and highlights some of the key sheet metal-specific features that can be created:

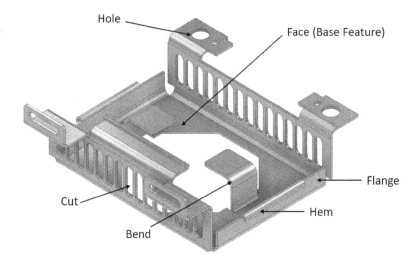

Figure 12.1: Sheet metal part highlighting specific sheet metal features

The sheet metal tools, upon selecting a sheet metal template in the part modeling environment, show the breadth of features that can be created:

Figure 12.2: Sheet metal environment active

In this chapter, you will learn about the following:

- Setting up and configuring your sheet metal template
- Creating a sheet metal part with faces, flanges, bends, hems, and cuts
- Applying advanced sheet metal base features – Contour Flange, Lofted Flange, and Contour Roll
- Converting a solid into a sheet metal part and performing a rip
- Creating and applying a custom sheet metal punch
- Designing sheet metal parts in assemblies and `.dxf` output
- Detailing sheet metal parts in the drawing environment

Technical requirements

To complete this chapter, you will need access to the practice files in the `Chapter 12` folder within the `Inventor Cookbook 2023` folder. It is important that you have a good understanding of the essential Inventor modeling techniques. An appreciation and understanding of sheet metal fabrication techniques is not essential but recommended.

Setting up and configuring your sheet metal template

This recipe will focus on the initial setup of a **sheet metal template**, prior to the creation of a sheet metal part. This process does not have to be carried out each time a sheet metal part is required. Once a template has been configured, it can be saved and used again, applying the settings and constraints to the next design. Multiple sheet metal templates can be configured, and in most cases will be required based on the materials and fabrication equipment available to the engineer.

Getting ready

To begin this recipe, you will not need any practice files.

How to do it...

We will start by creating a blank new sheet metal document, and then we will configure the desired sheet metal rules and settings within this. The last step will be to save this as an accessible template for future use:

1. From the Inventor **Home** screen, select **New** followed by **New**. Select **Templates | en-Us**, then select **Metric**. Select **Sheet Metal (mm).ipt** and then **Create**.

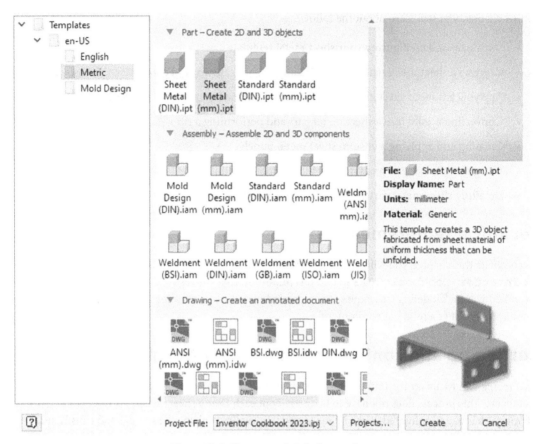

Figure 12.3: Sheet metal default template

2. A new sheet metal part file will open. Navigate to the far right of the ribbon, in the **Sheet Metal** tab, and select **Sheet Metal Defaults**.

Sheet Metal
Defaults

Setup ▼

Figure 12.4: Sheet Metal Defaults

This is where we can configure the template sheet metal rules.

3. Select **Edit Sheet Metal Rule** as shown in *Figure 12.5*.

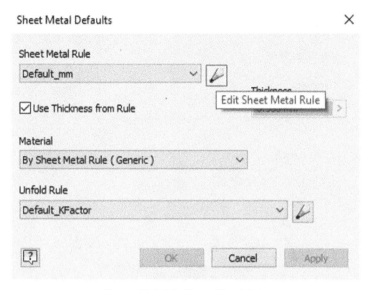

Figure 12.5: Edit Sheet Metal Rule

4. Select **Default**, then select **New…** to create a new local style.

Figure 12.6: New local style

5. Name your new local style `Sheet Metal TEST` and select **OK**.

6. The settings applied in reality will be dependent on the material and fabrication equipment. The following settings are not reflective of a real-world manufacturing facility. Under **Material**, select **Stainless Steel**. For **Thickness**, change this to a value from .500 mm to 1 mm. Leave the default value of **Report Bending Angle (A)**.

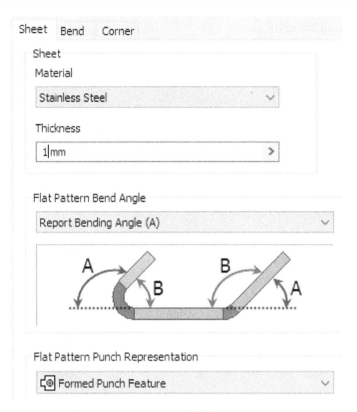

Figure 12.7: Material and Thickness settings

7. Select the **Bend** tab in the **Style and Standard Editor** dialog box.

8. Change **Relief Shape** to **Round**, and leave all other settings as the default. Select the **Corner** tab from **Style and Standard Editor**.

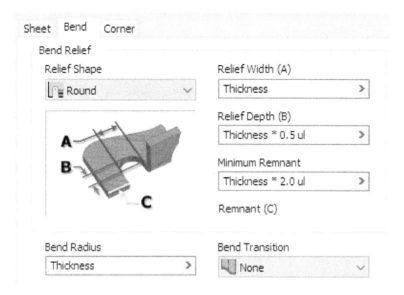

Figure 12.8: Bend relief options

9.   Define the **Corner** settings, as shown in *Figure 12.9*:

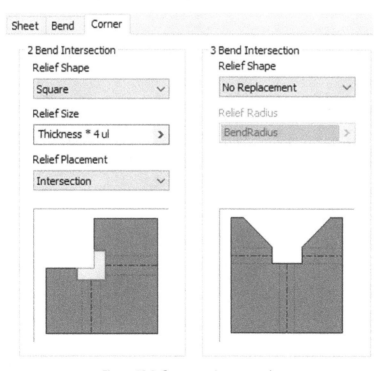

Figure 12.9: Corner settings to apply

10. Select **Save & Close**.

11. Select **Sheet Metal TEST** from the template list, then select **OK**.

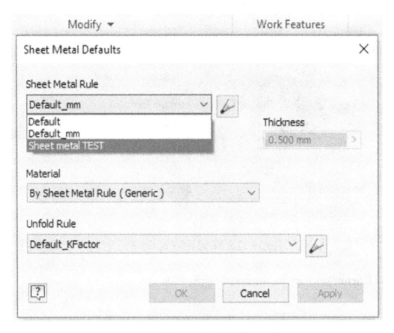

Figure 12.10: Sheet Metal TEST template

12. Select **File | Save As | Save Copy As Template**.

Figure 12.11: Save Copy As Template

13. Browse to the location shown in *Figure 12.12*. Select `Metric`. Save the template as `Sheet Metal TEST`.

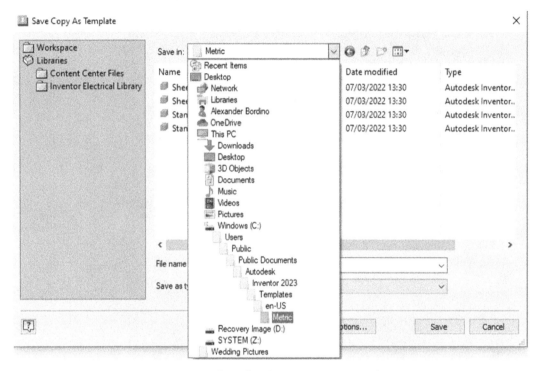

Figure 12.12: File path to browse to save a template

14. Select **Save**.

15. Now select **File | New**. The new `Sheet Metal TEST.ipt` template is available for selection with all the settings we previously applied in this recipe active.

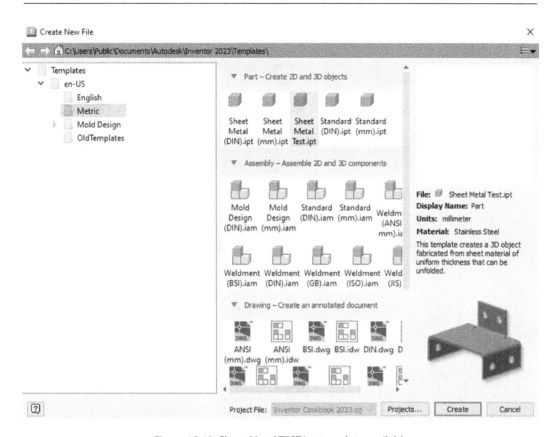

Figure 12.13: Sheet Metal TEST.ipt template available

By using the sheet metal defaults, you have defined and created a custom sheet metal template file, ready for future sheet metal component creation. This means that future parts created with this sheet metal template will follow the same sheet metal rules.

Creating a sheet metal part with faces, flanges, bends, hems, and cuts

You will create a basic sheet metal part using key feature commands such as **Face**, **Flange**, **Bend**, and **Hem** to create the final part. You will also learn how to unfold and refold a model.

Getting ready

You will not need any practice files for this recipe.

How to do it...

The first steps will involve creating a new sheet metal part and defining the base feature. From there, we will then start to use the sheet metal functionality to create the additional required features:

1. From the Inventor **Home** screen, select **New** followed by **Templates | en-US | Metric**. Then, select Sheet metal (mm).ipt and then **Create**.

2. Select **Start 2D Sketch**, then select the XZ plane. Select **Rectangle | Two Point Center Rectangle**, and then create the base sketch, as defined in *Figure 12.14*:

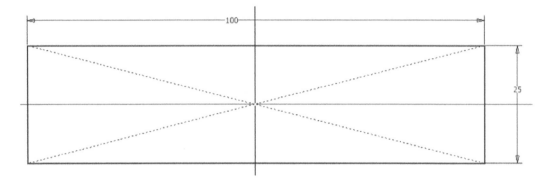

Figure 12.14: Initial sketch to create a 100 mm x 25 mm rectangle

Then, select **Finish Sketch**.

3. Select **Face** from the **Sheet metal** tab.

4. The profile is already selected to apply the default sheet metal and default thickness from the sheet metal rule in the template. Select **OK**.

5. Now, we will change the default thickness of the material. Select **Sheet Metal Defaults** and uncheck **Use Thickness from Rule**. Define a new thickness of .650 mm. Select **OK**.

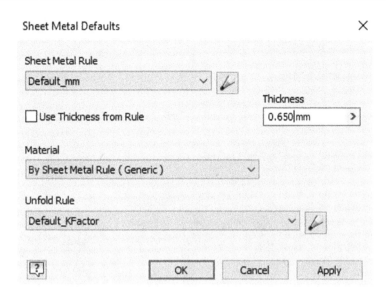

Figure 12.15: Sheet Metal Defaults to adjust the thickness

The face previously created in *steps 3* and *4* will update in thickness from .500 mm to .650 mm.

6. Create a new sketch on the top face of the sheet metal part. Create the sketch as per *Figure 12.16*. These sketch lines will form the basis of a fold operation.

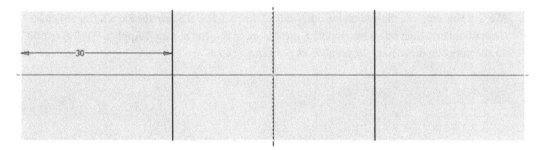

Figure 12.16: New sketch to define

7. From the sheet metal tools, select **Fold**. Select the sketch line shown in *Figure 12.17* as the reference. Use **Flip Controls** so that the direction of the bend is as in *Figure 12.17*. Leave **Fold Angle** at 90 degrees. Select **OK** to complete. A fold is applied to the sheet metal part.

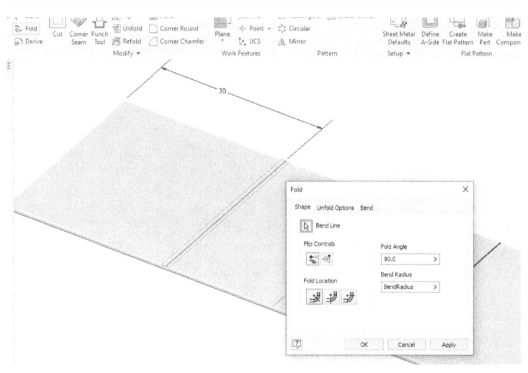

Figure 12.17: Fold to apply to a face

8. We will now replicate the fold on the other side of the part. Select **Share the Sketch** from **Sketch2**. Select **Fold** and then select the opposite sketch line to the one in *step 7*. Apply a 90-degree fold in the opposite direction. The result is as per *Figure 12.18*:

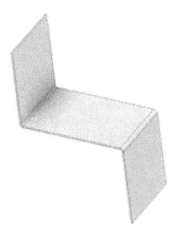

Figure 12.18: Second fold operation applied

9. As with all other features in Inventor, you can edit the sheet metal features using the **Model Browser**. Right-click on **Fold2** from the **Model Browser** and select **Edit Feature**. Adjust **Bend Angle** to 60 degrees.

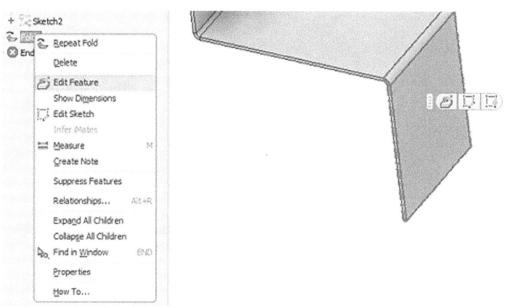

Figure 12.19: Edit Feature of a sheet metal feature

10. Select the **Flange** command. By selecting existing edges on a sheet metal part, you can specify and apply various **Flange** features. Multiple edges can be selected at once if the type of flange is to be the same. Select the four edges shown in *Figure 12.20*:

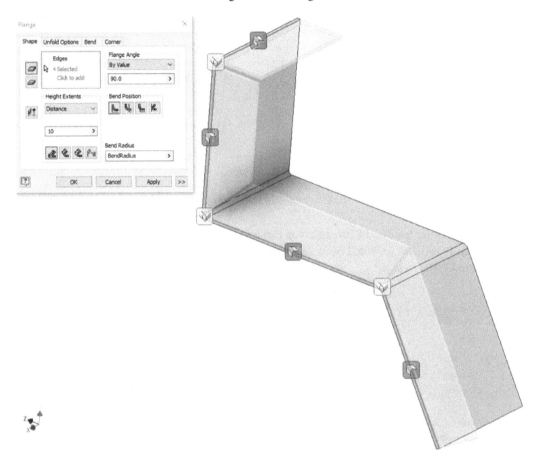

Figure 12.20: Edges of the flange selected

11. Change **Height Extents** to 10 mm. **Bend Rules, Bend Position**, and **Angle** can also all be adjusted. By default, Inventor will use values previously defined in the sheet metal rules in the template file. Select **Apply** to apply these flanges.

12. Select the edge shown in *Figure 12.21*, then change **Height Extents** to 20 mm and **Angle** to 70 degrees. Select **OK**.

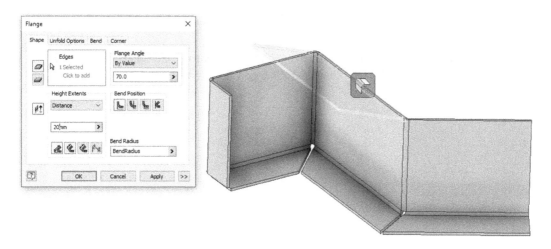

Figure 12.21: Second flange to apply

13. In the sheet metal create tools, in the ribbon, select **Hem**. This allows you to apply **Hem** features to sheet metal edges. Select the edge shown in *Figure 12.22*:

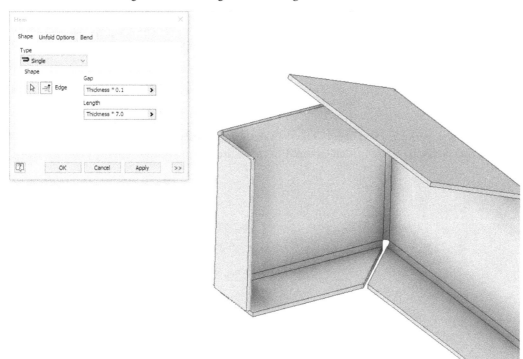

Figure 12.22: Hem to apply

14. Change **Type** to **Single**. Adjust **Thickness** to `Thickness * 0.1` and **Length** to `Thickness * 7.0`. Select **Apply** to complete.

15. Select the edge shown in *Figure 12.23*. Change **Type** to **Teardrop**. Select the >> button. Change **Width Type Extents** to **Width | Centered**. Change **Width** to `20` mm.

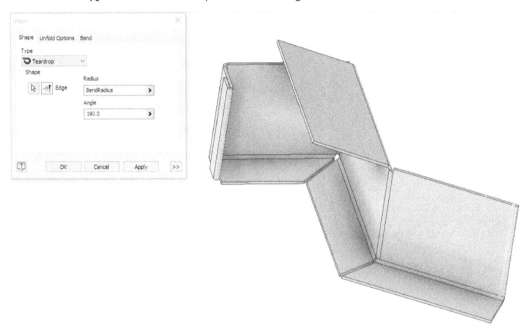

Figure 12.23: Second hem to apply

16. We will now perform a **Cut** operation on one of the faces. Select the face shown in *Figure 12.24* and then **Start 2D Sketch**. Then, proceed to create the sketch as defined in *Figure 12.25*.

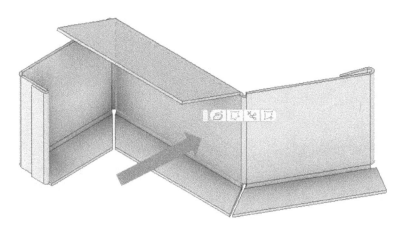

Figure 12.24: Face to start the 2D sketch

17. By pressing *F7* once, the sketch plane is selected, which will allow you to slice the graphics to gain a better view of the face.

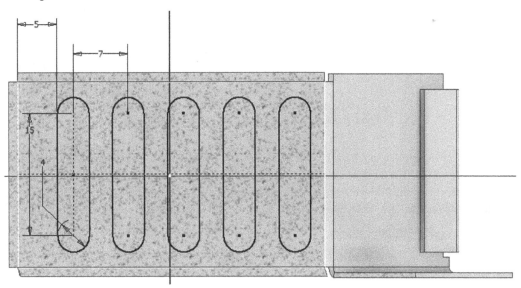

Figure 12.25: Sketch to create for the cut

18. Select **Finish Sketch**. Select **Cut**, then select the sketch slotted profiles and select **OK** to complete.

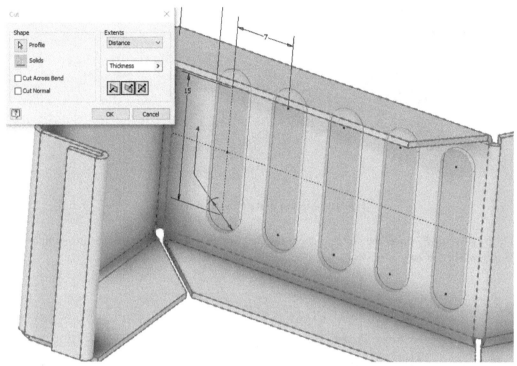

Figure 12.26: Cut operation to perform on sketched profiles

19. Now, we will unfold and refold the model. During the **Unfold** operation, Inventor will scan the part and ensure that no intersections are present. This could possibly indicate flaws that would mean aspects of the part would be unmanufacturable. Select **Unfold**, then select the **Stationary Reference** as the **Face**, as shown in *Figure 12.27*. Select **Add All Bends** and then select **OK**.

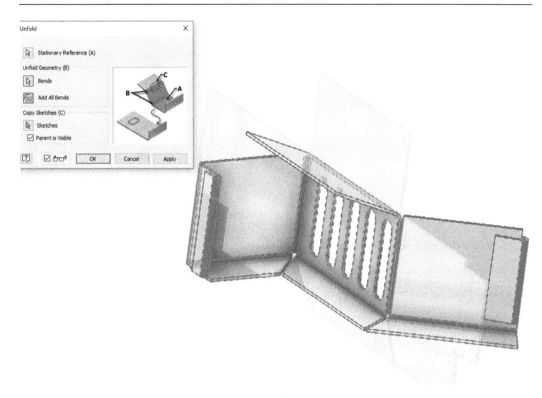

Figure 12.27: Unfold operation

A warning displays that informs you that there is an intersection of material in the part. Select **Accept** to proceed. The flat pattern is shown, as in *Figure 12.27*. The circled area in *Figure 12.28* is the possible intersection and problem area:

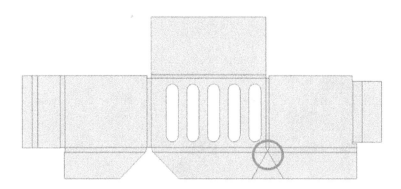

Figure 12.28: Part unfolded

A design change is required here prior to manufacture.

20. Select *Ctrl + Z* to cancel the **Unfold** operation. Edit **Flange1**.

21. Use *Ctrl* to deselect the edge shown in *Figure 12.29* and select **OK** to complete the edit to **Flange1**:

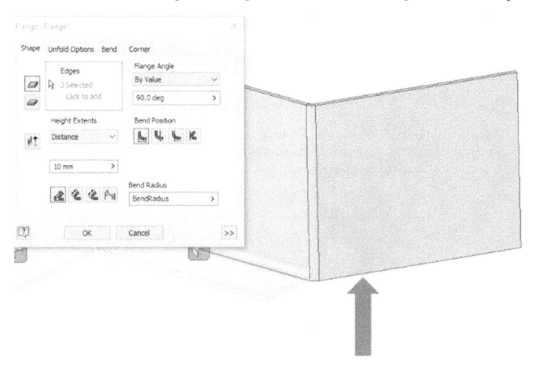

Figure 12.29: Edge to deselect from Flange1

22. Select **Unfold**. Select the same face for the stationary reference as before in *step 18*. Select **All Bends** to be included and select **OK**. The **Unfold** operation is performed without any warnings as the problem area has been rectified.

23. Select **Refold**, then select the same stationary reference as *step 22*. Select **All Bends** and then **OK**. The part is folded back.

You have successfully created a simple sheet metal part using a face base feature, flanges, bends, hems, and cuts. You have also used **Unfold/Refold** to identify problems prior to manufacture, and then addressed the issue.

Applying advanced sheet metal base features – Contour Flange, Lofted Flange, and Contour Roll

The face feature is not the only base feature available. **Contour Flange**, **Lofted Flange**, and **Contour Roll** can all be used as base features. In this recipe, you will learn how each is created at a basic level. All these operations require an open sketch to begin with.

Getting ready

To begin this recipe, you will need to navigate to `Inventor Cookbook 2023|Chapter 12`. Then, open `Contour Flange & Roll.ipt`.

With `Contour Flange & Roll.ipt` open, you will see that there is already an open sketch created. This sketch will provide the structure for the **Contour Flange** feature. The profile will be extruded, and the default sheet metal thickness value will be added.

All bend radii are added to the sketch.

Contour Flange can be used as a base feature or secondary feature. It is often used to create a rolled feature or multiple flanges on a sheet metal part.

How to do it...

The sketch profile we require for **Contour Flange** has already been created, so we will head straight into the sheet metal feature tools:

1. Select **Contour Flange**. The sketch profile is automatically picked up in the preview.

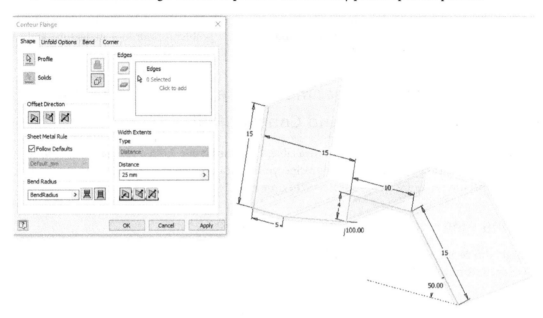

Figure 12.30: Contour Flange selected and a preview displayed

2. Change **Distance** from 25 mm to 15 mm. The preview will update.

3. Select **OK** to complete.

Figure 12.31: A contour flange is created from the sketch

Using **Contour Flange**, complex sheet metal forms can be made in a single operation, rather than multiple flanges and bends. Additional sheet metal features can now be added if required to the model.

4. Delete the contour flange feature but maintain the original sketch by deselecting the checkbox in the warning.

5. We will now perform a contour roll on the same sketch profile. This requires additional sketch geometry to act as an axis of revolution. Right-click **Sketch1** in the **Model Browser** and select **Edit Sketch**. Create the sketch lines as shown in *Figure 12.32*:

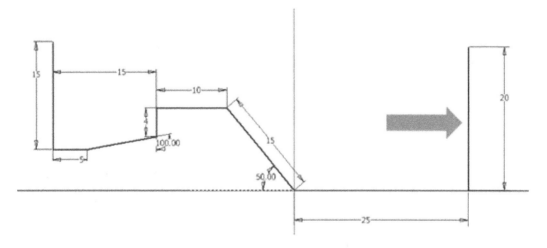

Figure 12.32: Sketch line added to act as an axis of revolution

6.   Select **Finish Sketch**. Select **Contour Roll**.

7.   Select the original sketch profile as the profile. Then, select the new sketch line created in step 5 as the axis. The **Contour Roll** preview is created as per *Figure 12.33*:

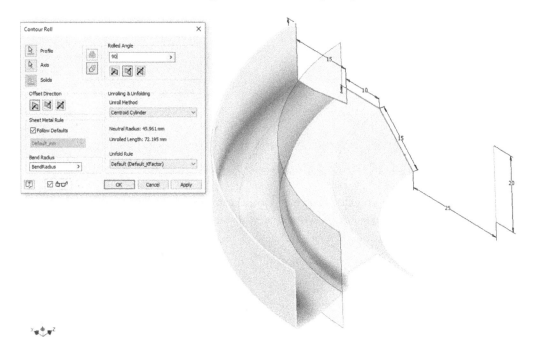

Figure 12.33: Contour Roll preview

8.   Now, we will adjust the contour roll parameters before completing the operation. Change **Rolled Angle** to 100 degrees. Then, select **OK** to complete.

The contour roll is completed.

9.   Close Contour Flange & Roll.ipt.

10.   Now, we will create a lofted flange using the sheet metal function. From the Chapter 12 folder, open Lofted Flange.ipt.

11.   The part contains two separate sketches of different profiles set on separate planes 50 mm from each other. We will create lofted flange between the two and create a sheet metal part that transitions from one profile to another.

12.   Select **Lofted Flange**.

13.   For **Profile 1**, pick the circle sketch in the **Graphics Window**. Select the ellipse for **Profile 2**. The preview of the lofted flange will generate the following:

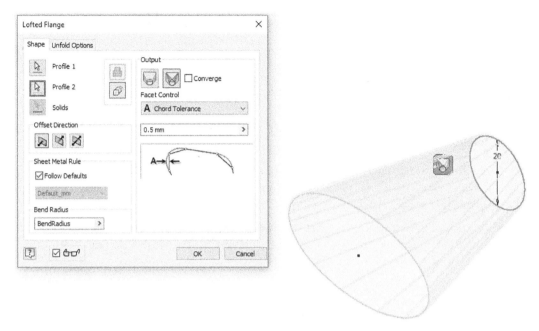

Figure 12.34: Lofted flange preview

14. We can now adjust the various options for the lofted flange. Output is the main factor – we can either switch from **Press Brake** as displayed or to a **Die Formed** solution, under **Output**. Leave this as **Press Brake**. This is shown in the **Output** section of **Lofted Flange**, as in *Figure 12.34*.

15. With **Press Brake**, **Facet Control** can be adjusted to either **Chord Tolerance**, **Facet Angle**, or **Facet Distance**. Select **Chord Tolerance** for this solution. Leave **Tolerance** as 0.5 mm and select **OK** to complete.

You have created several other base feature options for sheet metal parts in the form of **Contour Flange**, **Lofted Flange**, and **Contour Roll**.

Converting a solid into a sheet metal part and performing a rip

In this recipe, you will take an existing solid body part in Inventor, convert it into a sheet metal part, and then perform a **Rip** operation on the part.

Getting ready

To begin this recipe, you will need to navigate to `Inventor Cookbook 2023 | Chapter 12` and open `Rip.ipt`.

The part has been created via a lofted extrusion, followed by a **Shell** operation. This has created a solid body part. First, we need to convert this into a sheet metal component. As this has a uniform thickness of 2 mm, this is achievable.

How to do it...

We will start the process of converting the part into a sheet metal part, then we will define the reference points to perform the rip:

1. In the **3D Model** tab, navigate to the far right of the ribbon. Select **Convert to Sheet Metal**. Select one of the angled faces as the base feature. Select **OK** for **Sheet Metal Rules**. The part is converted into a sheet metal part.

2. Right-click on the top angled face and select **Start 2D Sketch**. Select **Project Geometry** and select the face to project the edges of the part to the sketch. Select **Finish Sketch**.

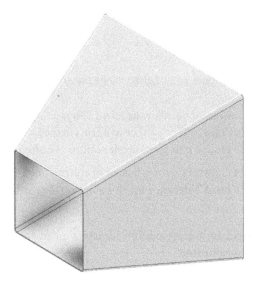

Figure 12.35: Projected sketch of top face

3. Select **Rip** from the **Sheet Metal** tab in the ribbon.

4. Keep **Type** as **Single Point** and select the face previously sketched on in *step 2* as the rip face. Then, select the midpoint of the edge of the shape, as shown in *Figure 12.36*. A preview of the rip will be shown:

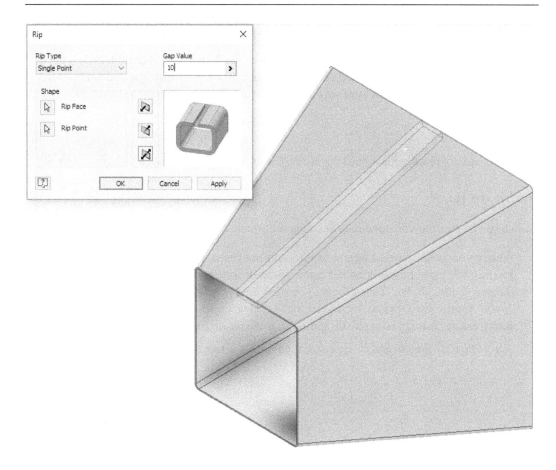

Figure 12.36: Rip face and rip point to select

5.   Overwrite **GapSize** to 10 mm. Then, select **OK**.

You have now converted a solid model into a sheet metal part and then performed a **Rip** operation. To ensure that converted solid models behave as expected when converted into a sheet metal, you must ensure the following:

- There is a consistent thickness to the model
- The sheet metal thickness parameter value matches that of the sheet metal rules
- There is some form of gap in the model and it is not a continuous face
- There is a rounded fillet on the outside edge

If these rules are not adhered to, converted sheet metal parts will not be able to be unfolded or have additional sheet metal features added.

Creating and applying a custom sheet metal punch

In this recipe, you will create a custom sheet metal punch file. You will then use this punch file in another sheet metal part. The punch tool can either remove or deform a material, and once created, can be reused in other files. Punch tools are extracted in a similar way to a standard iFeature.

Getting ready

You will not require a practice file to start this recipe. Create a new `Sheet metal (mm) .ipt` file.

How to do it...

We will begin by creating a face and then detailing the punch:

1. When creating punch tools, Inventor requires only one central point to locate this. Automatically, Inventor will project the origins and center points of sketch objects, which will make creating the punch difficult. For this reason, in your new `Sheet metal (mm) .ipt` file, select **Tools | Application Options**, then select the **Sketch** tab. Uncheck **Autoproject part origin on sketch create**. You can re-apply this setting after the punch has been created.

2. Select **Start 2D Sketch,** and create a sketch on the XY plane, as *Figure 12.37* shows:

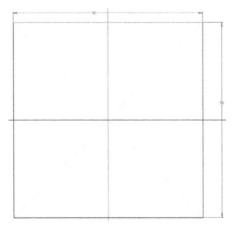

Figure 12.37: Initial sketch to create

This will be a 50 x 50 mm square at 0,0,0, the origin, as origin projection is temporarily turned off.

3. Select **Face**, and then select the sketch you created in *step 2*. Uncheck **Sheet Metal Defaults** for **Thickness** and change this from .500 mm to 1 mm. Select **OK** to complete.

4. Create a new 2D sketch on the face, as shown in *Figure 12.38*. This will become the punch tool detail.

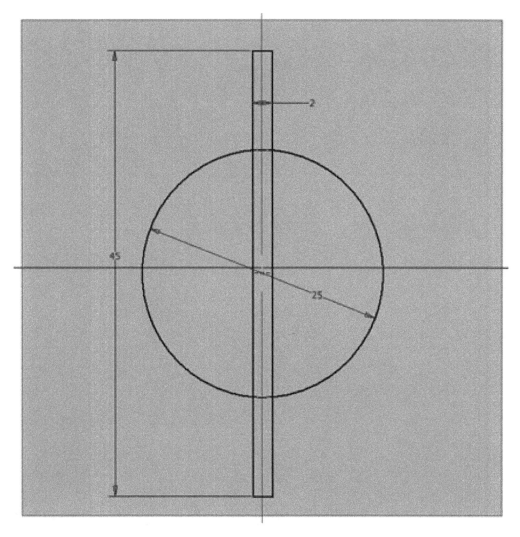

Figure 12.38: Sketch of punch tool to apply to the face

5. Once you have drawn the sketch as detailed in *Figure 12.38*, select **Point** from the sketch tools and select the center of the circle. This will be the locating point for the punch tool.

6. Select **Cut**, and then select the sketch previously created in *step 4*. Select the **Distance** and type **Thickness**. Select **OK** to complete.

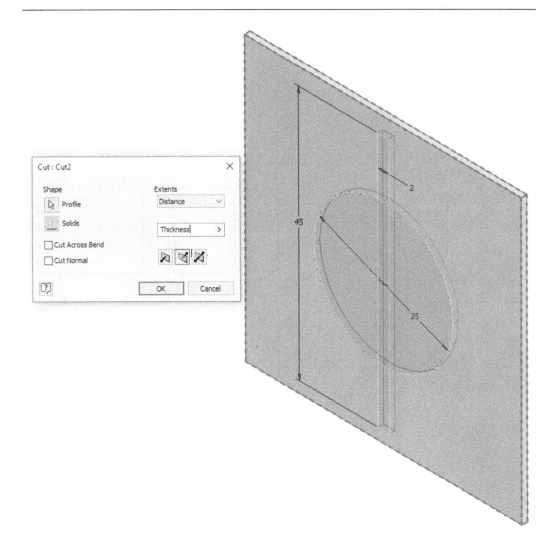

Figure 12.39: Cut applied to the sketch previously created

7. In the **Model Browser**, right-click **Sketch2** of the punch tool detail and select **Share Sketch**. This makes the sketch selectable when we export it as a punch file.

8. The punch is now complete. Now, we will define and extract it as a punch file. Select the **Manage** tab, then navigate right to the **Author** commands and select **Extract iFeature**.

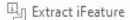

Figure 12.40: Extract iFeature

9. Select **Sheet Metal Punch iFeature** for **Type** in the **Extract iFeature** dialog.

10. Then, select **Cut1** from the **Model Browser** to bring this into the **Selected Features** section of **Extract iFeature**.

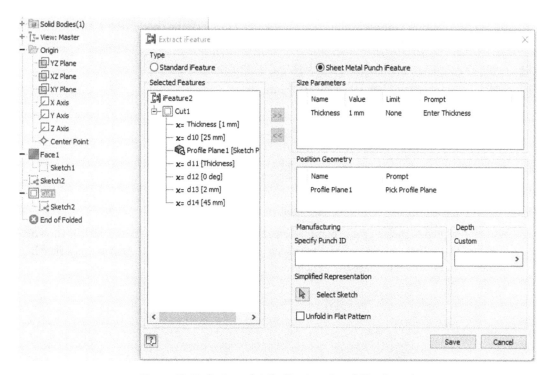

Figure 12.41: Options detailed in steps 9 and 10 selected

11. Under **Simplified Representation**, select the red cursor for sketch selection. Select the shared sketch, **Sketch2**, from the **Model Browser**.

12. Under **Specify Punch ID**, type `PunchTool01`. Select **Save**.

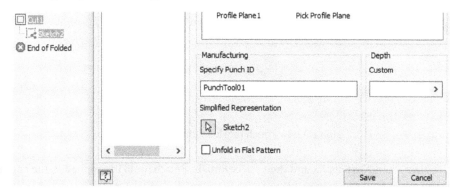

Figure 12.42: Punch ID specified and shared reference sketch selected

13. Inventor will now prompt you to save the file under a folder called `Catalog`. Rename the file `PunchTool01`, which is the same as your punch ID. Then, select **Save**. Select **Yes** on the warning.

14. The punch tool has now been successfully created and saved to an accessible library. Select **Tools | Application Options**, then select the **Sketch** tab. Click the checkbox for **Autoproject part origin on sketch create**. Save the Inventor file as `PunchTool01` in `Chapter 12`.

15. From the `Chapter 12` folder, open `Part to Punch.ipt`.

16. Select the **Manage** tab, then select **Insert iFeature**.

Figure 12.43: Insert iFeature

17. Select **PunchTool01** from the `Catalog` folder, then select **Open**.

18. When prompted to pick a profile plane, select the large rectangular face of the part, as shown in *Figure 12.44*. This will place the part centrally on the origin of the face. Alternatively, sketch points could be added to place the punch tool.

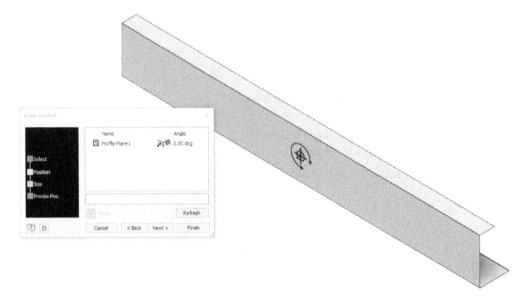

Figure 12.44: Punch tool location defined

19. Select **Next**, then **Next** again, and then select **Finish**. **PunchTool01** is applied to the part and the material is removed.

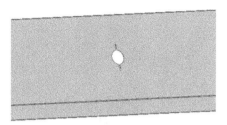

Figure 12.45: PunchTool01 applied

20. In this instance, we require multiple features to be punched into this face. Rather than importing the iFeature again to create more punch holes, we can instead apply a feature pattern of the punch cut to create the desired outcome in a more efficient way. Using the **Feature Pattern** tool, create the pattern shown in *Figure 12.46*:

Figure 12.46: PunchTool01 feature pattern to create more punch holes

You have created a custom sheet metal punch tool as a sheet metal punch tool iFeature, saved this to an accessible library, and then used and applied it to a punch material on an existing sheet metal part.

Designing sheet metal parts in assemblies and .dxf output

In many instances, it can be advantageous to design the sheet metal part required within the context of an existing assembly. Within Inventor, you can use multiple modeling techniques to achieve this with standard parts, and this also extends to sheet metal parts.

In this recipe, we will design and place an additional component for a wall-mounted TV bracket. Once complete, we will then produce flat patterns and export the final design to a .dxf file for manufacture.

Getting ready

To begin this recipe, you will need to navigate to Inventor Cookbook 2023 | Chapter 12 and open TV Bracket Assembly.iam.

The assembly file contains an incomplete TV wall-mounted bracket, as shown here:

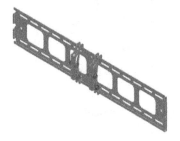

Figure 12.47: Incomplete TV wall-mounted bracket assembly – TV Bracket Assembly.iam

In this recipe, we will use sheet metal tools and standard modeling practices to create a new component in the context of the assembly. Once complete, we will generate a flat pattern of the component and export it as a .dxf file for manufacture.

How to do it...

We will begin by creating an in-place component of the additional bracket required. A face base feature will be defined, and then additional sheet metal features will be added. Once the part is complete, we will then duplicate and add another instance of this to the assembly:

1. Select the **Assemble** tab, then select **Create**.

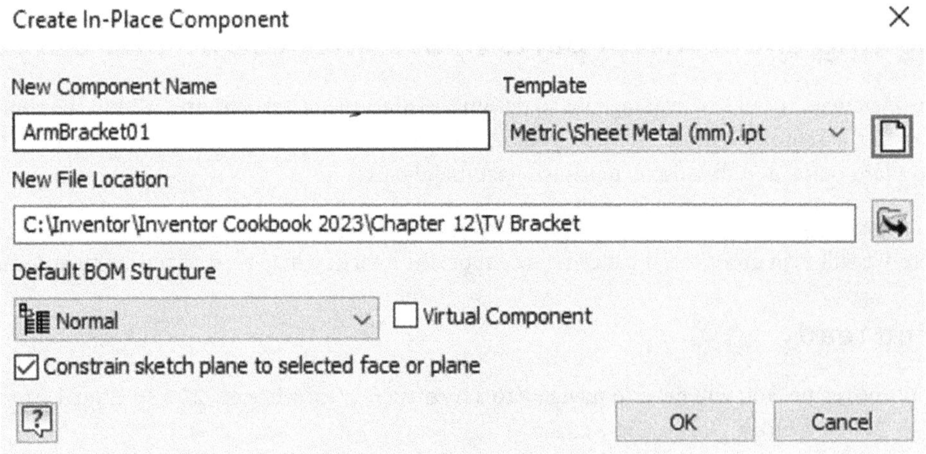

Figure 12.48: Component name, template, and file location

2. In the **Create In-Place Component** window, we now need to define **New Component Name**, **Template**, and **New File Location** for the new part:

   - For **New Component Name**, type `ArmBracket01`

   - For **Template**, select the **Browse** command, select **Metric**, and then select `Sheet Metal (mm).ipt`

   - Ensure **New File Location** is set to `Inventor Cookbook 2023\Chapter 12\ TV Bracket`

   - Select **OK** to complete

3. Select the face of the model highlighted in *Figure 12.49* as the reference:

Figure 12.49: Face to select as a reference

4. We are now creating an in-place component within the assembly. The sheet metal tools are now visible in the ribbon. Select **Start 2D Sketch**, then select the same face that is shown in *Figure 12.49* as the plane to sketch from. You may have to reorientate the model at this point.

5. In the new sketch, select **Project Geometry** and then select the two circles shown in *Figure 12.50*:

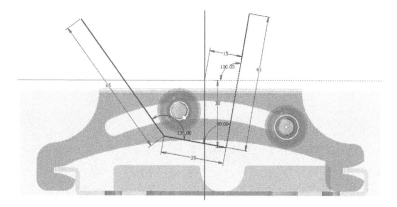

Figure 12.50: Existing geometry to project in the sketch

We will now define the first face of the sheet metal part.

6. Using the **Line | Construction Line** and **Dimension** commands, generate and define the sketch lines of the first face, as shown in *Figure 12.51*:

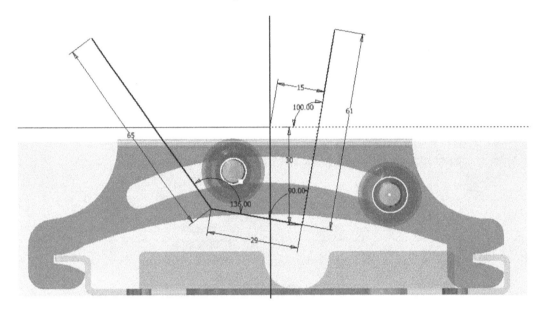

Figure 12.51: First sketch lines to generate

7. In the same sketch, add the remaining sketch geometry, as shown in *Figure 12.52*. Ensure that the sketch is fully defined before proceeding. Then, select **Finish Sketch**. You can also access the completed version of this file if required, with the completed final sketch. Go to Inventor Cookbook 2023 | Chapter 12 | TV Bracket | and open TV Bracket Assem Complete.iam.

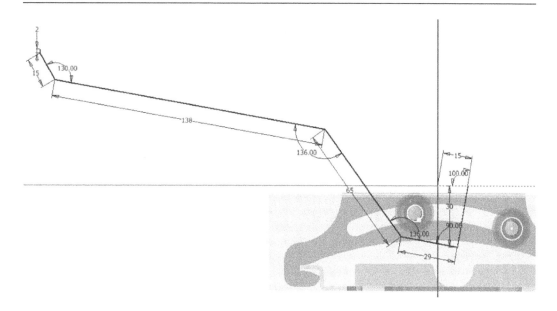

Figure 12.52: Remaining sketch geometry to apply

8. Select **Mirror**. Select all geometry to mirror and then select the line highlighted as the mirror line in *Figure 12.53*. Select **Apply** to complete.

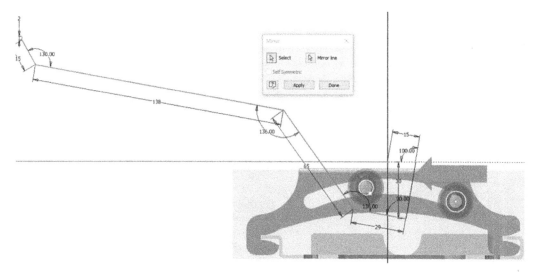

Figure 12.53: Mirror sketch to apply

9. Select **Line** and draw the final line, as shown in *Figure 12.54*, to complete a closed profile for the face. Then, select **Finish Sketch**.

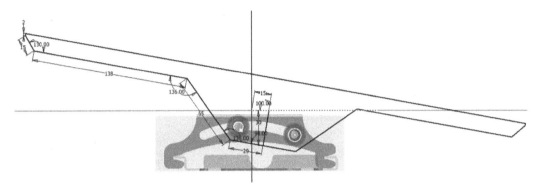

Figure 12.54: Final line to apply

10. Select **Sheet Metal Defaults**, then uncheck **Use Thickness from Rule** and type a new value of 3 mm for **Sheet Metal Thickness**. Select **Apply**.

11. Select **Face**, then select the sketch profile completed in *step 9*.

12. During the preview, move the view so you can see the bolted connection from the side. Ensure that the plate meets the washer. Then, select **OK**.

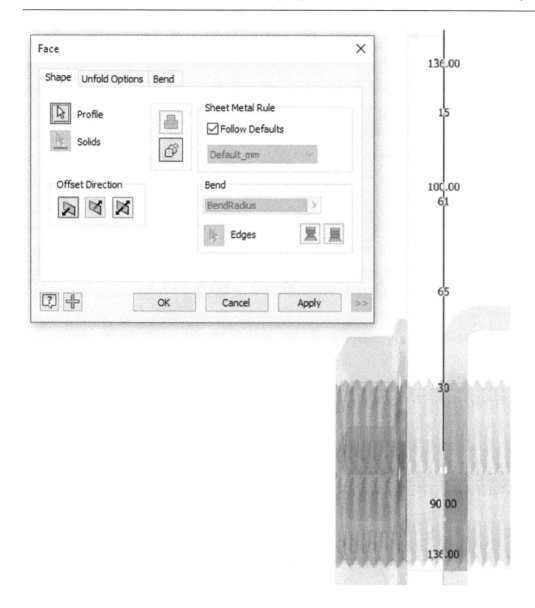

Figure 12.55: Face selected and 3 mm thickness applied

13. Select **Start 2D Sketch** and create a sketch on the face of the solid generated, as shown in *Figure 12.55*. Sketch the geometry shown in *Figure 12.56*, which will form a sheet metal cut in the next step:

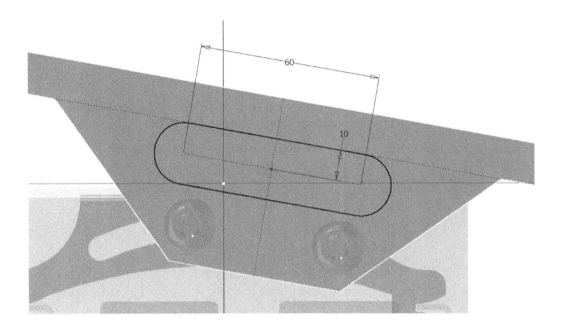

Figure 12.56: Slot geometry to sketch

14. Select **Finish Sketch**. Then, select **Cut**.

15. Select the sketch profile of the slot created in *step 13*. Ensure **Cut Distance** is set to **Thickness** and select **OK**.

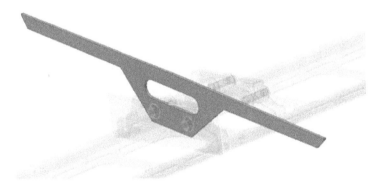

Figure 12.57: Cut performed

16. Select **Start 2D Sketch**. Select the same face as before for the plane.

17. Using **Project Geometry**, project the holes shown in *Figure 12.58*. Then, select **Finish Sketch**.

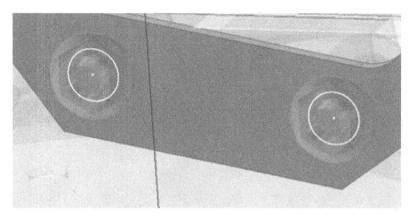

Figure 12.58: Holes to project in the sketch

18. Select the **Hole** command from the **Sheet Metal** tab. Select the two center points projected previously as the **Hole** centers. Under **Type**, select **Simple** for **Hole** and **None** for **Seat**. Then, enter 10 mm for the diameter. Select **OK** to generate the two cut holes.

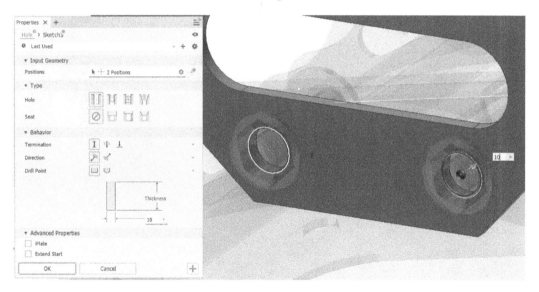

Figure 12.59: Hole command to generate two required holes

19. Select the **Flange** command from the **Sheet Metal** tab. Select the edge shown in *Figure 12.60*. Change **Flange Angle** to -90 degrees and **Distance** to 11 mm. Select **OK** to apply the changes.

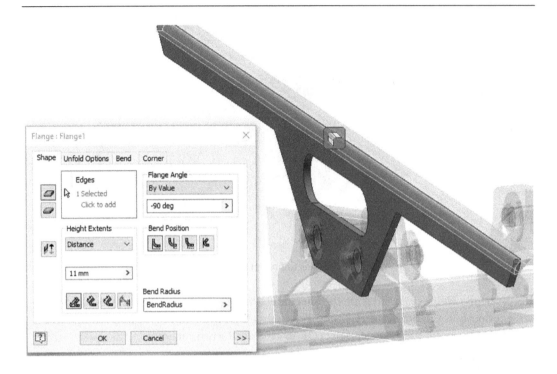

Figure 12.60: Flange applied to Face

20. Select **Mirror**. Ensure **Mirror features** is selected. Then, select the face, flange, bolt holes, and slot cut previously created. For **Mirror Plane**, select the face shown in *Figure 12.61*. Select **OK**.

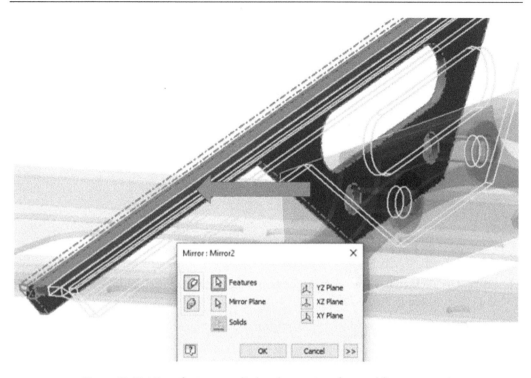

Figure 12.61: Mirror features applied to the previous face and flange created

21. Select **Corner Round** from the **Sheet Metal** tab. Select the 12 corners shown in *Figure 12.62*. Set **Radius** to 4 mm and select **OK** to complete.

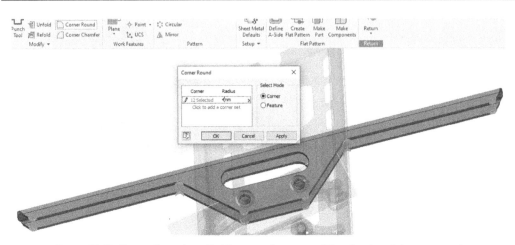

Figure 12.62: Corner Round applied to several corners of the sheet metal component

22. Select **Start 2D Sketch**. Then, select the top face of the bracket as the plane. Sketch the geometry and points shown in *Figure 12.63*:

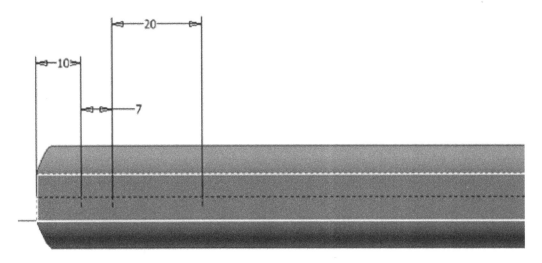

Figure 12.63: Top face of bracket and sketch to create

23. Select **Hole** and apply the following three hole sizes to each point, as detailed in *Figure 12.64*:

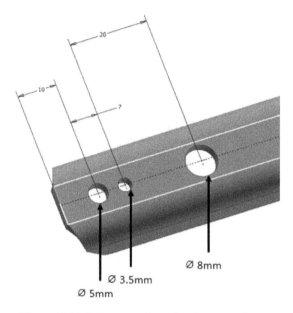

Figure 12.64: Holes to apply to the sheet metal part

Each hole has **Simple** set for the **Hole** type, and the **Seat** type is set to **None**. Only the diameter changes.

24. Select **Rectangular Pattern** from the **Sheet Metal** tab. Select the three holes created in *step 23* as the feature to pattern. Then, select the edge shown in *Figure 12.65*. For the spacing type, enter 50 mm. For the number type, enter 50 mm. Select **OK** to complete.

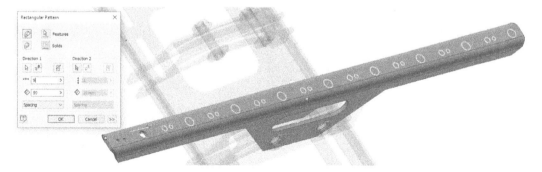

Figure 12.65: Edge to select for Rectangular Pattern, to create the desired number and spacing of holes

25. Select **Return** to return to the assembly. The additional bracket component required (**ArmBracket01**) is now complete. We now need to duplicate **ArmBracket01** for the other bolted connection in the assembly.

26. Select the **Assemble** tab, then select **Plane**. From the dropdown, select **Midplane between Two Planes**.

27. Select the face of the component shown in *Figure 12.66*. Select the same reverse face on the second instance of this identical component, also shown in *Figure 12.66*:

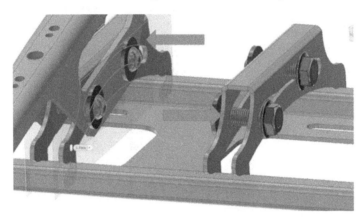

Figure 12.66: Mirror plane, reference planes to select

28. Now that a new workplane has been created, select **Mirror** from the **Assemble** tab.

29. Select **ArmBracket01** as the component to mirror, then select the workplane created in *steps 26* and *27* as the mirror plane. Select **Next**, then **OK**.

Figure 12.67: Completed TV wall-mounted bracket

30. We will now generate a flat pattern of the new **ArmBracket01** we created as part of the assembly and export this as a .dxf file for manufacture. In the **Model Browser**, right-click on **ArmBracket01**, then select **Open**.

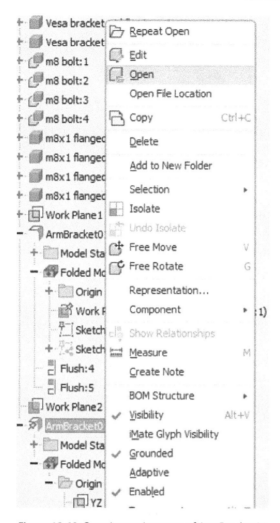

Figure 12.68: Opening an instance of ArmBracket01

31. Select **Create Flat Pattern** from the **Sheet Metal** tab. The sheet metal part is flattened into a flat pattern for manufacture:

- Select **Bend Order Annotation** to display the number of bends and the bend order

- Select **Go to Folded Part** to go back to the folded part

- Select **Flat Go To Flat Pattern**

Figure 12.69 shows the flat pattern of the bracket we created with the bend order displayed:

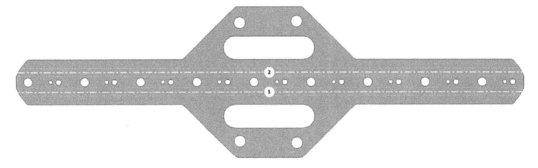

Figure 12.69: Flat pattern of ArmBracket01 with bend annotations

32. In the **Model Browser**, right-click **Flat Pattern** under **ArmBracket01**. Select **Save Copy As…**.

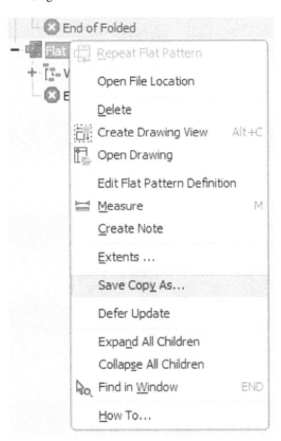

Figure 12.70: .dxf export of the Flat Pattern sheet metal part

33. Set **Save Type** to `.dxf`. Select **Save**. Select the desired **Export** options, then select **OK**. This exports the flat pattern of the part as a `.dxf` file.

You have created an in-place sheet metal part in an assembly, with numerous sheet metal features, such as holes, cuts, faces, and flanges. You have then applied duplication techniques to replicate the repeating features and components in the assembly, generated a flat pattern of the completed component, and exported this as a `.dxf` file for manufacture.

Detailing sheet metal parts in the drawing environment

Inventor has specific detailing tools for sheet metal parts, which include bend tables and hole tables. The standard 2D detailing tools can also be used to communicate design intent on the design. Models can also be displayed on a drawing as flat patterns and folded parts.

In this recipe, you will learn how to generate a hole table and bend table and place folded/unfolded views of an existing sheet metal component within the drawing environment.

Getting ready

To begin this recipe, you will need to create a new `ISO (mm) .dwg` file.

How to do it...

First, we will import the views of a ready-made component in both the **Folded** and **Flat Pattern** states:

1. Select **Base**.
2. Select the **Browse** command and search for and locate `Detail Sheet Metal.ipt` in the `Chapter 12` folder. Select **Open**.

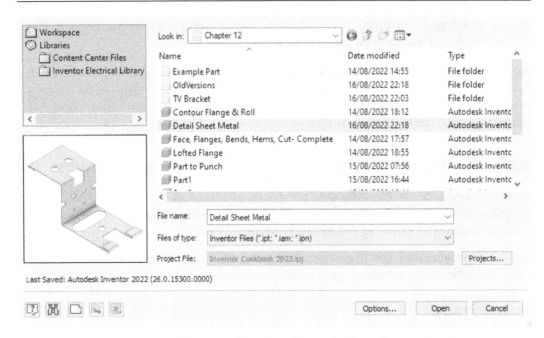

Figure 12.71: Detail Sheet Metal.ipt selected to apply a base view to a drawing

3. You will notice that the **Flat Pattern** option is not available. This is because the flat pattern has not yet been generated in the part file. We will first place a view of the part in its **Folded** state. Drag the view to the top-right corner of the drawing. Then, select the top left of the view cube to display an **Isometric** view of the part. Select **OK** to apply the changes.

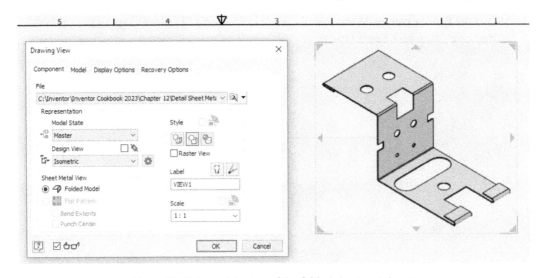

Figure 12.72: Isometric view of the folded sheet metal part

4. Right-click on the isometric view placed in *step 3*. Select **Open** to open .ipt for this view.

5. Select **Create Flat Pattern**. Then, select **Save**. Select **OK** and close .ipt.

6. Navigate back to the drawing file. Select **Base**.

7. Browse for Detail Sheet Metal.ipt. Select the file and then select **Flat Pattern**. Then, change **Scale** to 1:1. Place the view as per *Figure 12.73* and select **OK**:

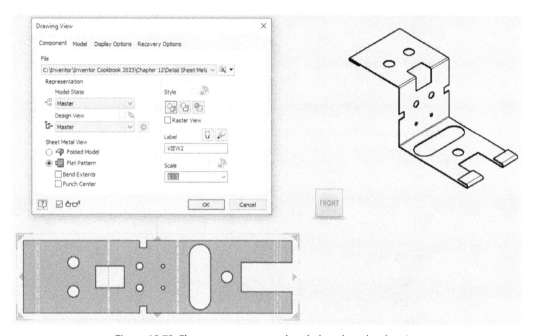

Figure 12.73: Flat pattern generated and placed on the drawing

8. Select the **Annotate** tab, then select **General** from the ribbon in the **Table** commands.

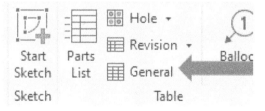

Figure 12.74: General table command

9. Select the **Flat Pattern** view placed in *step 7*. The dialog preview populates with the column headings that the bend table will have. You can select **Column Chooser** to alter the column types if required, by adding or removing items. The bend table is reading information directly from the model's iProperties, the same way a BOM table works.

Figure 12.75: Bend table preview

10. Select **OK** to place the bend table in the top left of the drawing.

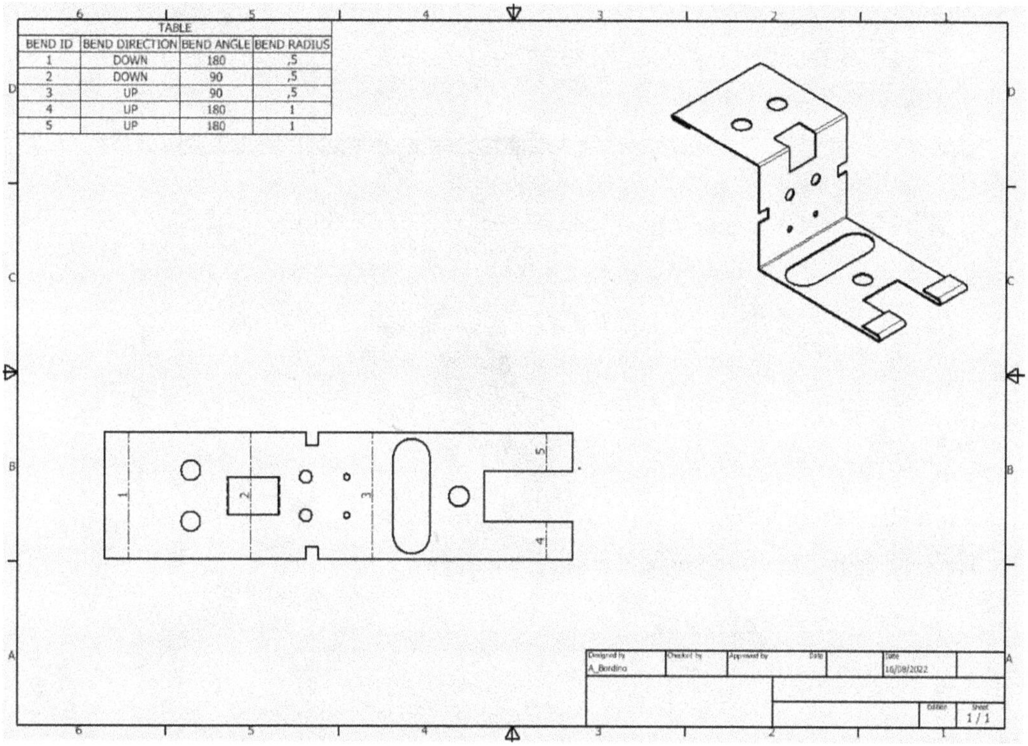
TABLE
BEND ID | BEND DIRECTION | BEND ANGLE | BEND RADIUS
1 | DOWN | 180 | ,5
2 | DOWN | 90 | ,5
3 | UP | 90 | ,5
4 | UP | 180 | 1
5 | UP | 180 | 1
A_Bordino
16/08/2022
1 / 1

Figure 12.76: The bend table is generated

11. The bend table is applied to the drawing. The **Flat Pattern** view is automatically edited to show corresponding **Bend ID** numbers that match the bend table. All information from the 3D model is displayed. We will now specify and generate a hole table. Select the dropdown next to **Hole** in the ribbon. Select **Hole View**.

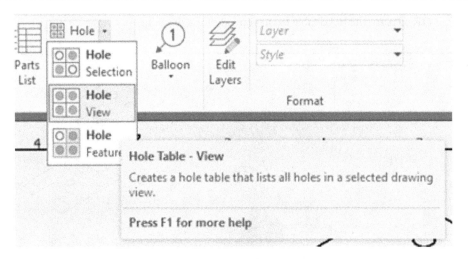

Figure 12.77: Hole View selected

12. Select **Flat Pattern View**.

13. Place the datum on the top right of the flat pattern view.

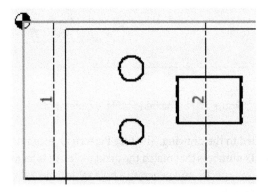

Figure 12.78: Datum placed

14. Click under the bend table to place the hole table.

15. Double-click on the generated hole table. Select the **Formatting** tab.

16. Select **Column Chooser**. Select **QUANTITY** from **Available Properties** and select **Add**. Select **OK**, followed by **OK** again.

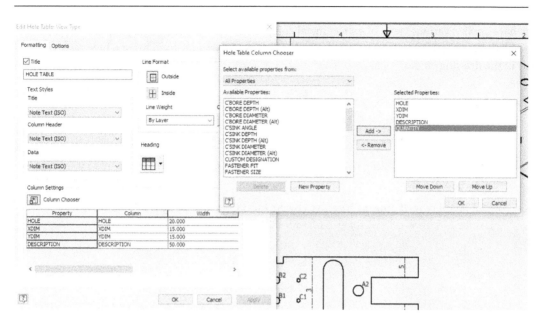

Figure 12.79: Hole Table Column Chooser

17. The hole table updates, and the flat pattern view also displays corresponding hole IDs.

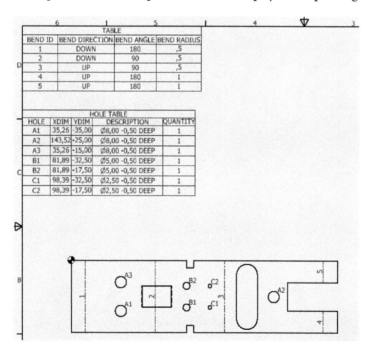

TABLE			
BEND ID	BEND DIRECTION	BEND ANGLE	BEND RADIUS
1	DOWN	180	,5
2	DOWN	90	,5
3	UP	90	,5
4	UP	180	1
5	UP	180	1

HOLE TABLE				
HOLE	XDIM	YDIM	DESCRIPTION	QUANTITY
A1	35,26	-35,00	Ø8,00 -0,50 DEEP	1
A2	143,52	-25,00	Ø8,00 -0,50 DEEP	1
A3	35,26	-15,00	Ø8,00 -0,50 DEEP	1
B1	81,89	-32,50	Ø5,00 -0,50 DEEP	1
B2	81,89	-17,50	Ø5,00 -0,50 DEEP	1
C1	98,39	-32,50	Ø2,50 -0,50 DEEP	1
C2	98,39	-17,50	Ø2,50 -0,50 DEEP	1

Figure 12.80: Sheet metal hole table and bend ID applied to the drawing

You have applied various views of a sheet metal component to a drawing, displaying both folded and unfolded states. You have also generated numbered bend tables and configured a hole table of the part in the drawing environment.

Model credits

The model credits for this chapter are as follows:

- **Sheet metal tray** (by Dileep Krishnadas K): `https://grabcad.com/library/sheet-metal-tray-3`

- **VESA display wall-mount bracket** (by Anindo Ghosh): `https://grabcad.com/library/vesa-display-wall-mount-bracket-1`

13

Inventor Professional 2023 – What's New?

Inventor 2023 was released in March 2022 and, as always, Autodesk has made many performance, automation, and core modeling workflow improvements on the previous release. Many of the additional features and enhancements have been driven by customer feedback and ideas from the Inventor forums community. The enhancements of version 2023 help streamline the design process and reduce repetition. Many of the key improvements have been made around connected workflows and interoperability with other products in the Autodesk portfolio, such as **Fusion 360** and **Revit** interoperability. The focus on this aspect shows that it is now more common for designers and engineers to work on more collaborative and complex projects that require seamless interaction and the interoperability of file types and workflows in CAD/CAM packages.

You can find the full Inventor 2023 release notes here: `https://help.autodesk.com/view/INVNTOR/2023/ENU/?guid=Inventor_ReleaseNotes_release_notes_html`

If you are thinking about moving to Inventor 2023, the **system requirements** can be found here: `https://knowledge.autodesk.com/support/inventor/learn-explore/caas/sfdcarticles/sfdcarticles/System-requirements-for-Autodesk-Inventor-2023.html`

In this chapter, you will learn about the following:

- General enhancements to Inventor 2023
- Interoperability between Fusion 360, Revit, and more
- Sketch enhancements
- Part enhancements
- Assembly and presentation enhancements
- Drawing enhancements

Technical requirements

You can access the practice files, where applicable, from the `Inventor Cookbook 2023 > Chapter 13` folder.

> **Important note**
>
> This chapter will not have a recipe for each enhancement/new feature or change, but wherever appropriate, these have been incorporated and practice files have been made available.
>
> Also, this chapter does not include future bug fixes or updates to the 2023 post-release version.

General enhancements to Inventor 2023

This section details general enhancements that have been made to version 2023.

New Home screen: Home replaces My Home

One of the most graphically dramatic differences between version 2022 and 2023, is the new **Home** screen. This has replaced the previous **My Home** screen. Upon opening Inventor 2023, you will be presented with a new graphical interface to start the creation of or open existing files.

The redesigned **Home** page has a design and layout that is more consistent with other Autodesk products, such as **AutoCAD** or **Fusion 360**, providing a more consistent and unified theme.

The new arrangement and features of the **Home** screen are as follows:

- The left panel offers access to select and create projects or to start or open files
- Recent documents can be displayed as thumbnails or as a list, similar to the 2022 release, but with a refreshed look and feel

Figure 13.1 shows the new **Home** screen:

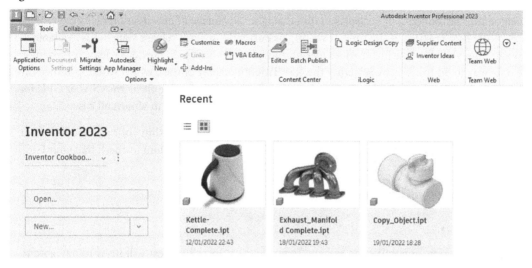

Figure 13.1: New Inventor 2023 Home screen

The new design of the **Home** page eliminates many of the duplicated functions found in the 2022 release and makes the process of accessing data and files much easier and more fluid.

Figure 13.2 shows the list option selected for preview files on the **Home** screen:

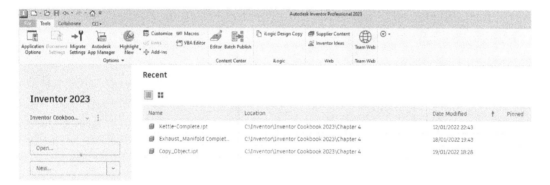

Figure 13.2: Home screen shown in list view

You can also pin files of interest, and refine the **Sort by** categories when viewing files.

GPU Ray Tracing display

Inventor 2023 introduces a new display option that takes advantage of new developments in graphics cards. **GPU Ray Tracing**, at the time of writing, is currently in a pre-release state but is available in Inventor now. GPU Ray Tracing provides real-time graphics rendering. In previous releases of Inventor, the only way to do this was by selecting the **Ray Tracing** display option, which would produce ray tracing via the CPU. The GPU Ray Tracing functionality allows you to use either the CPU or GPU for ray tracing. The use of the GPU should result in a higher-quality render in almost all cases.

The performance of GPU Ray Tracing depends heavily on the GPU within your machine, so your results may vary dramatically. Before using GPU Ray Tracing, it is important to check that your GPU is compatible with and supported by Inventor 2023.

The following recipe will demonstrate how to access GPU Ray Tracing.

How to do it

To begin this recipe, you will need to have Inventor 2023 installed. You will need to navigate to `Inventor Cookbook 2023` > `Chapter 13` and open `Bomba KSB ETA 80-20.ipt`. Then, follow these steps:

1. With the `Bomba KSB ETA 80-20.ipt` part open, select the **Tools** tab. Then, select **Application Options**.

2. Then select the **Hardware** tab, followed by **Enable Viewport GPU Ray Tracing (Pre-release)**. *Figure 13.3* shows these options:

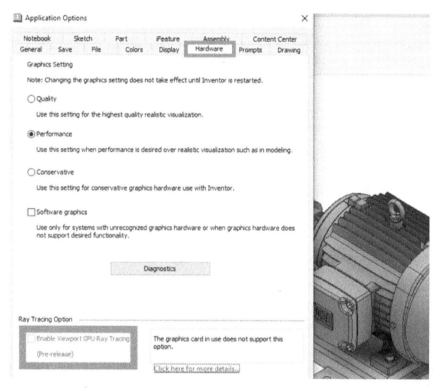

Figure 13.3: How to access GPU Ray Tracing

3. Close **Application Options**. Select the **View** tab. In the **Appearance** panel, change **Visual Style** to **Realistic**.

4. Select **Ray Tracing**. Configure your **Ray Tracing** requirements and begin **Ray Tracing**. If you don't have a compatible GPU, Inventor will still carry out ray tracing but using your CPU.

You have configured Inventor 2023 to utilize GPU Ray Tracing and create a real-time optimized render.

iLogic

A small **iLogic** enhancement has been made in this release. The custom command for an **External Rule** or **Global Form** can now be applied to the ribbon for easier access and execution.

Improved Free Orbit

As in previous releases of Inventor, you can hold *Shift* and scroll the *mouse wheel* to perform a free orbit around a model. In version 2023, this has been further enhanced by what is visible to the user in the **Graphics Window**, as follows:

- If the full model is in view, then the default **Pivot Point** is the model's center
- If the model is partially in view, then the central **Pivot Point** will snap to the nearest edge, face, or vertex
- If the model is outside the view, then the **Pivot Point** will default to the cursor location

These enhancements make **Free Orbit** more stable and easier to navigate.

[Primary] replaces Master in model states

For model states, view representations, and positional representations, the term **Master** has been removed. It has been replaced with **[Primary]**. This is to minimize conflict with potential part names featuring *Primary*.

Existing iLogic code that states that Master referring to master model states, for example, will not be affected by the change. The mechanics of model states, view representations, and positional representations will not be affected by this change.

For more on model states, see *Chapter 7, Model and Assembly Simplification with Simplify, Derive, and Model States*.

The Alt + Q new keyboard shortcut for Annotate

Holding *Alt + Q* is a new keyboard shortcut to the **Annotate** tab.

Performance enhancements and Express Mode

Inventor 2023 promises to bring substantial performance enhancements to both **Thread** display in assemblies, and **iFeature** performance with **Punch Tools**.

The **Express Mode** has been further improved – when component visibility is changed, performance is improved in **Express Mode**.

Express Mode is ideal for large assemblies and, when enabled, models will open much faster by loading only component-cached graphics into memory. **Express Mode** can be located via the **Application Options | Assembly** tab.

Interoperability between Fusion 360, Revit, and more

Inventor 2023 brings significant changes and enhancements to workflows and interoperability options, specifically with Fusion 360 and Revit. This makes for more seamless workflows when working with files in multiple products within the Autodesk portfolio and adds much more flexibility.

Fusion 360 tools

Within Inventor, there are now more options in the **Fusion 360** tab. These tools allow you to copy an Inventor design and upload it automatically to Fusion 360. Note that to use this feature, you must either have a standalone license of Fusion 360 installed and active or have Fusion 360 installed from your **Product Design & MFG Collection** license.

Figure 13.4 shows the new look and tools of the **Fusion 360** tab:

Figure 13.4: Fusion 360 tools in Inventor

When using these tools, you have the option of starting the **Fusion 360** workspace directly from Inventor. The **Fusion Modeling** command was previously known as **Send to Fusion 360**, and will take your Inventor model directly into the **Fusion 360 Design** workspace. Additionally, options in version 2023 enable a quick transition to other Fusion 360 workspaces, such as **Simulation**.

To use this new feature, simply open the Inventor file you wish to work with in Fusion 360, and then select the Fusion 360 workspace you want to work in. Inventor components selected are then derived to a single .ipt file, copied to **Fusion Team**, and then opened in the selected workspace within Fusion 360.

This makes transitioning from Inventor to Fusion 360 for specific tasks and workflows much easier and requires fewer steps than previous versions. The process is bi-directional to a degree – using the existing **AnyCAD** feature, you are also able to directly open Fusion 360 files within Inventor. This update enables Inventor users to get access to great cloud features such as **Simulation**, **Generative Design**, and cloud **Rendering** that are only found in Fusion 360.

Data exchange and interoperability enhancements for Revit and Inventor

Data exchange in Inventor is a new way to share specified subsets of Revit data with Inventor users. It is becoming increasingly common for manufacturers to work in close collaboration with the architectural stakeholders of a project. Data exchange makes this process much more efficient.

Data exchange enables you to view and filter data, and then make changes to the data in software such as Inventor.

It's important to note that Revit files can be huge in file size. Having a good understanding of the project and condensing the geometry in a data exchange to what is necessary for the Inventor user is essential.

The capability of referencing Revit data in Inventor utilizes a pre-installed Revit component. This must be the same release version as the Inventor version installed. Normally, the Revit component will install automatically when Inventor is installed. But if this was de-selected during the installation processl, it can also be manually added.

Once a Revit component is activated with a command, if the Revit component is not installed, then the Inventor command is canceled, and the Revit component is installed in the background automatically. Once installed, the Revit component allows the efficient importing of Inventor files.

Sketch enhancements

This release features a few sketch-based enhancements that focus on repairing broken projected geometry and additional options for accessing **iProperties** in sketch-based text or **3D Annotations**.

Fixing broken projected geometry

When modeling in Inventor, geometry can sometimes break due to associated sketches, dimensions, or geometry being deleted or changed. This often occurs when critical referenced geometry in the model browser has been changed or deleted. Associated sketches or geometry can then become broken or orphaned. Previously, this was a much more manual task to address, but now you can go to those sketches and repair or delete the broken projections. This is advantageous, as sometimes the geometry that needs to be changed is difficult to locate or identify.

The new workflow is to edit the sketch with the broken geometry – right-click the **Sketch Browser** node and select **Broken Projections**. This highlights the problematic geometry so that it can be deleted.

Model sketching and the annotating text enhancement

When applying sketch text or applying a general note with 3D annotations, **Standard iProperties** and **Custom iProperties**, if present in the model, standard and custom iProperties can now all be accessed from the active part file, within the **Format Text** box. This function is not available in the assembly environment. *Figure 13.5* shows the dropdown available in the **Format Text** options to access **iProperties**:

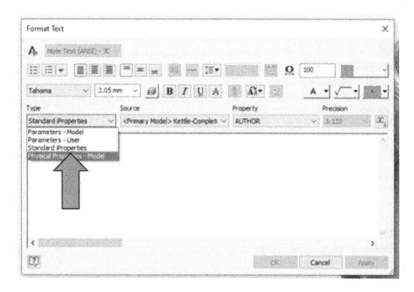

Figure 13.5: Standard iProperties accessed in either sketch text or 3D general annotation

This makes it easier to access and reference custom or standard iProperties from within sketch text or 3D general annotations.

The new link sketch format for sketch blocks

The new **Link sketch formatting from source component** option in **Derive**, enables the control of sketch blocks in the derived part, if the source sketch geometry is changed.

Part enhancements

3D annotations, sheet metal, and model-based definitions are areas that have had lots of improvements in the 2023 release. This section details what changes have been made to the Part environment of Inventor.

Sheet metal extended browser information

Version 2023 brings some new changes and enhancements to the Sheet Metal Part environment. The first is an **Extended Information in Browser Enabled for Multiple Commands**. This enables users to get more information about a sheet metal part or operation from the model browser.

To access the extended information, with a sheet metal part open, select **Application Options** | the **Parts** tab, then select **Show Extended Names**.

Figure 13.6 shows an example of the **Extended Information in Browser Enabled for Multiple Commands** active:

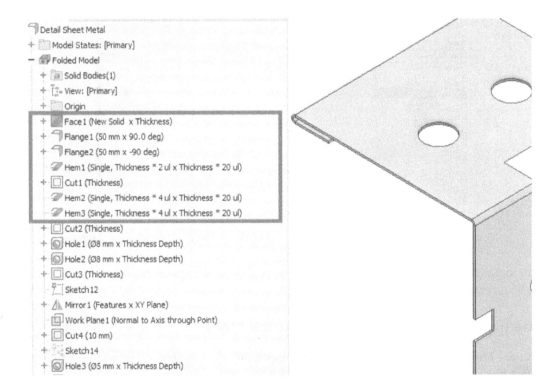

Figure 13.6: Extended information in browser enabled

The extended information in the browser feature makes it much easier to assess and differentiate between different types of sheet metal features within a part.

The Sheet Metal Template option

You can now access a new **Sheet Metal Template** option via the **Tools | Document Settings | Modeling** tab, then select **Make Components Options** to select or browse to the template used to create new **Sheet Metal** part files.

New Sheet Metal Mark command for engraving/etching features

This new command enables users to easily create and apply engraving or etching features to solid features of a model.

In this recipe, you will learn how to use the new **Mark** command.

How to do it

To begin this recipe, you will need to have Inventor 2023 installed. You will need to navigate to Inventor Cookbook 2023 | Chapter 13, open Mark.ipt, and follow these steps:

1. In the **Sheet Metal** tab, navigate to the **Modify** panel in the ribbon and select **Mark**:

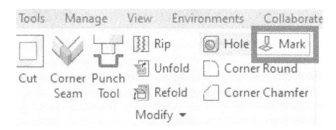

Figure 13.7: New Mark command

2. Select the **Inventor 2023** text on the face of the part.
3. **Mark Surface** or **Mark Through** can now be selected. Select **Mark Surface**.
4. Select **OK**. The text is marked on the face of the sheet metal part of Mark.ipt:

Figure 13.8: New Mark command used to etch text to the surface of a Sheet Metal part

You have now used the new **Mark** command to etch text into a sheet metal face.

3D annotation Datum Target command added

A **Datum Target** command has been added to the **3D General Annotations** and **Styles** commands, as shown in *Figure 13.9*:

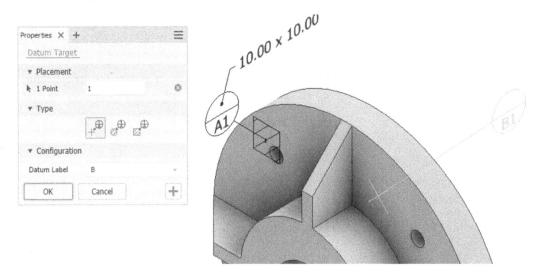

Figure 13.9: New Datum Target 3D annotation command

This enables you to establish datums on irregular surfaces. **Point**, **Circle**, and **Rectangle** datum targets can now be specified on a model as a **3D Model Based Definition**.

Tolerance feature enhancements

3D model based definitions of tolerance features can now be defined on more types of geometry, such as planes, sketch points, work points, user-defined axes, and existing datum targets.

Data reference frame label selection

Data reference frame labels have changed to allow quicker and easier access to 10 selectable frame labels and an additional **Custom** option, as shown here:

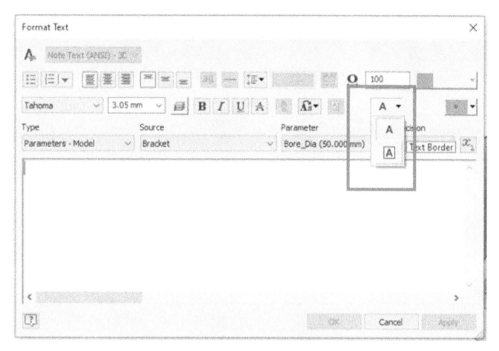

Figure 13.10: Data reference frame label selection

If you have already placed a data reference frame label, and are about to place a duplicate, Inventor will warn you that the data reference frame label is already present in the model.

Multiple selections for 3D annotation leaders

3D annotation leaders can now be placed by selecting multiple faces or edges of a model.

Rectangular borders for 3D annotation leaders, text, and general notes can now be applied using the option in the **Format Text** window, as shown here:

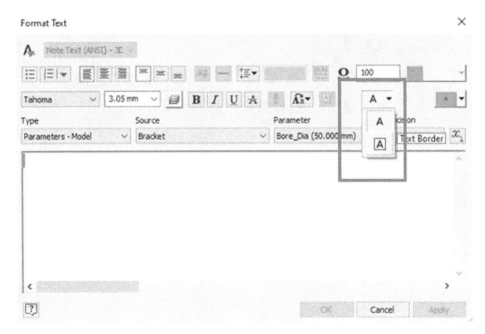

Figure 13.11: Rectangular border for 3D annotations, leaders, text, and general notes

Figure 13.11 shows where to access the rectangular border for 3D annotations, leaders, text, and general notes.

Tolerance and parameter updates to model states

The part and assembly model states now support unique tolerance settings for a model state. This means independent **Tolerance** values can be set in different model states of the same model.

Tolerance type options in Parameters

Within the **Parameters** dialogue, the **Tolerance** options in previous Inventor versions were limited to either **Upper**, **Median**, **Nominal**, or **Lower**. Now, by selecting a **Tolerance** row and selecting the pencil icon, you can access the full **Tolerance** dialogue:

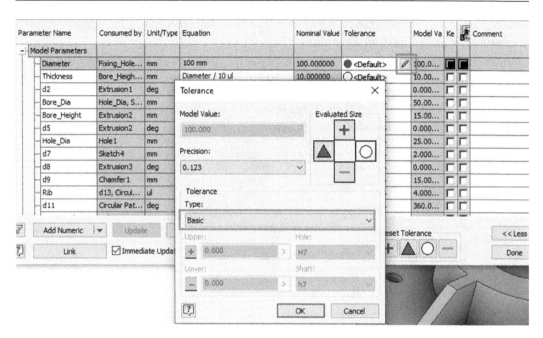

Figure 13.12: Basic tolerance type

The **Basic** tolerance type has been added to the **Type** selection options as shown here:

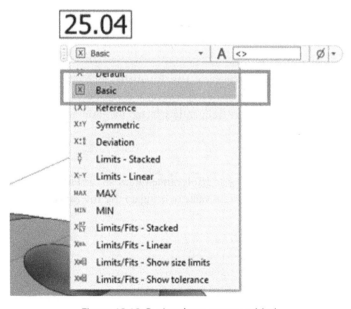

Figure 13.13: Basic tolerance type added

Tolerances can now be promoted and synchronized. The **Basic Tolerance** dimension is also available and accessible for **3D Annotation** dimensions.

Multi-Value sort order in Parameters

For user-defined parameters, when **Multi-Value** is selected, you can now choose to **Sort Order** for the values by selecting **Custom order** as shown here:

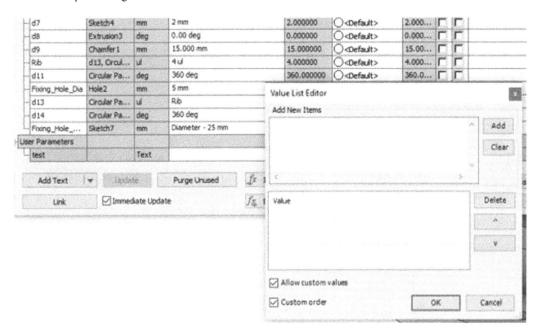

Figure 13.14: Custom order, available in User Parameters for Multi-Value options

This helps prioritize options that are more frequently selected for easier access.

Fillet tolerance

When placing a fillet on a part, you can now add **Tolerance** values. In the **Fillet** command, with an edge or edges selected, right-click on the dimension value to bring up the **Tolerance** options, as shown here:

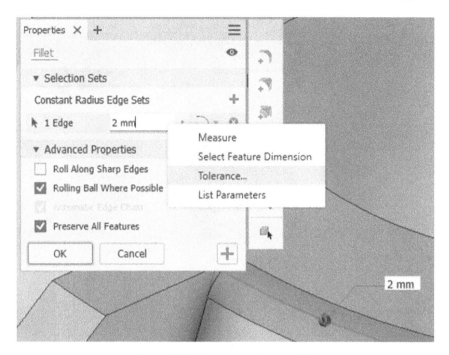

Figure 13.15: Tolerance options in Fillet

This reduces the number of clicks and operations previously required to do this.

The Relationships command for suppressed features

In the 2022 release, the **Relationships** command would only work for active features within a part. Now you can also run the command on suppressed features, to examine relationships between other features, as shown here:

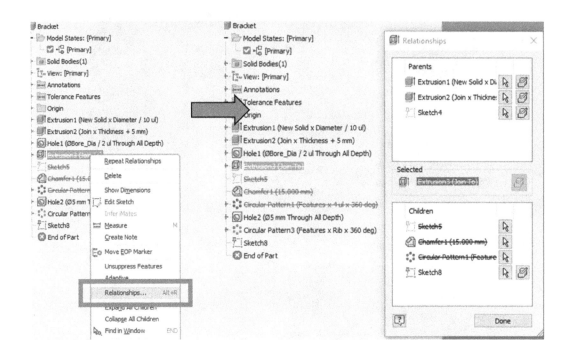

Figure 13.16: The Relationships command for suppressed features

The part enhancements are quite extensive within release 2023, particularly for the annotation and model-based definition features. The next section will detail changes, new **Bill Of Material** (**BOM**) settings, and functionality in the Assembly environment.

Assembly and presentation enhancements

This section details assembly enhancements that have been made to version 2023.

BOM setting updates

Allows for zero quantity visibility and the item number sequencing of components in **BOM**. Currently, if a component is suppressed in a model state, it will appear as 0 QTY in **BOM**. By selecting **BOM Settings** followed by the **Enable** or **Disable** options, **Hide Suppressed Components in BOM** or **Renumber Items Sequentially** can be selected as shown here:

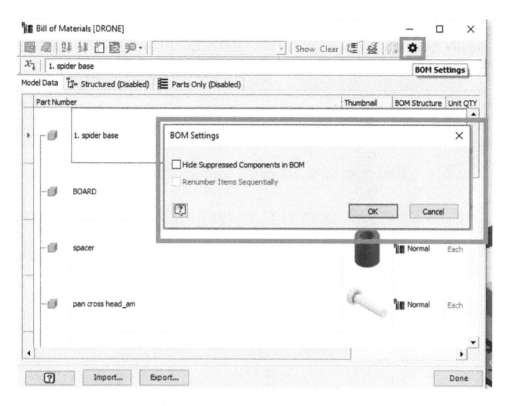

Figure 13.17: New BOM settings

Next, we will discuss changes to substitute model states.

Changes to substitute model states

Now, in version 2023, when a substitute model state is active, the following commands in the ribbon are disabled:

- **Place iLogic Component**
- **Analyze Interference**
- **Activate Contact Solver**
- **Convert to Weldment**
- **Component context menu**
- **Demote/Promote**
- **Replace from Content Center**

New options for Simplify

The **Simplify** command has been updated further to accommodate the following feature recognition options:

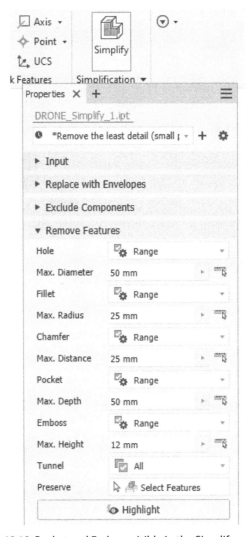

Figure 13.18: Pocket and Emboss visible in the Simplify command

This is to provide a clear classification of both additive and subtractive features. **Pocket** recognizes subtractive features, and **Emboss** recognizes additive features.

Constraint workflow command change

You can now suppress a constraint directly when editing it by selecting the **Suppress Constraint** checkbox in the **Place Constraint**, **Edit Constraint**, and **iMate** command dialogues.

Drawing enhancements

This section details drawing enhancements that have been made to version 2023.

Detail View drawing enhancement

When a **Detail Drawing View** is placed, selecting **Edit Detail Properties** on an existing detail section enables you to change the detail view **Fence Shape** option to either **Circular** or **Rectangular**, as shown here:

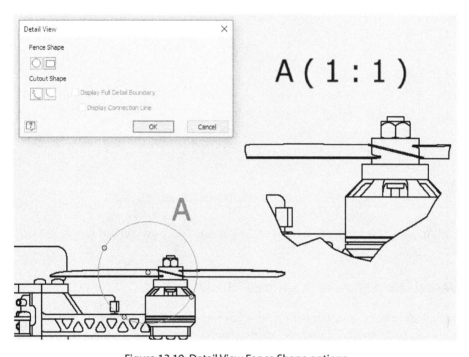

Figure 13.19: Detail View Fence Shape options

Figure 13.9 shows the **Detail View Fence Shape** options.

Model state name within drawing labels

You can now configure view labels to display a model state name in drawing views. If edits are made to the **Model State**, these are reflected in the view labels.

Within the **Overlay View** for parts, you can now also select the required model state to be shown in the view as shown here:

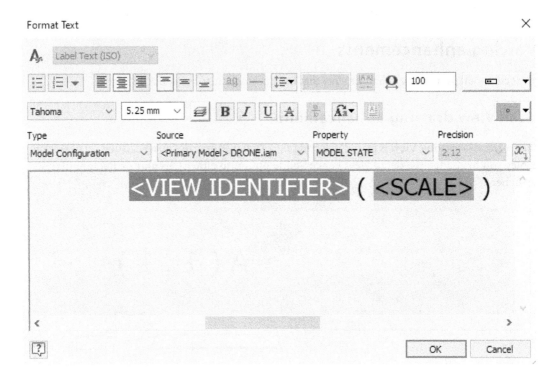

Figure 13.20: Model state name within drawing labels

In *Figure 13.20*, under the **Property** section of the **Format Text** box, you can see **MODEL STATE** is selected.

Remove existing sheets from new drawings

A new option, **Delete all existing template sheets**, has been added to the **Sheet** tab of the **Document Settings**. This will automatically remove existing sheets when a new drawing is created from a predefined sheet format.

Model credits

The model credits for this chapter are as follows:

Bomba KSB ETA 80-20 - 5CV - 1700rpm (by Jorge Omar and Ferreyra Libano): https://grabcad.com/library/bomba-ksb-eta-80-20-5cv-1700rpm-1

This brings us to the end of the book. I truly hope that you have had an enjoyable reading experience and that you have learned many new valuable skills that will aid you in your day-to-day usage of Inventor. It is my hope that these new advanced tools and techniques you have practiced using have accelerated your capabilities, broadened your knowledge, and raised your confidence in Inventor. You are now well on your way to becoming a pro!

Not every person uses Inventor the same way. Your processes and workflows are specific to you, but hopefully, the book has led you to ask questions about your workflows and techniques and has led to improvements. There is still much more to Inventor than this book covers, and I hope that some of the chapters' topics might have sparked your curiosity and interest in new areas of the software into which you can now "deep dive" and specialize.

May your sketches be fully defined and your load times swift. All the best to you with Autodesk Inventor!

Index

Hi!

I am Alexander Bordino, the author of Autodesk Inventor 2023 Cookbook. I really hope you enjoyed reading this book and found it useful for increasing your productivity and efficiency in Autodesk Inventor.

It would really help me (and other potential readers!) if you could leave a review on Amazon sharing your thoughts on Autodesk Inventor 2023 Cookbook.

Go to the link below or scan the QR code to leave your review:

`https://packt.link/r/1801810508`

Your review will help me to understand what's worked well in this book, and what could be improved upon for future editions, so it really is appreciated.

Best Wishes,

Alexander Bordino

Other Books You May Enjoy

If you enjoyed this book, you may be interested in these other books by Packt:

Practical Autodesk AutoCAD 2023 and AutoCAD LT 2023

Jaiprakash Pandey, Yasser Shoukry

ISBN: 978-1-80181-646-5

- Understand CAD fundamentals like functions, navigation, and components
- Create complex 3D objects using primitive shapes and editing tools
- Work with reusable objects like blocks and collaborate using xRef
- Explore advanced features like external references and dynamic blocks
- Discover surface and mesh modeling tools such as Fillet, Trim, and Extend
- Use the paper space layout to create plots for 2D and 3D models
- Convert your 2D drawings into 3D models

Mastering SOLIDWORKS Sheet Metal

Johno Ellison

ISBN: 978-1-80324-524-9

- Discover what Sheet Metal can be used for and how you can benefit from this skillset
- Create Sheet Metal parts, both from scratch and by converting existing 3D parts
- Select different Sheet Metal tools to be used in different situations
- Produce advanced shapes using Lofted Bends
- Relate the Sheet Metal techniques in the book to real-world manufacturing and design, including material selection and manufacturing limitations
- Practice Sheet Metal techniques using real-world examples

Packt is searching for authors like you

If you're interested in becoming an author for Packt, please visit authors.packtpub.com and apply today. We have worked with thousands of developers and tech professionals, just like you, to help them share their insight with the global tech community. You can make a general application, apply for a specific hot topic that we are recruiting an author for, or submit your own idea.

Download a free PDF copy of this book

Thanks for purchasing this book!

Do you like to read on the go but are unable to carry your print books everywhere?

Is your eBook purchase not compatible with the device of your choice?

Don't worry, now with every Packt book you get a DRM-free PDF version of that book at no cost.

Read anywhere, any place, on any device. Search, copy, and paste code from your favorite technical books directly into your application.

The perks don't stop there, you can get exclusive access to discounts, newsletters, and great free content in your inbox daily

Follow these simple steps to get the benefits:

1. Scan the QR code or visit the link below

https://packt.link/free-ebook/9781801810500

2. Submit your proof of purchase
3. That's it! We'll send your free PDF and other benefits to your email directly

www.ingramcontent.com/pod-product-compliance
Lightning Source LLC
Chambersburg PA
CBHW081448050326
40690CB00015B/2722